AF386151

J. Schulte am Esch · E. Kochs   (Eds.)

Central Nervous System Monitoring in Anesthesia and Intensive Care

With 127 Figures, Some in Color

Springer-Verlag
Berlin Heidelberg New York London Paris
Tokyo Hong Kong Barcelona Budapest

Prof. Dr. med. J. Schulte am Esch
Prof. Dr. med. E. Kochs

Department of Anesthesiology
University Hospital Eppendorf
University of Hamburg
Martinistrasse 52
20246 Hamburg
Germany

ISBN-13:978-3-642-78443-9   e-ISBN-13:978-3-642-78441-5
DOI: 10.1007/978-3-642-78441-5

Library of Congress Cataloging-in-Publication Data. Central nervous system monitoring in anesthesia and intensive care/J. Schulte am Esch, E. Kochs, eds.   p.   cm, Includes bibliographical references and index.ISBN-13:978-3-642-78443-9(U.S.)1.Electroencephalography.2.Evoked potential(Electrophysiology)3.Patient monitoring. 4. Intraoperative monitoring. 5. Critical care medicine. I. Schulte am Esch, J. (Jochen), 1939–   . II. Kochs, E. (Eberhard), 1943–   . RC386.6.E43C46   1994   616.8'047547--dc20   94-2610

The use of general descriptive names, registered names, trademarks, etc. in this publication does not imply, even in the absence of a specific statement, that such names are exempt from the relevant protective laws and regulations and therefore free for general use.

Product liability: The publishers cannot guarantee the accuracy of any information about dosage and application contained in this book. In every individual case the user must check such information by consulting the relevant literature.

Typesetting: Macmillan India Ltd., Bangalore-25
SPIN: 10099695   19/3130/SPS – 5 4 3 2 1 0 – Printed on acid-free paper

Preface

Research in electrophysiologic monitoring in anesthesia and intensive care has focussed mostly on questions pertinent for patient care: First how to quantitate drug effects on brain electrical activity and the degree of anesthetic-induced suppression of the central nervous system. Second, how to monitor functional impairment following cerebral ischemia and hypoxia. And third, how to differentiate between drug-induced effects on the central nervous system and deleterious events related to reductions in cerebral blood flow and/or oxygen delivery.

Even though progress has been achieved over the last 10 years in this field and fascinating new techniques have been developed, it is still not clear which monitor parameter will provide adequate information on the depth of anesthesia and the analgesic level. Because the central nervous system has been one of the main research areas in our department over the last 10 years, we organized a workshop to summarize the latest developments in central nervous system monitoring. This book comprises the topics of this workshop and is intended to provide insight into the current status of central nervous system monitoring, elucidating possible indications and delineating its limitations.

For more than 30 years a primary goal for intraoperative neurophysiological monitoring has been to get a warning when cerebral oxygen supply may be hampered. This, obviously, is of special interest during surgical procedures which may impair cerebral blood flow such as carotid endarterectomy, intracranial aneurysm, or spinal cord surgery. Many monitoring techniques have been developed to predict imminent neuronal damage. In this sense, intraoperative monitoring is used specifically to observe the response of the central nervous system to surgical interventions. The hope was that perioperative morbidity and mortality may be decreased by continuous or frequent assessment of neuronal function. Some monitoring techniques, such as EEG or evoked potential monitoring during carotid endarterectomies, have been shown to be very sensitive and specific for the prediction of postoperative neurologic outcome. For other surgical procedures the best monitoring parameter still has to be defined.

A specific anesthesiologic goal which also has gained interest over the past years is the assessment of the depth of anesthesia by neurophysiological monitoring. The hope that the EEG may provide unique parameters has not been substantiated because various anesthetic agents may produce different

EEG patterns. However, using computerized analysis methods, EEG frequency descriptors (i.e., median frequency and spectral edge frequency) have been advocated for the assessment of changes in depth of anesthesia. In closed-loop feedback systems the median frequency has been shown to be useful for supervising anesthetic drug administration. However, single parameters may not be helpful for assessing sedative/hypnotic as opposed to analgesic effects. It has become clear over the last years that the so-called intraoperative arousal phenomena may be of concern when there is an inadequate depth of anesthesia. In the EEG desynchronization with appearance of fast low wave activity has been considered to indicate arousal phenomena. Recent studies have shown that the shift to slower waves induced by surgical manipulations may also reflect electrophysiologic arousal reactions. This has made it clear that at the time being no single EEG parameter can be defined which unequivocally indicates intraoperative arousal. Furthermore, the EEG may not reflect the level of drug induced analgesic effects. Several studies suggest that evoked responses hold promise for intraoperative assessment of the inadequate depth of anesthesia. Early cortical auditory and somatosensory evoked responses change in a graded manner with changes in depth of anesthesia. With few exceptions these changes are not anesthetic specific. In addition, it has been shown that surgical stimulation during an inadequate depth of anesthesia will reverse the depressant effects of anesthetics. Further work has to show which anesthetic technique will be most useful for achieving this goal in intraoperative monitoring. There is evidence that auditory evoked responses may indicate intraoperative awareness. Specific cortical components of auditory evoked responses seem to be related to auditory signal processing. Further studies have to show if drug administration can be controlled by auditory evoked responses in order to impair transmission of sensory stimuli to implicit memory. From several studies on auditory evoked responses including the 40-Hz steady state response, it can be concluded that these measures hold promise for assessing the hypnotic effect of anesthesia. Somatosensory evoked responses have also been shown to change in relation to anesthetic drug concentration in a non-agent-specific manner (exception: etomidate). It has been demonstrated in a few studies that surgery may also offset the anesthetic induced depressant effects on somatosensory evoked responses. However, it is unclear if these changes are related to inadequate sedation. The stimulation of large mixed peripheral nerves, such as is used for conventional somatosensory evoked response monitoring, recruits different afferent pathways. In contrast, specific pathways have to be stimulated to assess changes in nociceptive signal transmission. Various studies have shown that somatosensory evoked responses following the stimulation of C- and Aδ- fibers correlate to drug induced changes in pain perception. However, these cortical evoked responses are very vulnerable to changes in psychophysiologic variables (i.e., attention, alertness, vigilance) and are suppressed by virtually all anesthetics. Few studies have shown that these evoked responses may be used for the assessment of analgesic treatment effects during anesthesia.

In contrast to somatosensory evoked responses, transcranial stimulation may be used for monitoring efferent pathways. This modality may be especially useful during surgical procedures when motor pathways are at risk. However, most anesthetics reduce the amplitudes of the evoked responses significantly. To interpret electrophysiologic data, changes in systemic variables have to be considered. Simultaneous recordings of brain function and cerebral blood flow velocities or venous jugular bulb oxygen saturation may help in the detection of deleterious effects.

Basic science and clinical applications complement each other. The synthesis of both, the exchange of scientific research and routine clinical practice, will lead to a concept which will bring about the maximal benefit for patient care. Therefore, one of today's pending challenges is to find monitoring techniques for clinical practice which allow the unequivocal assessment of central nervous system function during anesthesia and intensive care.

It was O. W. Holmes who, in 1840, stated that the great thing in this world is not so much where we stand, but in what direction we are going. In this sense we understand the current trends in the development of central nervous system monitoring techniques. It was the goal of this book to give an insight into the ever developing process on new specific monitoring techniques and how they may be used in the future. This in mind, we hope that this book will stimulate scientists and physicians to continued research in the field of central nervous system monitoring.

We are very grateful to the international group of distinguished speakers and chairmen as well as to the audience for the excellent contributions and discussions in this workshop. We are also much indebted to Mrs. L. Berger for her technical assistance in preparing the workshop and her help in organizing the publication of this book.

Hamburg, January 1994

JOCHEN SCHULTE AM ESCH
EBERHARD KOCHS

Contents

Part III   Monitoring of Stimulus Evoked Responses

Part IV   Evoked Response: Special Applications

Part V   Present and Future Trends in Cerebral Monitoring

Contributors

BISCHOFF, P., Dr.
Abteilung für Anästhesiologie, Universitäts-Krankenhaus Eppendorf,
Martinistr. 52, 20246 Hamburg, Germany

BLOOM, M.J., M.D., Ph.D. Assistant Professor
School of Medicine, Departments of Anesthesiology/CCM,
University of Pittsburgh, Pittsburgh, PA 15213, USA

BROMM, B., Prof. Dr. Dr.
Physiologisches Institut, Universitäts-Krankenhaus Eppendorf, Martinistr. 52,
20246 Hamburg, Germany

DEARDEN, N.M., M.D.
Dept. of Anaesthesia, Leeds General Infirmary, Leeds LS12 4AD, Scotland

DIMPFEL, W., Prof. Dr.
ProScience, Private Research Institute, Linden, 35390 Giessen, Germany

DINKEL, M., Dr.
Institut für Anästhesiologie der Universität Erlangen – Nürnberg,
Krankenhaus Str. 12, 91054 Erlangen, Germany

DOENICKE, A., Prof. Dr.
Anästhesiologische, Abteilung, Chirurgische Poliklinik der Universität,
Pettenkoferstr. 8a, 80336 München, Germany

FACCO, E., M.D.
Ist. Anestesiologia e Rianimazione, Università di Padova, Via C. Battisti 267,
35121 Padova, Italy

GURMAN, G., M.D., Prof.
Division of Anesthesiology, Soroka Medical Center, Beer-Sheva 84101, Israel

HEMPELMANN, G., Prof. Dr.
Institut für Anästhesiologie, Universitätskliniken, Klinikstr. 29, 35392 Giessen,
Germany

KALKMAN, C., M.D., Ph.D.
Department of Anesthesiology, University Hospital Amsterdam,
Meibergdreef 9, 1105 AZ Amsterdam, The Netherlands

KOCHS, E., Dipl. Phys., Prof. Dr.
Abteilung für Anästhesiologie, Universitäts-Krankenhaus Eppendorf,
Martinistr. 52, 20246 Hamburg, Germany

LEMMENS, H.J.M., M.D., Ph.D. Prof.
Veterans Administration Medical Center, School of Medicine,
Stanford University, Palo Alto, CA 94304, USA

LITSCHER, G., Dipl. Ing., Dr.
Univ.-Klinik für Anästhesiologie, Landeskrankenhaus, 8036 Graz, Austria

PLOURDE, G., M.D., M.Sc.
Department of Anaesthesia, Royal Victoria Hospital, 687 Pine Avenue West,
Suite 5.05, Montreal (Quebec), Canada H3A 1AB

ROM, G., Dipl. Ing.
Department of Medical Informatics, Institute of Biomedical Engineering,
Graz University of Technology, 8036 Graz, Austria

SCHAREIN, E., Dipl. Psych., Dr.
Physiologisches Institut, Universitäts-Krankenhaus Eppendorf,
Martinistr. 52, 20246 Hamburg, Germany

SCHULTE AM ESCH, J., Prof. Dr.
Abteilung für Anästhesiologie, Universitäts-Krankenhaus Eppendorf,
Martinistr. 52, 20246 Hamburg, Germany

SCHWENDER, D., PD Dr.
Institut für Anästhesiologie, Klinikum Grosshadern, Marchioninistr.,
81377 München, Germany

SCHWILDEN, H., Prof. Dr. Dr.
Institut für Anästhesiologie der Universität Bonn, Sigmund-Freud-Str. 25,
53127 Bonn, Germany

SEBEL, P.S., M.B., B.S., Ph.D., FFARC SI,
Department of Anesthesiology, Emory, University at Crawford Long Hospital,
Atlanta, GA 30308, USA

STOECKEL, H., Prof. Dr.
Institut für Anästhesiologie der Universität Bonn, Sigmund-Freud-Str. 25,
53127 Bonn, Germany

THORNTON, Ch., Ph.D.
Division of Anaesthesia, Clinical Research Centre,
Northwick Park Hospital, Harrow, Middlesex HAI 3UJ, UK

TREEDE, R.D., Prof. Dr.
Physiologisches Institut, Klinikum der Johannes-Gutenberg-Universität,
Saarstr. 21, 55122 Mainz, Germany

URBAN, W., Prof. Dr.
Institut für Anästhesiologie der Universität Bonn, Sigmund-Freud-Str. 25,
53127 Bonn, Germany

VAN AKEN, H., Prof. Dr.
Dept. of Anesthesiology, Universitaire Ziekenhuizen Katholieke Universiteit
Leuven, 3000 Leuven, Belgium

VERNON, J.M., Dr.
Department of Anesthesiology, Emory, University at Crawford Long Hospital,
Atlanta, GA 30308, USA

VILLEMURE, C., Dr.
Department of Anaesthesia, Royal Victoria Hospital, 687 Pine Avenue West,
Suite 5.05, Montreal (Quebec), Canada H3A 1AB

WERNER, C., Dr.
Abteilung für Anästhesiologie, Universitäts-Krankenhaus Eppendorf,
Martinistr. 52, 20246 Hamburg, Germany

ZENTNER, J., PD Dr.
Neurochirurgische Klinik der Universität Bonn, Sigmund-Freud-Str. 25,
53127 Bonn, Germany

Part I

Monitoring of the Electroencephalogram – Fundamentals

Interactions of Anesthetics at Different Levels of the Central Nervous System

B. W. Urban

Components of the Anesthetic State

Even today there is no agreement as to which molecular and higher level events are responsible for the state commonly called anesthesia [7–10, 40–41]. This is not only because of the many physiological changes that are observed during anesthesia [34], but also because of the fact that there is no agreement as to an unambiguous measurement procedure that would allow a quantitation of anesthesia [45]. This situation may not change as long as neuroscience is not able to quantitate consciousness, since the loss of consciousness is the goal of anesthesia [4, 27, 51]. However, not only is consciousness lost during anesthesia; perception, memory, pain and muscle relaxation and other physiological regulatory mechanisms are also affected [34]. Neuroscience today is not in a position to furnish a complete and molecular description of the various forms of sleep, memory and pain [27]. However, this list of our lack of knowledge suggests in itself that there are probably several and different molecular mechanisms that give rise to the state of anesthesia.

Although anesthetics primarily act on molecular structures, a complete understanding of the mechanisms of anesthetic action is not possible without considering the different levels of the nervous system, from the molecular level up to the intact brain (Fig. 1). While doing so, it is important to keep in mind the integrative aspect of the central nervous system organization. On each of its levels, anesthetic actions on single components have to be studied, followed by a consideration of the integration of these components into a higher level system. Since each level of the central nervous system constitutes more than the sum of its individual components, it does not suffice to only pay attention to the individual components. Even small anesthetic effects on components of a certain level may have much more far reaching consequences when these components are integrated into a new functional unit. It becomes clear that anesthesia cannot be explained simply by molecular studies without a knowledge of the network topology of the central nervous system at all its levels. On the other hand, a knowledge of its network topology alone (still little known) will not explain anesthesia without an understanding of anesthetic actions at the molecular level.

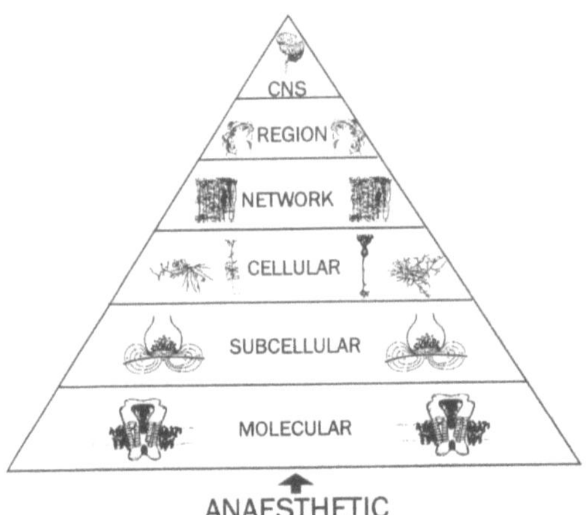

Fig. 1. The various levels of the central nervous system (*CNS*). Anesthetics interact with various molecular structures at the molecular level. Molecular structures are integrated at the next level into subcellular structures; therefore, the effect of anesthetics at this level results from the integration of different individual effects. This scheme continues throughout all levels of the central nervous system

Molecular Level

Properties of Anesthetics

At the molecular level, the molecular properties of anesthetics themselves as well as their target sites have to be investigated. A great number of mostly small organic molecules have anesthetic potency, without sharing a common chemical or physical structure [6, 38, 44]. In a more restricted sense, one can only call those substances anesthetics that produce anesthesia in the absence of any other drug. In this context, anesthesia may be defined as follows: "Anesthesia describes the condition of a patient corresponding to the one produced by diethyl-ether, which permits surgery to be performed without the patient moving, reacting to pain, or remembering the surgical intervention after recovery from exposure to the anesthetic drug [50]. A first extension of this definition includes all substances which could produce anesthesia in principle, while in clinical practice they are used in conjunction with other drugs, for example, muscle relaxants. A further extension also accepts substances as anesthetics that are used by basic scientists in animals or during in vitro experiments and that display comparable anesthetic effects. This extended definition is useful because it permits a systematic structure – function analysis of anesthetic actions with the help of deliberate changes in the physicochemical and chemical properties of an anesthetic molecule. For volatile and small organic anesthetics, it has proven useful to divide them into different groups, depending on whether they are lipophilic, surface-active, ionized or inhalation anesthetics [50]. Regarding intravenous anesthetics, various groups are distinguished clinically, for example: hypnotics, sedatives, neuroleptics, dissociative anesthetics and narcotics (opiates).

Since anesthetics differ so much in their physicochemical properties, it is unlikely that they all interact with only a few, highly specific receptor sites. The

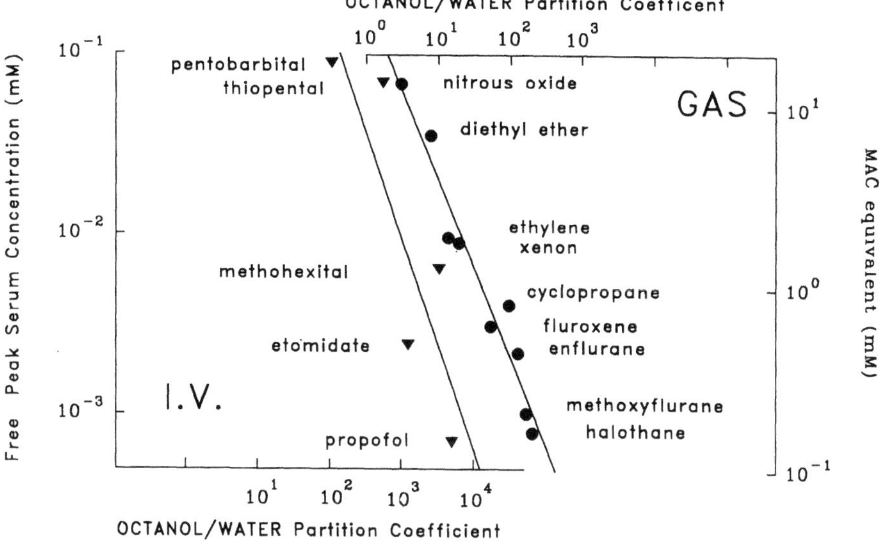

Fig. 2. Meyer–Overton correlation for volatile anesthetics and for intravenous hypnotics. To facilitate comparisons, the MAC values (minimum alveolar concentration to prevent movement at incision in 50% of patients) for volatile anesthetics [47] have been transformed into molar concentrations of aqueous solutions [50]. The slope of the straight line for intravenous anesthetics is − 1.18, with a correlation coefficient of 0.86. These values compare with a slope of − 0.99 and a correlation coefficient of 0.97 for inhalation anesthetics

double logarithmic Meyer–Overton correlation (Fig. 2) shows a linear dependence of anesthetic potency on the anesthetic's partition coefficient between a lipophilic phase and a gas phase or a buffer [32]. This correlation points to membranes as an important molecular site of action. Lipophilic interactions are non-specific [48]. Therefore, it seems quite likely that anesthetics act at different target sites, including not only the lipids of cell membranes, but also the lipophilic domains of their membrane proteins.

The Electrically Excitable Membrane

A pure lipid bilayer (Fig. 3) is an excellent insulator and does not permit any current or ion flow across the membrane, which, however, would be required for signal transmission to occur. Specialized membrane proteins, called ion channels, have to be incorporated into lipid bilayers in order to turn them into electrically excitable membranes [24]. Membranes adsorb and absorb anesthetics, which can result in quite a few changes of their properties [50]. Lipophilic anesthetics reside preferentially in the membrane interior and lead to a thickening of the lipid bilayer [16,17]. In addition, they can increase

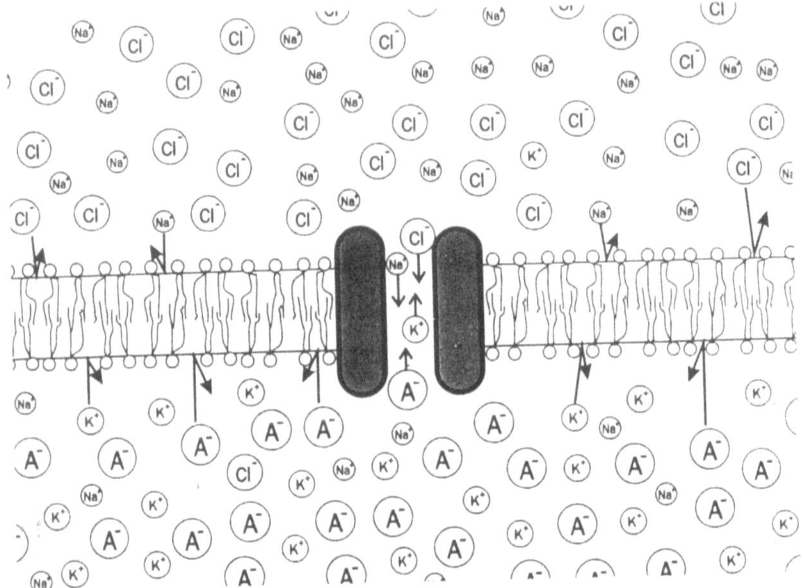

Fig. 3. The lipid bilayer consists of a bimolecular leaflet of lipid molecules which forms an almost impenetrable barrier for ions. The penetration of the lipid bilayer is made possible by specialized membrane proteins. An ion channel molecule is schematically shown containing an aqueous channel in its interior. Through this channel, ions can cross the membrane

membrane surface tension. They also affect membrane fluidity by changing the mobility of the fatty acid chains of the lipids, often as a function of bilayer depth. Surface-active substances such as alcohols are rarely found in the bilayer interior [18]. Instead, they adsorb to the membrane surface where they alter the surface tension and the electrical surface potential. The change in the surface potential results from an effect on the surface dipoles which reside close to the membrane interface. Inhalation anesthetics are mostly less polar than alcohols, and they possess properties that lie inbetween the two extremes discussed above [19]. Therefore, it is likely that inhalation anesthetics may be found both in the membrane interior as well as at its interface. The effects of these anesthetic substances on lipid bilayers do not normally lead to biologically significant bilayer conductance, with the exception of anesthetic concentrations that result in membrane breakdown.

Ion Channels

In order to render a lipid bilayer electrically excitable, specialized membrane proteins are needed. Nature uses ion channel proteins, which span the membrane and contain an aqueous channel through which ions can move from one side of the membrane to the other [24]. If each ion species could flow through these channels, the result would be an electrical short circuit. The membrane could not maintain an ion gradient and electrical excitability would be lost. For this reason, different types of ion channels have evolved that allow only certain ions, e.g. sodium ions, to flow through them. Similarly, there are potassium channels, calcium channels and chloride channels. These channels do not remain open all of the time; they contain molecular switches (Fig. 4) that are operated either by the membrane potential or by agonists. There are voltage-dependent sodium channels, potassium channels, calcium channels and a multitude of agonist-operative channels as well as channels that are dependent on both membrane potential and agonists. Important agonist-activated ion channels are the acetylcholine receptor channel, the glutamate receptor channel and the γ-aminobutyric acid (GABA) receptor channel. These channels all have in common the property that they do not permit ion flow in the resting state. Ion

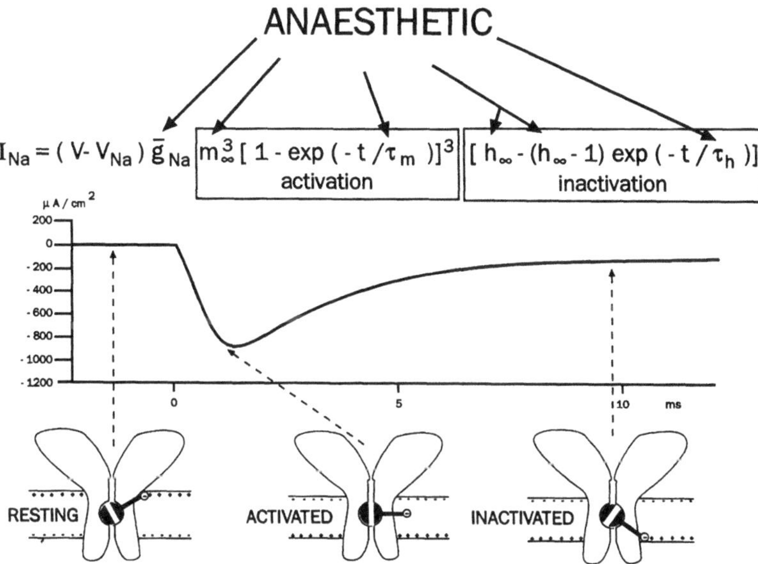

Fig. 4. The voltage-dependent sodium channel and sodium current. The *bottom* part of the figure shows the three important conformations of the sodium channel, which exists either in the nonconducting, *resting* state, in a conducting, *activated* state, or in another nonconducting, *inactivated* state. The molecular switch registers changes in membrane potential (shown by a reversed charge distribution in the activated and inactivated state). The *middle* part shows the time course of the sodium current from a typical voltage-clamp experiment. The *dashed arrows* indicate which conformation of the sodium channel is predominantly associated with a particular phase of the current. At the *top* is the Hodgkin–Huxley equation (see text), which describes the current course mathematically. The *thick arrows* indicate which parameters of this equation are affected by anesthetics

flow is possible during the activation phase and is terminated again in many cases by an inactivated or desensitized state, which differs from the resting state (Fig. 4). Each ion channel type consists of several subtypes [24].

Sodium Channels

Considering the great number of different ion channels, it becomes important to choose some as model channels on which to study the actions of anesthetics. The sodium channel is a suitable model channel as it probably constitutes the voltage-dependent ion channel that has been electrophysiologically characterized the most [1,25,31]. Its structural properties have also been thoroughly investigated using techniques from the fields of biochemistry, spectroscopy and molecular biology [5,28,36]. Similarly, the pharmacology of many other compounds has been described [6,35,39,44]. The voltage-dependent sodium channel is responsible for the generation of fast-action potentials and it therefore plays an important role in nerve impulse initiation, conduction and integration [27]. The sodium channel has been examined electrophysiologically using the voltage-clamp technique. This technique uses an amplifier which keeps the membrane potential constant and allows it to be almost instantaneously changed to a new and constant level. The resulting current is monitored and can be described by mathematical formalae, e.g. the Hodgkin–Huxley formula [25] (Fig. 4). When the membrane potential is suddenly changed in a voltage-clamp experiment, there is an inward current into the cell which first rises (activates), reaches a maximum and then declines again (inactivates). This current is described by an equation containing five separate parameters, which determine the maximal current flow through the channel ($\bar{g}_{Na}$), the time constant of activation (τ_m) and its steady state value (m_∞) and the time constant of inactivation (τ_h) and its steady state value (h_∞). Studies on volatile anesthetics have shown that not just one, but all of these five parameters are changed by anesthetics [50]. Consequently, at the molecular level, current through sodium channels is not only reduced by one type of action, but by several summating ones. Not all of these actions are depressant. For example, exclusively lipophilic substances such as n-pentane have an excitatory effect on the activation system, but the other inhibitory effects dominate in the steady state, resulting in an overall suppression [16,17]. Their kinetic behaviour is also of interest. When a nerve axon is exposed to n-pentane, spontaneous action potentials are initially observed, i.e. an excitation, which disappears with time. It seems that activation is affected before the other effects are manifested. This observation could parallel clinical observations where an excitation phase often follows anesthetic induction [2]. In conclusion, it seems clear that an anesthetic can have several molecular sites of action on ion channels. The overall depression of the sodium channel by anesthetics correlates with their lipophilicity and results from a summation of individual excitatory and inhibitory anesthetic actions.

Mechanisms of Anesthetic Actions on Ion Channels

As far as the modes of actions of anesthetics on the maximal conductance, activation and inactivation of sodium channels are concerned hypotheses which are consistent with the data do exist, although their absolute validity has not been proven yet [50]. According to these hypotheses, lipophilic substances produce their anesthetic effects by changing lipid bilayer properties such as thickness, surface tensions and fluidity. While these changes do not significantly alter the ionic conductance of a pure lipid bilayer, significant effects result because of the interaction of the ion channels with the lipid bilayer. Thus, the integration of these two different molecular components results in considerable anesthetic effects. This was demonstrated directly in experiments using artificial bilayers and the ion channel gramicidin A [15, 23]. Gramicidin A is an antibiotic that forms cation-selective ion channels which do not contain a molecular switch. These channels are not normally found in excitable membranes, but because of their simple structure, they have been very useful as model channels. Using this model, the "thickness–tension " hypothesis has been developed, which is the only molecular theory of anesthesia that has been quantitatively verified [15]. This theory states that two molecularly effects, i.e. membrane thickening and an increase in membrane surface tension, explain the anesthetic effect of purely lipophilic substances on the gramicidin A ion channel. Only the summation of both effects quantitatively accounts for the overall depression. The direct effect of lipophilic substances on the gramicidin A channel is negligibly small, so that the anesthetic effect manifests itself only in the integrated system of lipid bilayer and ion channel.

Subcellular Level

At the next level of the central nervous system, the subcellular level, different ion channels and other membrane proteins are integrated into the lipid bilayer to form electrically excitable membranes, the properties of which change depending on whether they are membranes of synapses, axons, dendrites or the cell soma [27]. Different functional components are combined into a new system, for example, different types of ion channels contribute to the generation of the fast-action potentials (Fig. 5). In the squid giant axon, a standard model of an excitable membrane [3], there are three different ion channels involved. The resting potential of this axon is determined largely by potassium channels that have not yet been fully characterized [22]. In conjunction with sodium channels, these potassium channels determine the threshold at which an action potential is elicited. The rising phase of the action potential results predominantly from an opening of sodium channels, while membrane repolarization and the refractory period are determined by sodium channel inactivation and the activation of potassium channels of the delayed rectifier type (Fig. 5). This potassium channel

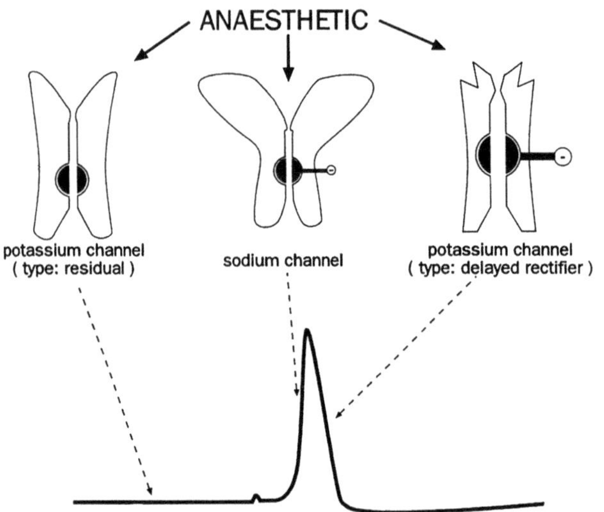

Fig. 5. The molecular components of action potentials. The action potential results from a combined action of different types of ion channels. The *dashed arrows* indicate which phases of the action potential are predominantly affected by a particular channel type (see text). As indicated by the *thick arrows*, anesthetics act on all three types of ion channels, thereby changing various aspects of the action potential

differs from that already mentioned in its activation properties and pharmacology [22]. All three channels are altered in their functions by anesthetics. These effects occur at clinically relevant anesthetic concentrations [22]. It is interesting to compare the voltage-dependent sodium and potassium (delayed rectifier) channels. Many volatile anesthetics have a larger impact on these sodium channels than on the potassium (delayed rectifier) channels [20]. This observation can be partly explained by the fact that potassium channels of the delayed rectifier type in squid giant axons do not become inactivated. Since the anesthetic effect on sodium channel inactivation leads to current depression, it can be easily shown mathematically that sodium currents would be suppressed less by anesthetics in the absence of inactivation mechanisms [49, 50]. Indeed, as far as the non-specific actions of anesthetics are concerned, a good correlation can be observed between sodium and potassium current suppression [20], which is consistent with an action mediated through the lipid bilayer. As was the case for sodium channels, anesthetics have more than one effect on potassium (delayed rectifier) channels. Not only do these effects have to be summed, but also the overall suppressions of sodium channels and the two types of potassium channels have to be integrated before the anesthetic effects on action potentials can be fully accounted for [22]. These effects are an alteration in firing threshold, action potential rise time and the refractory period, which determines the maximal impulse firing rate.

Cellular Level

At the next level of integration, different excitable membranes are combined within a neurone. Action potentials spread from one neurone to the next, and they are integrated with impulses from other cells via a system of sometimes widely distributed synapses. There are various possibilities how incoming signals which have not been suppressed much can be depressed much more at the next level of integration. It is not common for incoming signals to produce an action potential in the postsynaptic neuron, unless there has been some temporal or spatial summation (particularly at the axon hillock region) of incoming excitatory inputs [27] (Fig. 6). If this integration is compromised by an upset in the temporal correlation of incoming signals, then this could result in no action potential being elicited in the postsynaptic neurone (Fig. 6). A partial blockade at one level can lead to a complete blockade at the next higher level. On the other hand, a normally occurring blockade of an action potential by simultaneous inhibitory inputs may be lifted due to inhibitory and excitatory signals being shifted in their temporal correlation as a result of anesthetic action (Fig. 6). In this case, anesthetics could have an excitatory effect, although their effect on the lower level of integration of the central nervous system would have

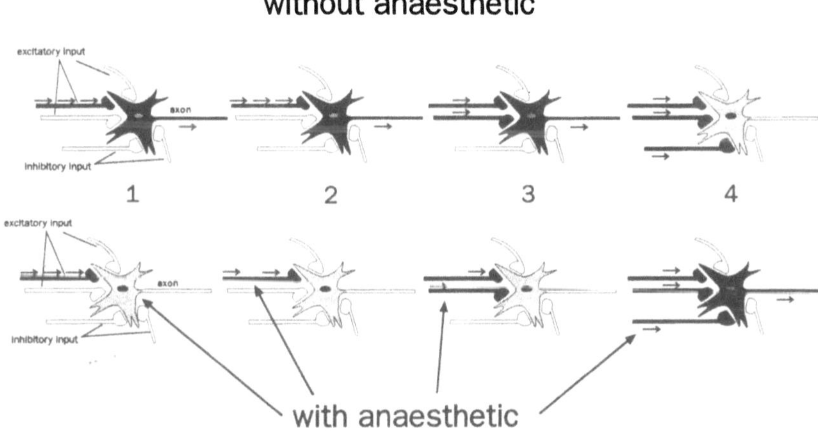

Fig. 6. Spatial and temporal integration of excitation and inhibition. *Top row;* only the temporal (*1, 2*) or spatial (*3*) summation of incoming signals leads to an action potential being elicited in the postsynaptic neurone, unless this is prevented by simultaneous inhibitory inputs (*4*). *Bottom row,* Possible alterations of signal integration through the effects of anesthetics on a neurone, indicated by *thick arrows*. A direct anesthetic action on the integration properties of the neurone, for example, by changing the properties of the axon hillock region (*1*), may lead to impulse blockade as well as a temporal change (frequency reduction or shift) in the incoming excitatory signals (*2, 3*). If the inhibitory and excitatory input signals no longer arrive simultaneously (*4*), then a previously blocked action potential may now be elicitcd

been an inhibitory one. From these considerations it can be concluded that the effect of an anesthetic cannot be predicted by only knowing its molecular properties; the manner in which these components are integrated into a higher-level system is also of great consequence. The complexity of this system may be appreciated by considering that a typical neurone in the brain cortex does not only contain a few synapses, as shown here (Fig. 6), it contains thousands of synapses.

Network Level

The next higher level of central nervous system integration is characterized by the networking of neurones. At this level at least feedback loops, reflex arcs and oscillators appear [29]. Little is known about anesthetic mechanisms of action at this level. Theories of brain function stress the importance of oscillations and resonances within the central nervous system, which are thought to be very important for global functional states such as the sleep–wake cycle and conscious perception [29].

Oscillators are specific systems which can be easily upset by small and non-specific perturbations. This is another example of how non-specific actions can have specific results. Hypotheses concerning the importance of altered or suppressed oscillators within the central nervous system are consistent with observations of electrical brain activity during anesthesia and corresponding changes in the frequency patterns of electroencephalogram (EEG) power spectra [26, 37].

Meyer–Overton Correlation and Lipophilic Interactions

Before the molecular actions of anesthetics reach the highest level of central nervous system integration, they have to pass through various intermediate levels. In the process, the original effect is altered more and more because it has been combined with an ever-increasing number of other effects. Are any of the original properties left? Do different anesthetic effects share common properties? The Meyer–Overton correlation states that anesthetic potency correlates with lipophilicity. This correlation holds for volatile and small organic substances [47] (Fig. 2). If anesthetic potency is plotted against lipid solubility for the different classes of clinically used intravenous anesthetic agents, only a weak correlation is observed [13]. A better correlation results when this is restricted to intravenous hypnotics such as propofol and barbiturates (Fig. 2). Contrary to benzodiazepines and opiates, these molecules are still relatively simple and undifferentiated. The Meyer–Overton correlation also holds at the molecular level. This is shown in Fig. 7 for the sodium channel, but it can also be demonstrated for potassium channels [50] or acetylcholine receptor channels

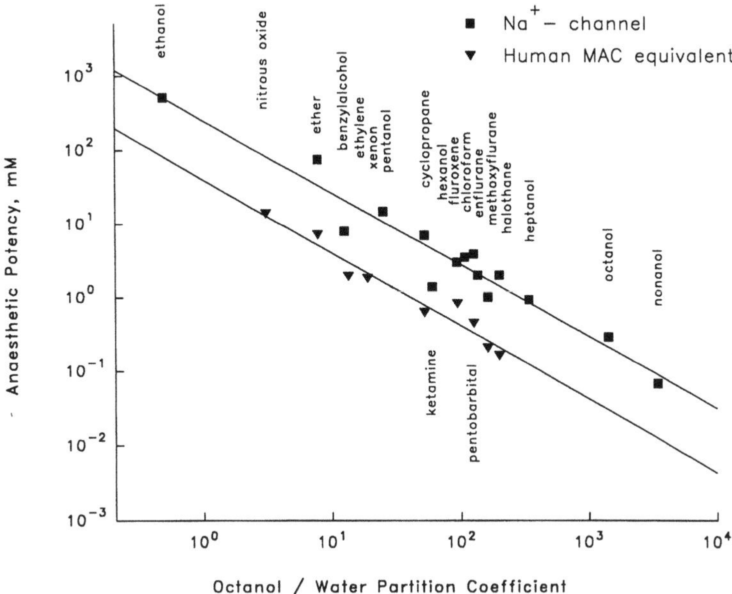

Fig. 7. Comparison of Meyer–Overton correlations for the molecular and the highest level of the central nervous system. MAC (minimum alveolar concentration to prevent movement at incision in 50% of patients)-equivalent data as in Fig. 2; slope -0.991, correlation coefficient 0.973, The sodium channel data were measured at the squid giant axon [50]; slope -0.975, correlation coefficient 0.967

[14]. The correlations run parallel to each other. However, compared with human anesthesia, a ten-fold higher concentration of anesthetics is required to reduce sodium currents by 50%. This does not necessarily mean that sodium channel suppression does not contribute to anesthesia. Rather, it seems to be important that the suppression of sodium currents correlates with lipophilicity equally as well as does human anesthesia. It is with knowledge of network topology that statements can be made concerning the percentage of sodium current suppression required for anesthesia to occur. In addition, the possibility that many other molecular components may contribute to anesthesia by being integrated into the overall effect must be considered. On the other hand, there are molecular entities that are affected at subanesthetic doses that do not produce anesthesia. Since anesthesia is presumably the result of the integration of many different molecular actions of anesthetics, there is no reason to expect that anesthesia and the individual molecular effects all display the same dose–response curve.

However, it might be expected that as far as simple and relatively non-specific anesthetics are concerned, the majority of the molecular structures within the central nervous system that are important in anesthesia show a similar Meyer–Overton correlation to that observed in human anesthesia.

Hypnotic intravenous anesthetics conform to the Meyer–Overton correlation (Fig. 2). Even the effects of the opiates fentanyl, alfentanil and sufentanil correlate in their effects with lipid solubility [13] when compared with each other; however, the correlation is shifted towards much lower concentrations than those shown here (Fig. 7). Lipophilic properties of anesthetics are obviously very important for anesthesia; they are quite non-specific [48], which may explain why so many different molecular structures are affected by anesthetics. These observations suggest that the type of anesthesia that can be produced through the action of a single substance without the addition of any other drug results from non-specific actions on the central nervous system. The more specific the physicochemical properties of an anesthetic substance, the less it is likely to produce clinical anesthesia by itself without the presence of any other drug.

Figure 8 attempts to classify anesthetics on the basis of their physicochemical properties. With regard to the anesthetics shown in Fig. 8, the importance of lipophilic interactions within the overall anesthetic response decreases from left to right, while the proportion of polar interactions increases. A presently still incomplete list of different lipophilic and polar interactions and effects is also shown.

Lipophilic interactions are characterized by van der Waal's interactions and entropic considerations, and they are by their very nature non-specific and non-directed [48]. They influence membrane properties such as thickness, surface tension and fluidity, and through absorption into lipophilic domains of proteins they can alter their conformations. Polar interactions are specific and

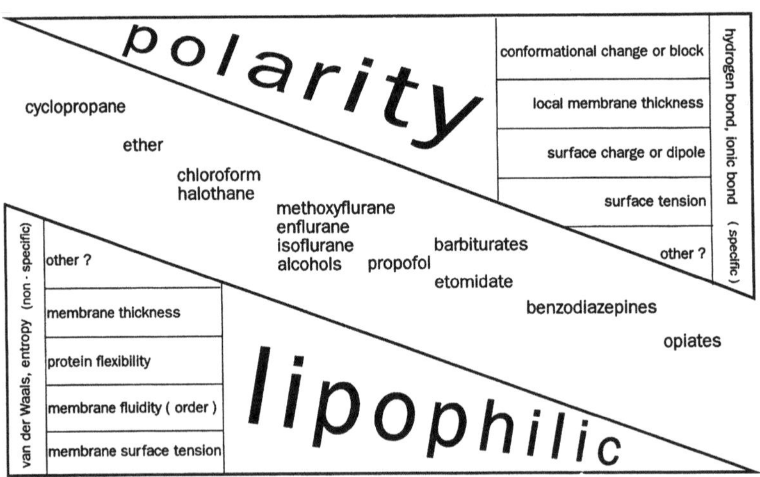

Fig. 8. Attempt at grouping anesthetics according to their physicochemical properties. With regard to the anesthetics shown, the influence of lipophilic interactions within the overall anesthetic suppression decreases from *left* to *right*, while the contribution of polar interactions to the overall effect increases. A presently incomplete list of various lipophilic and polar interactions and effects is also shown

directed; they lead to receptor binding and the formation of hydrogen and ionic bonds. Their influence on membrane properties comprises a change in surface potentials and surface tensions as well as a local thickening of membranes, resulting from receptor binding. It is even conceivable that highly specific substances undergo lipophilic interactions. This mechanism has been postulated for anesthetics with ester groups [21]: the specific binding to a receptor raises the local anesthetic concentration, which in turn leads to a thickening of the lipophilic interior of the membrane (lipophilic effect). Other results of receptor binding could be various conformational changes in membrane molecules. The question has not yet been settled as to whether anesthetics act primarily through interaction with proteins [11, 38] or indirectly by changing lipid bilayer properties [6, 33] or whether both types of molecular target sites are important [49, 50].

Coordinating Measurements on Human Subjects

In order for results collected at various levels of the nervous system to be quantitated and compared, it is important that research – whenever possible – is conducted on comparable species. Experimental progress in the neurosciences has enabled molecular in vitro studies of human cell and tissue material as well as the non-invasive monitoring of human central nervous system activity. Therefore, it becomes meaningful and possible to concentrate anesthesia research more than before on humans. Comparisons between molecular in vitro and systematic in vivo monitoring should be tried in order to gain a better understanding of both areas. Molecular electrophysiology, for example, uses the patch-clamp method on cultured human cells [42] or investigates human brain membrane proteins in lipid bilayers [13], while non-invasive system electrophysiology may be achieved by applying EEG and evoked potential methods to human subjects [37], as is already done in today's neuromonitoring in anesthesiology and intensive care medicine. In future, the scope of monitoring may be broadened. In vitro biochemistry with human brain proteins [43] can be complemented by non-invasive biochemical investigation of the living human brain using nuclear magnetic resonance (NMR) techniques [30]. Molecular in vitro pharmacology of human brain receptors [12] can be compared with an in vivo pharmacology of the human central nervous system using positron emission tomography [46]. In order for these different pieces of information to complement each other, it is important that the various research projects are coordinated, so that comparable conditions can be defined and created. Such coordination requires collaboration between clinicians and basic scientists.

Acknowledgements. Part of this work was supported by NIH grant NS22602. The author wishes to thank Ms. Sabine Schmitz for expert assistance in producing the figures.

References

1. Armstrong CM (1981) Sodium channels and gating currents. Physiol Rev 61:644–683
2. Artusio JF (1955) Ether analgesia during major surgery. J Am Med Assoc 157:33–36
3. Baker PF (1984) The squid axon. Academic, Orlando (Current topics in membranes and transport, vol 22)
4. Bennett HL (1987) Learning and memory in anaesthesia. In: Rosen M, Lunn JN (eds) Consciousness, awareness and pain in general anaesthesia. Butterworths, London, pp 132–139
5. Catterall WA (1988) Structure and function of voltage-sensitive ion channels. Science 242:50–61
6. Elliott JR, Haydon DA (1989) The actions of neutral anaesthetics on ion conductances of nerve membranes. Biochim Biophys Acta 988:257–286
7. Fink BR (1967) Conference on neurophysiology in relation to anesthesiology. Anesthesiology 28:1–200
8. Fink BR (1972) Cellular biology and toxicity of anesthetics. Williams and Wilkins, Baltimore
9. Fink BR (1975) Molecular mechanisms of anesthesia, vol 1. Raven, New York
10. Fink BR (1980) Molecular mechanisms of anesthesia, vol 2. Raven, New York
11. Franks NP, Lieb WR (1987) What is the molecular nature of general anaesthetic target sites. Trends Pharmacol Sci 8:169–174
12. Frenkel C, Duch DS, Urban BW (1990) Molecular actions of pentobarbital isomers on sodium channels from human brain cortex Anesthesiology 72:640–649
13. Frenkel C, Duch DS, Urban BW (1993) Effects of intravenous anesthetics on human brain sodium channels. Br J Anesth 71:15–24
14. Gage PW, Hamill OP (1976) Effects of several inhalation anaesthetics on the kinetics of postsynaptic conductance changes in mouse diaphragm. Br J Pharmacol 57:263–272
15. Haydon DA, Hendry BM, Levinson SR, Requena J (1977) The molecular mechanisms of anaesthesia. Nature 268:356–358
16. Haydon DA, Requena J, Urban BW (1980) Some effects of aliphatic hydrocarbons on the electrical capacity and ionic currents of the squid giant axon membrane. J Physiol (Lond) 309:229–245
17. Haydon DA, Urban BW (1983) The action of hydrocarbons and carbon tetrachloride on the sodium current of the squid giant axon. J Physiol (Lond) 338:435–450
18. Haydon DA, Urban BW (1983) The action of alcohols and other non-ionic surface active substances on the sodium current of the squid giant axon. J Physiol (Lond) 341:411–428
19. Haydon DA, Urban BW (1983) The effects of some inhalation anaesthetics on the sodium current of the squid giant axon. J Physiol (Lond) 341:429–440
20. Haydon DA, Urban BW (1986) The action of hydrophobic, polar, and some inhalation anaesthetic substances on the potassium current of the squid giant axon. J Physiol (Lond) 373:311–327
21. Haydon DA, Elliott JR, Hendry BM, Urban BW (1986) The action of nonionic anesthetic substances on voltage-gated ion conductances in squid giant axons. In: Roth SH, Miller KW (eds) Molecular and cellular mechanisms of anesthetics. Plenum, New York, pp 267–277

22. Haydon DA, Simon AJB (1988) Excitation of the squid giant axon by general anaesthetics. J Physiol (Lond) 402:375–389

23. Hendry BM, Urban BW, Haydon DA (1978) The blockage of the electrical conductance in a pore-containing membrane by the n-alkanes. Biochim Biophys Acta 513:106–116

24. Hille B (1992) Ionic channels of excitable membranes, 2nd edn. Sinauer Associates, Sunderland

25. Hodgkin AL, Huxley AF (1952) A quantitative description of membrane current and its application to conduction and excitation in nerve. J Physiol (Lond) 117:500–544

26. Jones JG (1987) Use of evoked responses in the EEG to measure depth of anaesthesia. In: Rosen M, Lunn JN (eds) Consciousness, awareness and pain in general anaesthesia. Butterworths, London, pp 99–111

27. Kandel ER, Schwartz JH, Jessell TM (1991) Principles of neural science, 3rd edn. Elsevier, New York

28. Levinson SR, Thornhill WB, Duch DS, Recio-Pinto E, Urban BW (1990) The role of nonprotein domains in the function and synthesis of voltage-gated sodium channels. In: Narahashi T (ed) Ion channels, vol 2. Plenum, New York, pp 33–64

29. Llinas RR (1988) The intrinsic electrophysiological properties of mammalian neurons: insights into central nervous system function. Science 242:1654–1664

30. Merboldt KD, Bruhn H, Hanicke W, Michaelis T, Frahm J (1992) Decrease of glucose in the human visual cortex during photic stimulation. Magn Reson Med 25:187–194

31. Meves H (1984) Hodgkin–Huxley: thirty years after. Curr Top Membr Transp 22:279–329

32. Meyer KH (1937) Contributions to the theory of narcosis. Trans Faraday Soc 33:1062–1068

33. Miller KW (1985) The nature of the site of general anaesthesia. Int Rev Neurobiol 27:1–61

34. Miller RD (1990) Anesthesia, 3rd edn. Churchill–Livingstone, New York

35. Narahashi T (1984) Pharmacology of nerve membrane sodium channels. Curr Top Membr Transp 22:483–516

36. Noda M, Numa S (1987) Structure and function of sodium channel. J Recept Res 7:467–497

37. Pichlmayr I, Lips U, Kuenkel H (1984) The electroencephalogram in anesthesia. Fundamentals, practical applications, examples. Springer, Berlin Heidelberg New York

38. Richards CD (1980) The mechanisms of general anaesthesia. In: Norman J, Whitwam JG (eds) Topical reviews in anesthesia. Wright, Bristol, pp 1–84

39. Rosenberg P (1981) The squid giant axon: methods and applications. In: Lahue R (ed) Methods in neurobiology, vol 1. Plenum, New York, pp 1–134

40. Roth SH, Miller KW (1986) Molecular and cellular mechanisms of anesthetics. Plenum, New York

41. Rubin E, Miller KW, Roth SH (1991) Molecular and cellular mechanisms of alcohol and anesthetics. Ann NY Acad Sci 625:1–848

42. Ruppersberg JP, Ruedel R (1988) Differential effects of halothane on adult and juvenile sodium channels in human muscle. Pflugers Arch 412:17–21

43. de Rycker C, Grandfils C, Bettendorff L, Schoffeniels E (1989) Solubilization of sodium channel from human brain. J Neurochem 52:349–353

44. Seeman P (1972) The membrane action of anesthetics and tranquilizers. Pharmacol Rev 24:583–655
45. Stanski DR (1990) Monitoring depth of anesthesia. In: Miller RD (ed) Anesthesia, vol, 1 3rd edn. Churchill–Livinstone, New York, pp 1001–1029
46. Stöcklin G (1992) Applications of positron emission tomography (PET) to the measurement of regional cerebral pharmacokinetics. Anasth Intensivmed Notfallmed Schmerzther 27:84–92
47. Taheri S, Halsey MJ, Liu J, Eger EI, Koblin DD, Laster MJ (1991) What solvent best represents the site of action of inhaled anesthetics in humans, rats, and dogs. Anesth Analg 72:627–634
48. Tanford C (1980) The hydrophobic effect: formation of micelles and biological membranes, 2nd edn. Wiley, New York
49. Urban BW (1985) Modifications of excitable membranes by volatile and gaseous anesthetics. In: Covino BG, Fozzard HA, Rehder K, Strichartz G (eds) Effects of anesthesia. American Physiological Society, Bethesda, pp 13–28
50. Urban BW (1993) Differential effects of gaseous and volatile anaesthetics on sodium and potassium channels. Br J Anaesth 71:25–38
51. White DC (1987) Anesthesia: a privation of the senses. A historical introduction and some definitions. In: Rosen M, Lunn JN (eds) Consciousness, awareness and pain in general anaesthesia. Butterworths, London, pp 1–9

Principles of Central Nervous System Monitoring in Humans

R. -D. Treede

Introduction

Central nervous system (CNS) monitoring now plays an increasingly important role in anesthesiology. As illustrated in Table 1, this is due to two major questions that are of interest intra-operatively: (1) How does anesthesia affect the brain? and (2) How does surgery affect the brain? In addition to the effects of drugs and surgical lesions, those caused by temperature and ischemia are also assessed in CNS monitoring.

The brain is the primary target organ for drugs used in general anesthesia. It is therefore evident that depth of anesthesia should be measured by assessing the functional status of the CNS. For the sedative and hypnotic effects of general anesthetics, this aim has been accomplished using electroencephalographic techniques that are described in detail in the subsequent chapters. An important component of balanced anesthesia techniques is muscle relaxation, which can be monitored using electromyography. Heart rate and blood pressure are still the most commonly used parameters for intraoperative monitoring. The absence of increases in these autonomic nervous system parameters is also thought to reflect adequate analgesia. The increasing use of drugs with a direct effect on the autonomic nervous system has created a need to differentiate between analgesia and sympatholysis. In addition, stability of the autonomic nervous system has been stated as an aim in its own right. The intraoperative monitoring of pain-related evoked potentials (Bromm 1989; Kochs et al. 1990; Bromm and Treede 1991) promises to provide useful information on the adequacy of analgesia. This new approach will be discussed in detail elsewhere in this volume.

Another application of intraoperative CNS monitoring derives from the danger that surgery can pose for the nervous system; examples are spinal surgery or carotid endarterectomy. Here, modern electrophysiological methods can largely replace the wake-up test. Somatosensory, auditory, and visual evoked potentials are used instead of psychophysical methods, i.e., asking the patient which stimuli he can perceive. Instead of having the patient perform voluntary movements, transcranial stimulation of the motor cortex is combined with electromyography to test the motor pathways.

Almost all methods of intraoperative CNS monitoring use noninvasive electro-physiological techniques. In the next few paragraphs, therefore, we will

Table 1. Aims and methods of central nervous system monitoring in the operating room

Aim of monitoring	Depth of anesthesia	Affection of neuronal pathways
Methods used	Electroencephalogram	Somatosensory evoked potential
	Electromyogram	Auditory evoked potential
	Heart rate and blood pressure	Visual evoked potential
	Pain-related evoked potential	Transcranial stimulation of motor cortex

The measured parameters can be affected by drugs, temperature, cerebral ischemia, and surgical lesions.

review some principles of electrophysiology. In contrast to the peripheral nervous system, where action potentials of muscle or nerve fibers are recorded, the slow synaptic potentials are more important for CNS monitoring.

Generator Mechanisms of the Electroencephalogram

For the ongoing spontaneous electroencephalogram (EEG), the classic work by Creutzfeldt and coworkers (for review, see Creutzfeld 1983) has shown that action potentials of cortical neurons do not contribute to the electrical fields recorded on the scalp. In intracellular recordings from cortical pyramidal cells, they found that only the low-amplitude synaptic potential fluctuations, but not the high-amplitude action potentials, correlated with the EEG. This was corroborated by phase-locked superimposition of the EEG and the intracellular recording. The amplitude of the fields generated by subcortical neurons is too small to be measured in the EEG.

Rhythmic EEG activity, therefore, reflects rhythmic excitation of cortical neurons. In this sense, the EEG monitors only cortical activity. The rhythmicity, however, is not generated by the cortical neurons themselves: if a small island of cortex is deafferented by cutting the underlying white matter, its rhythmic activity stops. The rhythmic potential fluctuations in cortical cells are driven by excitatory input from the specific thalamic nuclei. The rhythmic discharge of thalamic neurons can be explained by negative feedback circuits within the thalamus. The duration of the inhibitory postsynaptic potentials (70–150 ms) matches the period of the EEG oscillations. The thalamic generators in turn are influenced by the midbrain reticular formation (inhibition of the inhibitory interneurons). Activation of this midbrain region leads to interruption of the rhythmic EEG oscillations, i.e., a desynchronized EEG (see Pfurtscheller and Aranibar 1977). Thus, although the EEG is a correlate of the electrical activity of cortical neurons, it may indirectly indicate changes in the afferent input to the cortex from the thalamus and from deeper brain regions.

Generator Mechanisms and Scalp Recording
of the Primary Evoked Response

Direct assessment of subcortical electrical activity is possible with evoked potential methods, because the signal-to-noise ratio can be improved by stimulus-locked averaging, such that low-amplitude signals are also detectable. Figure 1 illustrates how the polarity of the signal recorded on the cortical surface is explained by the location of the synapse on the dendritic tree of a pyramidal neuron. Excitatory input from specific thalamic afferents to layer 4 creates a current sink at that layer. Completion of the current loop leads to a current source in the apical dendrites, i.e., a surface-positive, primary evoked potential. Inhibitory input to layer 4 or excitatory input to apical dendrites create the inverse pattern (surface-negative evoked potential). These small local fields are only recordable on the scalp, several centimeters from their source, when many neural elements that are arranged in parallel are excited simultaneously. This condition is fulfilled by the large pyramidal cells in the cortex. The resulting electrical field distribution has been called "open field" (Lorente de Nó 1947). In this case, all dipoles add up and, for a far-field recording, can be considered equivalent to one integral dipole vector, similar to the Einthoven recording of the electrocardiogram (EKG). Closed electrical fields are generated when the individual dipoles show no preferred direction of orientation. This arrangement is typical for many brain stem nuclei, basal ganglia, some thalamic

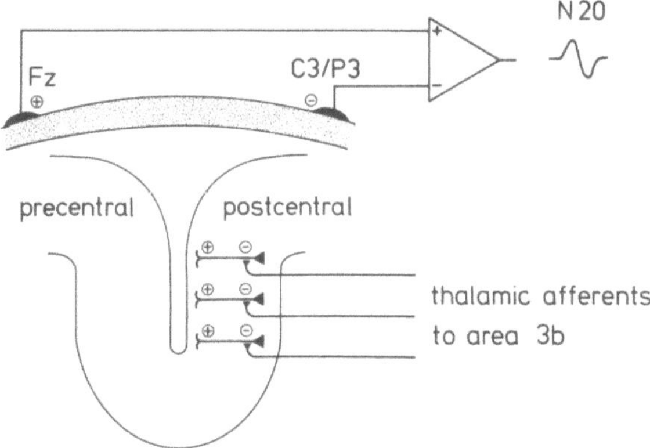

Fig. 1. The primary evoked response from area 3b of the somatosensory cortex. Activation of the pyramidal cells in layer 4 by specific thalamic afferents generates a deep current sink and a corresponding current source at the apical dendrites. The resulting electrical dipole is perpendicular to the cortical surface, but parallel to the surface of the head. The scalp potential N20–P20 is best picked up from a parietal electrode with a frontal reference (modified from Creutzfeld 1983 and Allison et al. 1989). *Fz*, midline frontal area

nuclei, and for the stellate cells in the cortex (Vaughan and Arezzo 1988). The individual dipoles therefore cancel each other, and with electrodes outside the cell ensemble, no potential difference can be recorded. A mixed arrangement with a preponderance of parallel dipoles is typical for cerebellar cortex, hippocampus, and cerebral cortex.

The orientation of the resulting integral dipole relative to the recording electrodes is important for the amplitude of the signal. Due to the folded structure of the cerebral cortex, the cortical surface is perpendicular to the surface of the head in many areas, for example, in area 3b of the primary somatosensory cortex. As a consequence, a primary potential in area 3b will lead to a tangential dipole projected onto the scalp (Fig. 1). The N20 of the somatosensory evoked potential corresponds to the negativity of this primary evoked potential dipole (Allison et al. 1989). The positivity projects to the midline frontal area (Fz). This example illustrates the fact that the optimal reference depends very much on the purpose of the measurements. If, for monitoring purposes, one is interested only in activity originating from area 3b, a derivation between C3 and Fz would be optimal, because it picks up the largest potential difference on the scalp. A potential originating in area 1 near the top of the postcentral gyrus would project its positivity directly to the overlaying scalp (P25). The corresponding negativity would project to the base of the skull. In this case, an extracephalic reference is optimal.

Stationary Fields from Propagated Action Potentials

The propagating action potential creates spatial voltage gradients that are measurable within the volume conductor or – in the case of noninvasive recordings – on its surface. Figure 2 gives a schematic representation of this. The action potential propagates along the axon within the volume conductor. The top part symbolizes unipolar measurements from different sites in the volume conductor. As the action potential propagates, current enters the axon in the region that is being depolarized. This moving current sink leads to a negative voltage drop over the extracellular space, depicted here as an upward deflection. Current leaves the axon mostly in front of the depolarized area, but also behind it, leading to the classic triphasic action potential form, which is recorded with increasing latencies along the course of the axon. This phenomenon forms the basis of peripheral conduction velocity estimates.

A less well known phenomenon is the generation of stationary far-field potentials due to inhomogeneities of the volume conductor. The downward deflection at the center of Fig. 2 is an example of such a stationary component: it is recorded with the same latency from all recording positions. This model consists of an axon that runs through two joined cylinders of different diameters (Stegeman et al. 1987). The different size of the volume conductors on the left and right creates a change in overall conductivity at the junction. When the

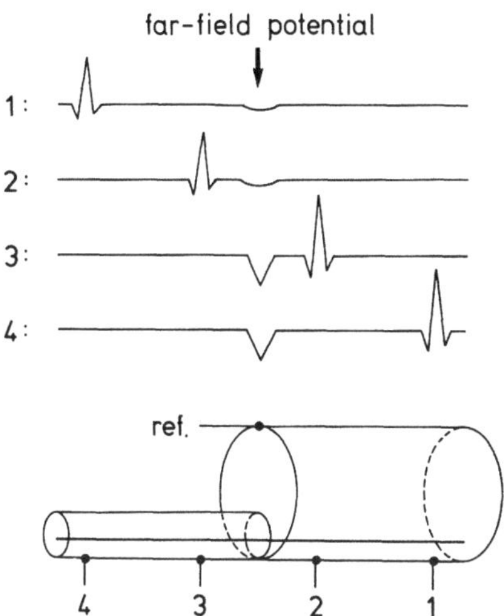

Fig. 2. A far-field potential such as the P9 of the somatosensory evoked potential. The stationary potential is due to a virtual surface charge generated by an action potential that propagates across a change in volume conductor geometry (modified from Stegeman et al. 1987 and Kimura 1989). *ref.*, reference electrode

action potential passes through this conductivity change, it creates a virtual, stationary generator at the boundary between the two volume conductors, which is equivalent to surface charges on that boundary. This virtual, secondary generator gives rise to a far-field component. It creates signals that can be recorded with the same latency independent of the recording location. It should be noted that polarity and location of the nonmoving far-field depend on the choice of the reference site and not on the direction of action potential propagation.

A typical example of such a change in geometry is the shoulder region where the arm joins the larger trunk. The P9 far-field component of the median nerve somatosensory evoked potential is thought to be generated by such a mechanism (Kimura 1989). Changes in the geometry or conductivity of the volume conductor or abrupt changes in direction of action potential propagation may cause a stationary far-field component in the evoked potential. It is conceivable that such a phenomenon may occur in the brain stem and may thus contribute to other early evoked potential components. Whereas most peaks in far-field recordings are from fixed neural activity such as synaptic discharges, some peaks may arise from an advancing front of axonal depolarization such as a junctional or intercompartmental potential (Kimura 1989). Therefore, for CNS monitoring with evoked potentials, action potentials may play some role in addition to synaptic potentials.

Alternate Methods to Assess Neuronal Activity

For reasons outlined above, we have so far focused exclusively on EEG and evoked potentials. Are there other methods for CNS monitoring that can be used either at present or in the future? Table 2 is an attempt at summarizing the

Table 2. Methods to assess neuronal activity

	Single cells	Cell ensembles
Ionic currents	Patch-clamp techniques	Ion-selective electrodes (e.g., K^+),
	Ion-sensitive Dyes (e.g., Ca^{++})	Ion-sensitive dyes (e.g., Ca^{++})
Synaptic potentials	Intracellular recordings, Voltage-sensitive dyes	**Field potentials, EEG, MEG,** voltage-sensitive dyes
Action potentials	**Extracellular recordings**	**ENG, EMG**
Metabolism	Autoradiography	**PET, SPECT, Phosphorus MRI**

EEG, electroencephalogram; MEG, magnetoencephalogram; ENG, electroneurogram; EMG, electromyogram; PET, positron emission tomography; SPECT, single photon emission computed tomography; MRI, magnetic resonance imaging.

techniques that are used on the single cell level and for cell groups, in both humans and in animals.

The ionic currents underlying nervous activity are nowadays characterized in detail by using patch-clamp techniques of single channels. This approach is obviously limited to individual cells or membrane patches. Indirect assessment of currents is possible by monitoring the concentration of ions. A well-known example is the measurement of extracellular potassium concentrations by using ion-selective microelectrodes. Calcium-sensitive dyes are widely used for individual cells and for cell ensembles as well. None of these approaches has been useful for CNS monitoring in humans so far. There is a possibility that the optical recording techniques might eventually become available.

Intracellular recordings have been used to characterize postsynaptic potentials. Evidently, they are not useful for human studies, but voltage-sensitive dyes may eventually be used for this purpose, especially since they are also useful for studying cell ensembles (Grinvald et al. 1982; Orbach et al. 1985). Voltage-sensitive dyes are also fast enough for action potential recording. As we discussed above, synaptic potentials are the basis for the field potentials that are recorded as EEG and evoked potentials. The magnetic fields that accompany intra- and extracellular currents are measured in neurology as the magneto-encephalogram (for a recent review see Williamson and Kaufman 1990). Current technology does not allow these devices to be used in the electrically noisy operating room environment. We should, however, keep in mind the fact that originally EEG recordings were only possible inside a Faraday cage.

Action potentials of single cells are large enough to be picked up with extra-cellular microelectrodes. Such recordings have been done in humans, especially in the peripheral nerve and in the thalamus (Lenz et al. 1988; Vallbo et al. 1979), but they are not used for monitoring because of their invasive nature. Noninvasive action potential recording is the basis of conventional electroneurography and electromyography studies.

In addition to these electrical signals, neuronal activity causes metabolic changes. These have been studied in great detail with deoxyglucose and autoradiographic techniques. Positron emission tomography and single photon emission computed tomography have made it possible to map metabolic activity noninvasively in humans. Again, as with magnetoencephalography, it is not yet technically feasible to do these measurements intra-operatively. However, aortic blood flow has already been assessed with a single detector pair. Last but not least, magnetic resonance imaging can also yield information on the metabolic state of tissue.

In summary, we have a core repertoire of electrophysiological methods that are used for intraoperative CNS monitoring (EEG, evoked potentials, ENG, electromyography). This is the topic of several other chapters in this volume. In addition, there is a fringe of methods that are already used in humans, but not yet for monitoring purposes (the methods printed in boldface in Table 2); finally, there are a number of techniques that have not been used in humans so far. One message from this list may be that whenever there is question left open – a parameter that we think is useful to monitor but that cannot be assessed yet – we should keep our minds open to look beyond electrophysiology.

Acknowledgements. The author appreciates critical reading of the manuscript by Dr. E. Scharein and the technical assistance of G. Günther and B. Müller.

References

Allison T, McCarthy G, Wood CC, Darcey TM, Spencer DD, Williamson PD (1989) Human cortical potentials evoked by stimulation of the median nerve. I. Cytoarchitectonic areas generating short-latency activity. J Neurophysiol 62:694–710

Bromm B (1989) Laboratory animal and human volunteer in the assessment of analgetic efficacy. In: Chapman CR, Loeser JD (eds) Issues in pain measurement. Raven, New York, pp 117–143

Bromm B, Treede RD (1991) Laser evoked cerebral potentials in the assessment of cutaneous pain sensitivity in normal subjects and in patients. Rev Neurol (Paris) 147:625–643

Creutzfeldt OD (1983) Cortex cerebri. Springer, Berlin Heidelberg New York

Grinvald A, Manker A, Segal M (1982) Visualization of the spread of electrical activity in rat hippocampal slices by voltage-sensitive optical probes. J Physiol (Lond) 333:269–291

Kimura J (1989) Electrodiagnosis in diseases of nerve and muscle: principles and practice. Davis, Philadelphia

Kochs E, Treede RD, Schulte am Esch J, Bromm B (1990) Modulation of pain-related somatosensory evoked potentials by general anesthesia. Anesth Analg 71:225–230

Lenz FA, Dostrovsky JO, Kwan HC, Tasker RR, Yamashiro K, Murphy JT (1988) Methods for microstimulation and recording of single neurons and evoked potentials in the human central nervous system. J Neurosurg 68:630–634

Lorente de Nó R (1947) Action potential of the moto-neurons of the hypoglossus nucleus. J Cell Comp Physiol 29:207–287
Orbach HS, Cohen LB, Grinvald A (1985) Optical mapping of electrical activity in rat somatosensory and visual cortex. J Neurosci 5:1886–1895
Pfurtscheller G, Aranibar A (1977) Event-related cortical desynchronization detected by power measurements of scalp EEG. Electroencephalogr Clin Neurophysiol 42:817–826
Stegeman DF, Van Oosterom A, Colon EJ (1987) Far-field evoked potential components induced by a propagating generator: computational evidence. Electroencephalogr Clin Neurophysiol 67:176–187
Vallbo AB, Hagbarth KE, Torebjörk HE, Wallin BG (1979) Somatosensory, proprioceptive, and sympathetic activity in human peripheral nerves. Physiol Rev 59:919–957
Vaughan HG, Arezzo JC (1988) The neural basis of event-related potentials. In: Picton TW (ed) Human event-related potentials. Elsevier, Amsterdam, pp 45–96 (Handbook of electroencephalography and clinical neurophysiology, vol 3)
Williamson SJ, Kaufman L (1990) Evolution of neuromagnetic topographic mapping. Brain Topogr 3:113–127

Part II

Techniques and Applications of Electroencephalogram – Monitoring

Pharmacokinetic and Pharmacodynamic Interactions Relevant to Cerebral Monitoring

H. J. M. Lemmens

Introduction

Many drugs acting on the central nervous system produce characteristic changes on the electroencephalogram (EEG). As early as the third decade of this century, Berger described in several papers that morphine, cocaine, barbiturates, and scopolamine altered the EEG [1, 2]. In the last 10 years, drug-induced EEG changes have been described in detail for many intravenous drugs used in anesthesia. It has now generally been established that drugs acting on the central nervous system cause EEG changes that are specific, continuous, measurable, objective, sensitive, and reproducible. These properties made the EEG a powerful tool to measure drug effects. However, the correlation between EEG drug effects and clinical drug effects such as analgesia, hypnosis, and sedation has not yet been unraveled and remains to be definitively established. Therefore, the EEG must be considered as a "surrogate" effect measurement. A surrogate measure of effect can be defined as a secondary, drug-induced, measurable change in body physiology.

Electroencephalogram Drug Effects

In 1984, Stanski [3] described the characteristic EEG changes caused by thiopental administered as a constant infusion over several minutes (Fig. 1). He observed four different phases. In phase 1, activation of the EEG occurred, characterized by fast-frequency, high-amplitude activity. This was followed in phase 2 by an increase in slow wave activity. At this stage, loss of consciousness occurs. A further increase in dose causes a reduction of activity in all frequency ranges until burst suppression occurs at plasma concentrations above 40 μg/ml. Etomidate and propofol cause basically the same EEG changes as thiopental. In 1987, Schutler described the EEG changes caused by ketamine. Compared to the baseline, he observed first a loss of alpha rhythm. With increasing dose, this was followed by a predominantly theta rhythm. The maximal EEG effect was characterized by an intermittent delta rhythm. Benzodiazepines cause an increase in beta activity and a decrease in alpha activity. Figure 2 shows an

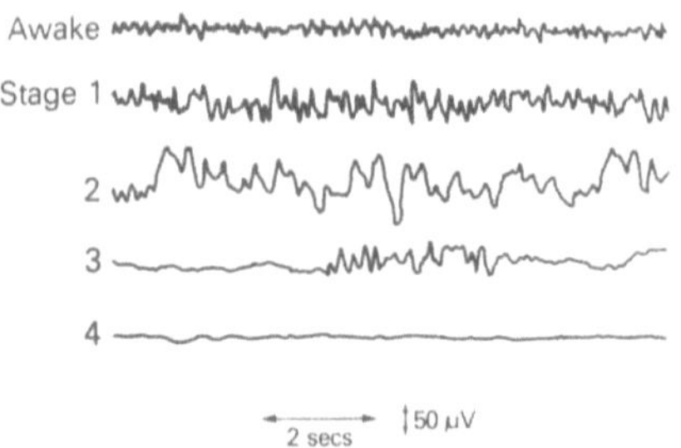

Fig. 1. Thiopental electroencephalogram effects

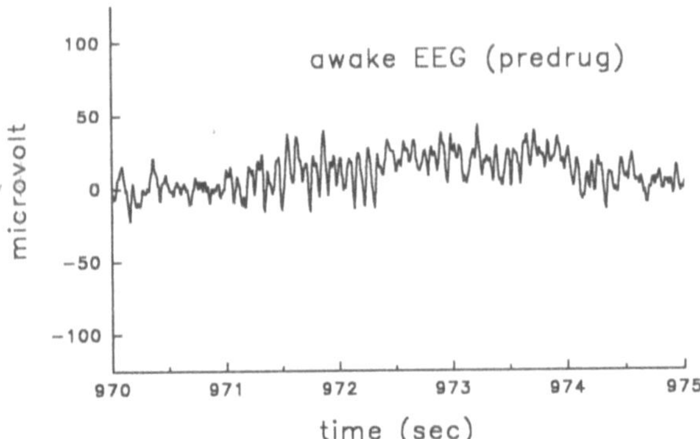

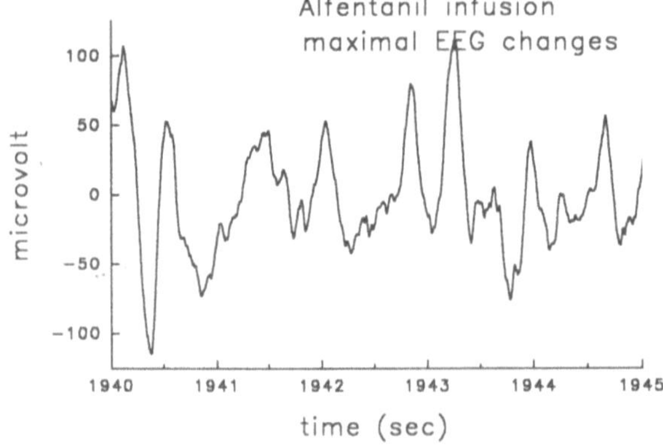

Fig. 2. Alfentanil electroencephalogram effects

example of the EEG effects of the pure mu opioid agonist alfentanil. Compared to the baseline, we see an increase in amplitude and a decrease in frequency. These changes are identical to those caused by fentanyl and sufentanil. The EEG changes caused by butorphanol, a mixed agonist–antagonist opiate are

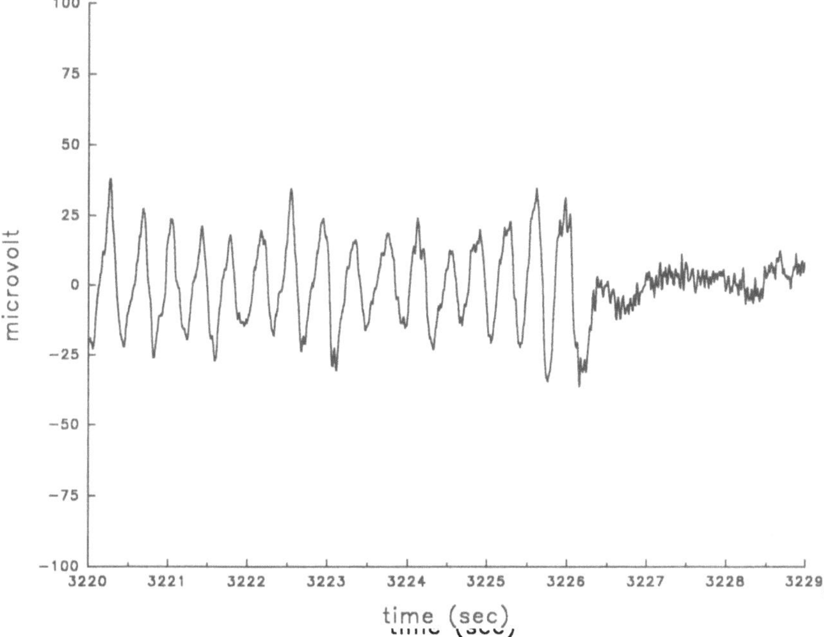

Fig. 3. Butorphanol electroencephalogram effects

characterized by intermittent periods of slowing in frequency and an increase in amplitude (Fig. 3).

Quantitation of Electroencephalogram Drug Effect and Relation to Plasma Concentration

To quantitate the drug-induced EEG changes of intravenous anesthetics, two methods are commonly used, Fourier analysis and aperiodic analysis.

In Fourier analysis, the EEG is considered to be represented by the summation of sine and cosine waves of different frequencies, and amplitudes. Parameters such as spectral edge frequency, median frequency, and total power can be computed. Aperiodic analysis determines simply the amplitude and frequency of each wave. Parameters such as number of waves and amplitude can be derived. The quantitation of drug-induced EEG changes makes it possible to determine the relationship to the plasma concentration. Figure 4 shows plasma concentration and EEG response of the investigational analgesic drug A-3665. (Note the lag or hysteresis between changes in drug plasma concentration and changes in EEG effect.) The time delay between changes in plasma concentration and EEG effect are shown in Fig. 5 in a different way. The consecutive effect

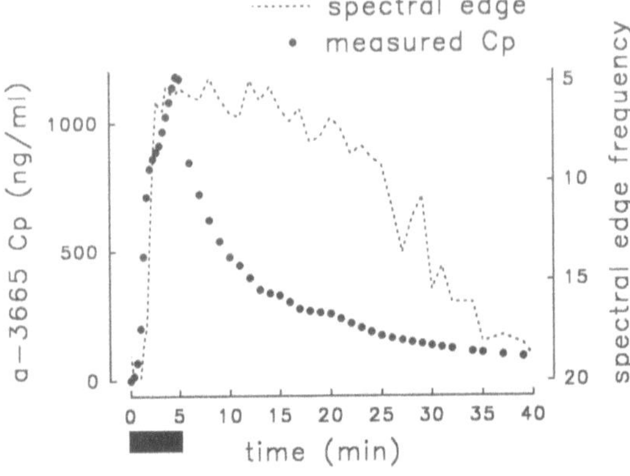

Fig. 4. Plasma concentration and electroencephalogram (EEG) drug effect of the new investigational opioid A-3665. The *x axis* shows the time; the drug infusion time is indicated by the *solid bar*. The *left y axis* shows the plasma concentration of A-3665. The *right y-axis* shows the EEG effect, in this case the spectral edge parameter derived from fourier analysis. *Dashed line*, spectral edge; *dots*, measured CP

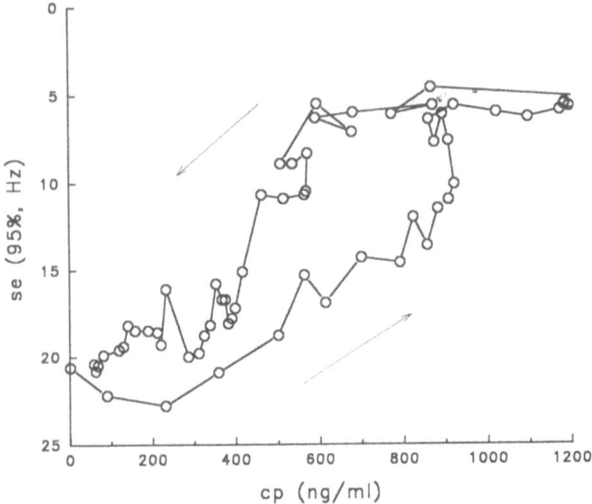

Fig. 5. The *x axis* shows the plasma concentration, and the *y axis* shows the electroencephalogram effect of A-3665. Note the counterclockwise hysteresis *loop* starting in the *lower left corner*

measurements are plotted against plasma concentration. The counterclockwise hysteresis loop starting in the lower left corner of Fig. 5 can be clearly seen. The occurrence of hysteresis makes it impossible to fit the plasma concentration–effect data. Therefore, we have to remove the hysteresis first. This can done thus by assuming that there must be an effect site compartment that has to equilibrate with the plasma or biological fluid concentration. The equilibration between plasma and effect site concentration is characterized by a first-order rate constant, Ke0, which is the rate constant for elimination of drug from the effect compartment [5]. We find the optimal Ke0 by minimizing the area in the

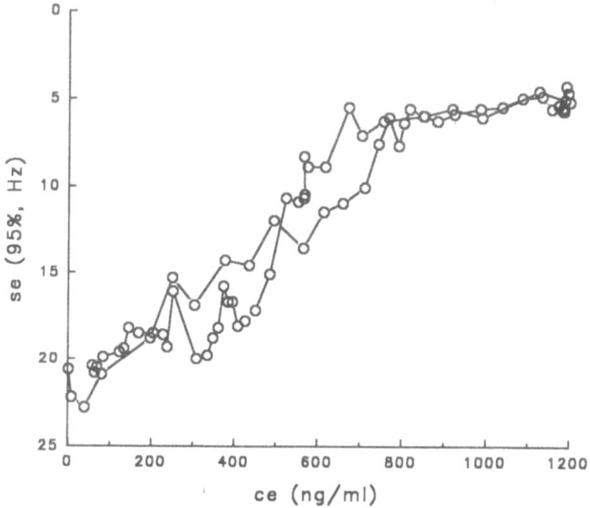

Fig. 6. Collapsed hysteresis loop

hysteresis loop of Fig. 5. Figure 6 shows the collapsed hysteresis loop based on the Ke0 value that minimizes the area in the hysteresis loop. The x axis shows the calculated effect site concentration, and the y axis shows the effect. Once we have removed hysteresis, we can use the effect site concentration data to fit pharmacodynamic models and to derive parameters such as potency and maximal effect. This concept has been used for the pharmacodynamic modeling of many intravenous anesthetics in man [3, 4, 6–8].

Electroencephalogram Effect and Benzodiapine Receptor Complex

It was recently demonstrated by Mandema [9] that the EEG effects we measure really reflect interaction of the drug with the receptor complex. He determined the relation in rats between concentration and EEG effects for four benzodiazepines, flunitrazepam, midazolam, oxazepam, and clobazam. Figure 7 shows the relationship between plasma concentration and EEG effect for these drugs. No hysteresis was observed between plasma concentration and EEG effect. Therefore, concentrations could be directly related to the EEG effect. The solid line is the fitted function to a sigmoid E_{max} model for each drug. Table 1 shows the calculated pharmacodynamic parameters for each benzodiazepine. The receptor affinity K_i was measured by displacement of labeled flumazanil. The ratio of unbound CP_{50} to receptor affinity is more or less constant for these four drugs. This indicates that potency of benzodiazepines as measured by the EEG is an appropriate pharmacodynamic measure of benzodiazepine drug effect.

　　　　　　　　　　　　　　　　　　　　　　　　　　　　H. J. M. Lemmens

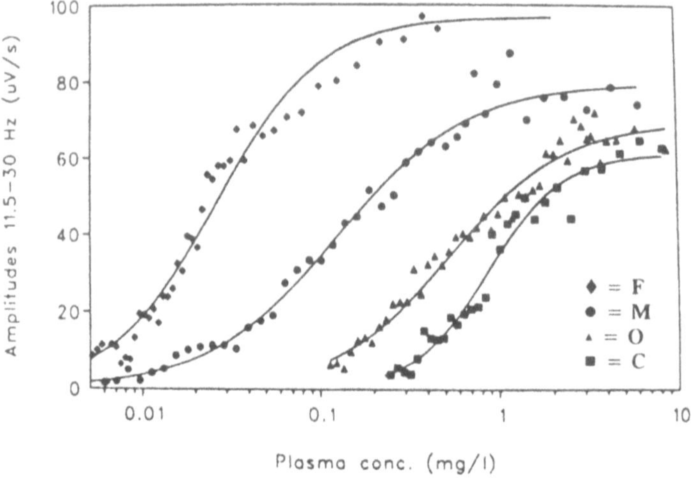

Fig. 7. Relationship between concentration and electroencephalogram in rats for flunitrazepam (*F*), midazolam (*M*), oxazepam (*O*), and clobazam (*C*) (adapted from [9])

Table 1. Comparison of electroencephalogram activity and benzodiazepine receptor affinity

	E_{max} (μV/s)	Total drug CP_{50}	Unbound CP_{50}	k_i ng/ml	Ratio $\dfrac{CP_{50}}{\bar{k}_i}$
Flunitrazepam	90 ± 5	26 ± 3	4.2 ± 0.7	7.0 ± 0.8	0.60
Midazolam	73 ± 2	105 ± 10	3.7 ± 0.5	4.9 ± 0.5	0.76
Oxazepam	65 ± 6	559 ± 37	49 ± 4	86 ± 15	0.57
Clobazam	63 ± 8	859 ± 98	277 ± 34	350 ± 61	0.79

Mean $\pm$ SE Adapted from [9].

The Application of the Electroencephalogram in Clinical Pharmacology

The fact that different classes of central nervous-acting drugs produce characteristic EEG changes and that adequate pharmacodynamic and pharmacokinetic modeling concepts have been developed makes it possible to examine the comparative clinical pharmacology of drugs and the effect of variables such as ageing and disease. The importance and clinical relevance of this conceptual approach (precise pharmacokinetic and pharmacodynamic modeling using the EEG as a measure of effect) has been demonstrated by Buhrer [10]. He compared the pharmacology of the benzodiazepines midazolam and diazepam.

He found that the plasma effect site equilibration half-life of diazepam was 1.6 min, more than three times faster than that of midazolam (5.4 min). He also demonstrated that midazolam is four times more potent than diazepam. Before this study was done, when midazolam was introduced on the market in America in 1987, midazolam was considered to be only twice as potent as diazepam and to have a more rapid onset of action than diazepam. Therefore, it was in retrospect no surprise that soon after the release of midazolam, more than 70 cases of significant overdosage with major clinical consequences were reported. This illustrates how incomplete clinical pharmacological characterization of a drug with traditional studies can result in significant patient morbidity and mortality.

Presently, pharmacokinetic and pharmacodynamic modeling concepts are being more and more systematically applied in the development and regulatory approval of new drugs. This will result in more scientific, efficient, and cost-effective drug development.

References

1. Berger H (1931) Über das Elektroenkephalogramm des Menschen.III. Mitteilung. Arch Psychiatr Nervenkr 94:16–60
2. Berger H (1933) Über das Elektroenkephalogramm des Menschen. VIII. Mitteilung. Arch Psychiatr Nervenkr 101:453–469
3. Stanski DR, Hudson RJ, Homer TD, Saidman LJ, Math, E (1984) Pharmacodynamic modeling of thiopental anesthesia. J Pharmacokinet Biopharm 12:223–240
4. Schuttler J, Stanski DR, White PF, Trevor AJ, Horai Y, Verotta D, Sheiner LB (1987) Pharmacodynamic modeling of the EEG effects of ketamine and its enantiomers in man. J Pharmacokinet Biopharm 15:241–253
5. Verotta D, Sheiner LB (1987) Simultaneous modeling of pharmacokinetics and pharmacodynamics:an improved algorithm. CABIOS 3:345–349
6. Scott JC, Ponganis KV, Stanski DR (1985) EEG quantitation of narcotic effect:the comparative pharmacodynamics of fentanyl and alfentanil Anesthesiology 62:234–241
7. Schwilden HJ, Schuttler J, Stoeckel H (1985) Quantitation of EEG and pharmacodynamic modeling of hypnotic drugs etomidate as an example. Eur J Anaesthesiol 2:121
8. Scott JC, Cooke JE, Stanski DR (1991) Electroencephalographic quantitation of opioid effect:comparative pharmacodynamics of fentanyl and sufentanil. Anesthesiology 74:34–42
9. Mandema JW (1991) EEG effect measures and relationships between pharmacokinetics and pharmacodynamics of psychotropic drugs. Thesis, State University Leiden, the Netherlands
10. Buhrer M, Maitre P, Crevoisier C, Stanski DR (1990) Electroencephalographic effects of benzodiazepines.II. Pharmacodynamic modeling of the electroencephalographic effects of midazolam and diazepam. Clin Pharmacol Ther 48:544–554

Electroencephalographic Feedback Control of Anesthetic Drug Administration

H. Schwilden

Introduction

The relationship between drug dosing and the induced time course of effect is of major interest for the anesthetist when aiming at inducing and maintaining an anesthetic state. Especially when the patient is treated with neuromuscular blocking agents, there are only very limited apparent clinical signs for detecting awareness. A reliable estimation and prediction of drug dosing to prevent intra-operative awareness is therefore required.

Gibbs et al. [13] were presumably the first to recognize the possible value of the electroencephalogram (EEG) for estimating depth of anesthesia. In 1937, in their article "Effect on the electro-encephalogram of certain drugs which influence nervous activity", they wrote, "A practical application of these observations might be the use of the electro-encephalogram as a measure of the depth of anesthesia during surgical operations. The anesthetist and surgeon could have before them on tape or screen a continuous record of the electrical activity of both heart and brain".

The main reasoning was thus that the EEG gives information about the functional state of the brain. Later on, Brazier and Finesinger [8] demonstrated the dose dependence of drug-induced EEG changes. In 1950, Bickford [6] described an apparatus for the automatic electroencephalographic control of ether anesthesia, which used the integrated EEG power (being equivalent to mean EEG amplitude squared) as a pharmacodynamic signal. Later investigations by Bickford [7] and Bellville et al. [4, 5] took into consideration the change in the EEG frequency distribution.

In more recent years, the pharmaco-EEG [15] has been established as a subspecialty to study the electrically recordable changes in cerebral functional states and behaviour. Fink stated the underlying hypothesis [11] as follows: "1. EEG changes are directly related to the biomedical changes each compound induces in the brain, and 2. the behavioral effects are directly related to the biomedical effects." This approach uses diverse techniques of EEG processing to identify those EEG characteristics indicating certain behavioural effects. In general, the method uses a plethora of EEG parameters derived from the raw EEG. For the practice of anesthesia, especially for the purpose of monitoring depth of anesthesia, it would be desirable to have only a few EEG parameters to be monitored, to prevent the anesthetist being overburdened with too much data.

This paper examines the use of median EEG frequency for the feedback control of i.v. anesthetic drug delivery.

Spectral Electroencephalogram derivations

The Fourier transformation transforms a given EEG epoch into a sum of sinus waves of given frequency which are characterized by their corresponding amplitudes and phases. In most cases, the phases behave like white noise and thus may not be considered. Because in general the phases do not give any significant additional information, one most often restricts the quantitation of an EEG to the power spectrum. More recent methods of so-called chaos analysis open, however, the possibility that there might be a certain degree of determinism in the EEG time series. Grassberger and Procaccia [14] have developed theoretical and computational tools to examine deterministic chaotic behavior in terms of so-called correlation dimensions. Most recently, direct methods for the assessment of the degree of determinism have been developed [26], which, however, have so far not been applied to EEG analysis. The power spectrum is nothing but a distribution; thus, the derivation of spectral EEG parameters is equivalent to the derivation of descriptors of a distribution. It should, however, be mentioned here that the estimation of the power spectrum is not a unique procedure. The transformation from the time domain into the frequency domain requires a signal extending from $-\infty$ to $+\infty$ in time. Because an EEG epoch is finite in time, methods to extrapolate the signal to the future and into the past are required. Fourier transformation assumes that the epoch under consideration is repeated indefinitely to $\pm\infty$, while other methods of power spectrum estimation, such as maximum entropy analysis [49], assume other methods of extrapolation. For the purpose of EEG monitoring during anesthesia, these differences in estimating the power spectrum are of minor importance if the EEG epoch is long enough. Figure 1 gives scheme of an EEG power spectrum and

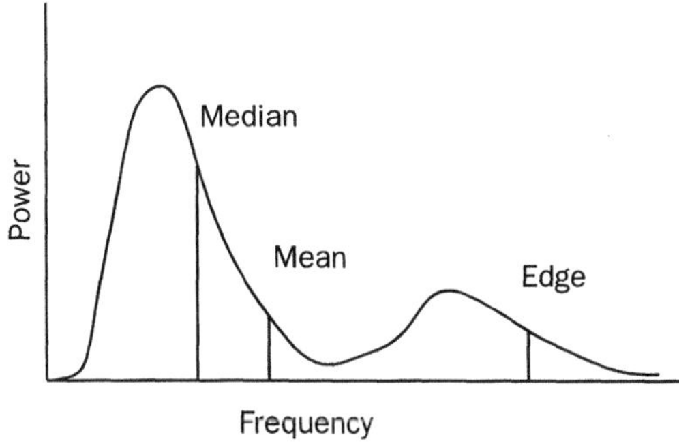

Fig. 1. Electroencephalogram (EEG) power spectrum regarded as a distribution and three parameters which have been used to describe EEG frequency shifts during anesthesia

the three monoparametric descriptors of distribution which were used. The simplest one is mean EEG frequency, being defined as the mean of the power spectrum regarded as a distribution. As is known from ordinary statistics, the mean can give inappropriate measures of the centre in the case of skew symmetric distributions and is in addition rather sensitive to outliers. Another descriptor which has been used is edge frequency [25], defined as that frequency below which 95% of total power (95% quantile) is located. The 50% quantile, i.e. median EEG frequency [32], is generally recommended as the most stable quantile with respect to outliers for the description of the centre of a distribution.

Median Electroencephalogram Frequency

Median EEG frequency has been studied in various groups of surgical patients and volunteers [30, 34, 36–40, 42, 45, 46] treated with different agents and drug combinations. In a group of 60 patients who were treated with a variety of drugs and drug combination, the dosing of which was left to the discretion of the anesthetist, we observed a distribution of median EEG frequency which peaked at 2.5 Hz [46]. The 75% quantile of all observed values was below 5 Hz.

The relationship between median EEG frequency and clinical signs has been studied in volunteers treated with intravenous hypnotics such as etomidate [29, 39], methohexitone and propofol [30]. The methodological approach was to generate linearly increasing drug concentration, the slope of which was chosen such that the onset of drug action and diverse clinical effects could be followed with reasonable precision. Such a method, which has already been described else where [28, 29, 33, 34], requires a computer-controlled infusion for the calculation of the infusion scheme as well as transmission of the steadily changing infusion rates to the pump. The linear increase of blood concentrations was maintained as long as burst suppression pattern occurred in the EEG trace. At this point, the infusion was stopped and the recovery phase observed. After having regained consciousness with respect to time and location, the infusion was restarted. This cycle was performed three times for each volunteer. Figure 2 depicts a typical time course of median EEG frequency, edge frequency and mean amplitude for a volunteer treated with propofol. During onset and offset of drug action, the following clinical events and associated values of spectral EEG derivations were recorded: baseline (A), falling asleep (ES), loss of eyelash reflex (øL), loss of corneal reflex (øC) and occurrence of burst suppression (BS). During recovery, the following signs were documented: disappearance of burst suppression (øBS), recurrence of corneal reflex (+ C), recurrence of eyelash reflex (+ L) reaction to verbal commands (+ R) early orientation (FO) and full orientation (VO).

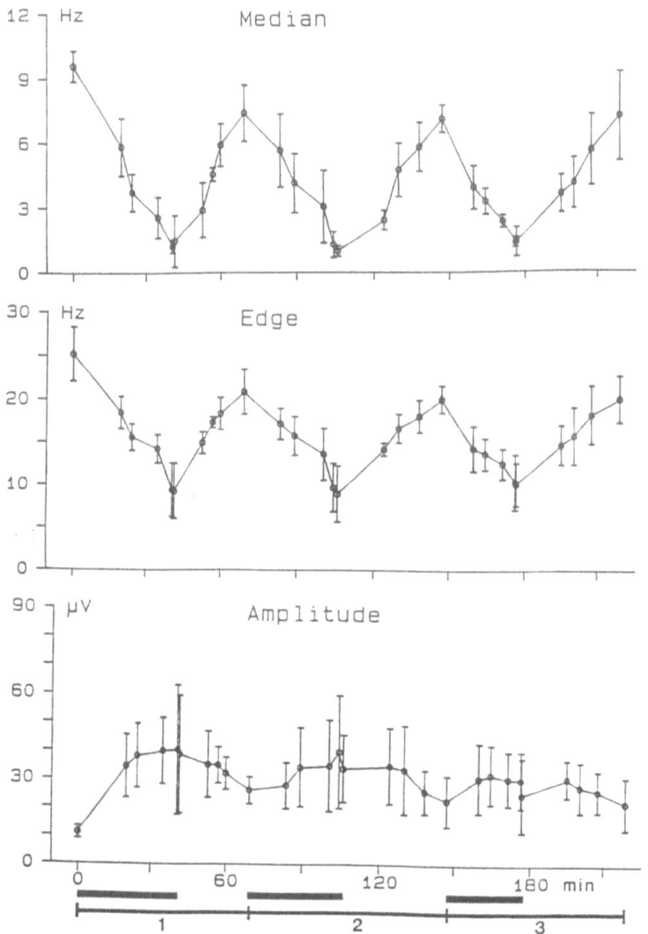

Fig. 2. Mean ± SD of median electroencephalogram (EEG) frequency, edge frequency and mean amplitudes at defined clinical endpoints (see text) in a group of five volunteers during a triple slope infusion of methohexitone

Figure 3 depicts two cumulative distributions as a function of median EEG frequency. The right curve gives the empirical probability derived from the above-mentioned volunteer studies for the occurrence of signs of undue light planes of anesthesia as a function of median EEG frequency. Around a value of 6 Hz, the probability is approximately 50%. The curve on the left gives the empirical probability of occurrence or disappearance of burst suppression pattern as a function of median EEG frequency. If one allows the optimal range of median EEG frequency to be defined as having probabilities of 5% or less for the occurrence of signs of undue anesthesia, Fig. 3 defines the interval between 2 Hz and 4 Hz as the optimal range.

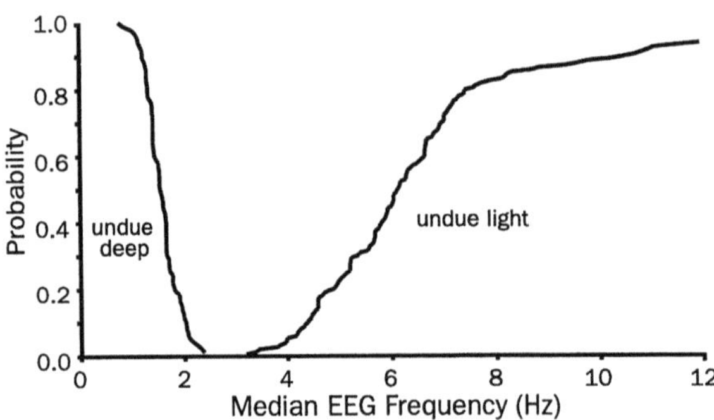

Fig. 3. Empirical probability of the occurrence of clinical signs which were considered as an indication of too light an anesthesia (*right*) and too deep an anesthesia (*left*) as a function of median electroencephalogram (EEG) frequency

Median Electroencephalogram Frequency as a Function of Drug Delivery – Theoretical Considerations

The classic dose–response relationship (in vivo) relating a given dose to the effect at a certain moment in time and the static concentration–response relationship of isolated organs or tissues in a bath of defined drug concentration (in vitro) have been generalized by the development of pharmacokinetics, allowing us to measure simultaneously drug concentration in the blood and effect in humans. The pair pharmacokinetics–pharmacodynamics [16, 17, 44] summarizes our present understanding of the dose–effect relationship as a function of time. Formalizing the aspects of pharmacokinetics and pharmacodynamics in mathematical terms [35] in order to establish quantitative models of both, pharmacokinetics has to be regarded as a functional K mapping time courses of drug input functions $I(t')$ to time courses of concentrations $c(t)$. Correspondingly, pharmacodynamics has to be regarded as a functional D mapping time occurs of concentrations $c(t')$ to time courses of effects $E(t)$:

$$c(t) = K[I(t')] \tag{1}$$

$$E(t) = D[c(t')] \tag{2}$$

In general, the relationship between drug dosing $I(t)$ and effect $E(t)$ allows no further specification of the functions K and D without additional assumptions. A vast literature on the pharmacokinetics of anesthetic agents has proven, however, that in nearly all cases pharmacokinetics can be assumed to be linear, at least if the concentrations are in the therapeutic range. Hence, the entire non-linearity of the dose–effect relationship is focused in the pharmacodynamic functional D. Moreover, it is commonly accepted that the concentration at the site of drug action is immediately related to effect. This may, not apply to situations where receptor association and receptor dissociation form a slow process (in the order of some 10 s to hours).

Under this condition, the pharmacokinetic functional K has the form

$$c(t) = \int_{-\infty}^{t} dt'\, G(t, t')\, I(t') \tag{3}$$

If one assumes in addition that the state variables of the pharmacokinetic model are independent of time, one can prove that the function $G(t, t')$ depends only on the difference $t - t'$:

$$G(t, t') = G(t - t') \tag{4}$$

and Eq. 3 thus reads

$$c(t) = \int_{0}^{t} dt'\, G(t - t')\, I(t') \tag{5}$$

Equation 5 is the mathematical formulation of the superposition principal of linear pharmacokinetics. $G(t)$ is nothing but the drug concentration after a bolus of unit dose, which most often is modelled by bi- or tri-exponential function:

$$G(t) = A e^{-\alpha t} + B e^{-\beta t} \tag{6}$$

Such functions are often interpreted in terms of compartment models. On the condition that changes in concentration at the site of drug action immediately cause changes in effect, one can regard the functional D as a simple function $D(c)$:

$$E = D(c) \tag{7}$$

It can be shown that under fairly general conditions the formula

$$E = E_{\max} \frac{c^{\gamma}}{c_0^{\gamma} + c^{\gamma}} \tag{8}$$

represents a reasonable approach to model the concentration effect relationship whereby $E_{\max}$, c_0 and γ are fixed constants. Equation 8 describes a concentration–effect relationship which has zero effect at zero concentration, an effect $E_{\max}/2$ at $c = c_0$ and maximum concentration at $c \gg c_0$.

As median EEG frequency generally decreases as anesthetic drug concentration increases, one has to modify Eq. 8 slightly to

$$E = E_0 - E_1 \frac{c^{\gamma}}{c_0^{\gamma} + c^{\gamma}} \tag{9}$$

where E_0 denotes baseline median EEG frequency and the difference $E_0 - E_1$ is the smallest median EEG frequency achievable.

Measuring drug concentration and effect simultaneously the parameters E_0, E_1, c_0 and γ can be determined by least square fitting. Figure 4 shows for a volunteer treated with a triple slope infusion of etomidate the measured

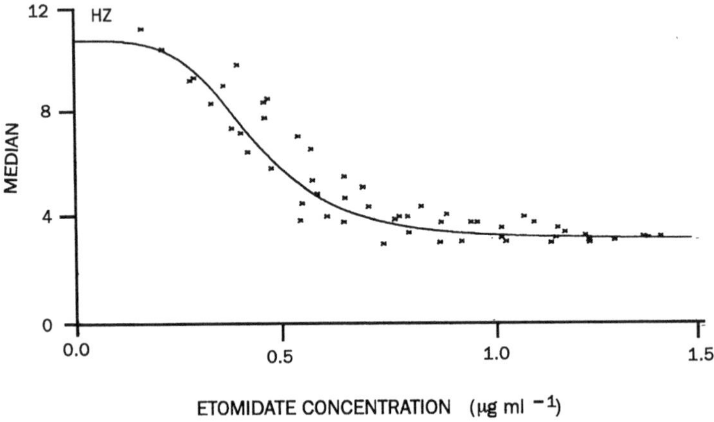

Fig. 4. Concentration–response curve for etomidate. The *solid line* represents the best fit to the measured data

concentration–effect relationship and the fitted curve. In a group of six volunteers, the following values for E_0, E_1, c_0 and γ were found:

$$E_0,\ 9.3 \pm 0.3\ \text{Hz};\ E_1,\ 7.9 \pm 0.5\ \text{Hz},\ c_0,\ 0.31 \pm 0.08\ \text{mg/l};\ \gamma:\ 3.3 \pm 2.1.$$

Input–Output Modelling

A full characterization of the pharmacokinetic and pharmacodynamic relationship requires three pieces of information: the drug input function, the time course of drug concentration and the time course of effect. The relationship between drug dosing and concentration determines the pharmacokinetic model, and the relationship between drug concentration and effect determines the pharmacodynamic model. A remarkable property of Eq. 8 is its invariance under scale transformation of the type $(c, c_0) \rightarrow (\lambda c, \lambda c_0)$:

$$E(c, c_0) = E(\lambda c, \lambda c_0) \tag{10}$$

This invariance has immediate practical consequences for the derivation of the pharmacodynamic model parameters. It is obviously not necessary to measure the concentration c and c_0 at the site of drug action, but instead any other site of drug concentration measurement, e.g. blood or plasma, can be used, as long as the site of drug action and the site of concentration measurement are in equilibrium.

Going one step further, one may choose a non-dimensionless scaling factor such as volume of distribution or clearance. In so doing, the concentrations in Eq. 8 are transformed into the corresponding amounts of drug (M, M_0)

or steady state infusion rates (I_{ss}, I_{ss0}). The corresponding formula is then given by

$$E = E_{max}\frac{I_{ss}^{\gamma}}{I_{ss0}^{\gamma} + I_{ss}^{\gamma}} \tag{11}$$

If c_i and E_i denote measured concentrations and effect at time t_i, the pharmacokinetic dynamic modelling procedure regards these measured quantities as a function of the model parameters A, α, B, β, $\ldots$, c_0, γ:

$$c_i = c(t_i, A, \alpha, B, \beta, \ldots; I(t)) + \varepsilon_i \tag{12}$$

$$E_i = E(c_i, c_0, \gamma, \ldots) + \eta_i \tag{13}$$

whereby ε_i and η_i are considered as random variables. Least square fitting procedures identify that set of parameters A, α, B, β, $\ldots$ and c_0, γ, $\ldots$ which minimizes objective functions such as $\Sigma\, \varepsilon_i^2$ and $\Sigma\eta_i^2$.

The input–output modelling procedure inserts Eq. 12 into Eq. 13. Thus, the effect E_i is expressed as a function of all model parameters:

$$E_i = E(c(t_i, A, \alpha, B, \beta, \ldots), c_0, \gamma, \ldots) + \eta_i \tag{14}$$

Because of the scale invariance of Eq. 8, the parameters A, B and c_0 do not appear independently in Eq. 14, but only their ratios A/c_0 and B/c_0, etc.

Thus, if one waives the measurements of blood concentrations, the input–output modelling procedure can identify all parameters of the pharmacokinetic dynamic model except for one, such as the initial volume of distribution. The knowledge of this variable is, however, not necessary for the determination of drug dosing, because the time course of concentration at the site of drug action only needs to be known up to a scaling factor. Input–output modelling thus replaces common pharmacokinetic quantities such as initial volume of distribution, volume of distribution at steady state or clearance by the corresponding amount of drugs, e.g. bolus dose, to achieve initially a certain effect, or the total amount of drug in the body at a steady state or the steady state infusion rate to maintain a given effect.

For some drugs, one observes that EEG frequency does not decrease at low concentrations, but initially leads to increases in frequency, e.g. barbiturates, propofol or benzodiazepines. This behaviour can be expressed by the following formula:

$$E = E_1 - E_2\frac{c^{\gamma}}{c_1^{\gamma} + c^{\gamma}} + E_2\frac{c^{\gamma}}{c_2^{\gamma} + c^{\gamma}} \tag{15}$$

Model-Based Adaptive Feedback Control of Anesthesia

If the pharmacodynamic effect of drug administration can be measured on line, e.g. median EEG frequency or neuromuscular blockade, one can take into consideration the possibility of automatically delivering the drug such as to induce and maintain a given value of pharmacodynamic response.

Control theory distinguishes between open-loop control and closed-loop control. In open-loop control, the input (e.g. drug dosage) is independent of the output (e.g depth of anesthesia). In closed-loop control systems, the input at any particular time depends on the previous output. Both control systems require a controller to determine the optimum dosage strategy. This might be the anesthesiologist and/or a model of the process to be controlled. When the input to the system is controlled by a model, this control is commonly referred to as being model based. Model-based closed-loop control systems may use the measured output of the system not only to determine the next input, but also to update the model describing the relationship between input and output. This method is referred to as model based and adaptive. Among the models used, one can distinguish between heuristic and deterministic models. PID (proportional–integral–differential) control is a very often used heuristic model for feedback control. In this case, it is assumed that the input to the system needed to correct for a difference between measured output set-point is related to the difference between the set-point and the output itself, the integral of the output as well as the derivative. Pharmacokinetic dynamic models are examples of deterministic models used to control drug dosage. Figure 5 compares the schematic structures of closed-loop feedback control. The upper part depicts the most simple configuration, the bottom part, the schematics of a model-based adaptive feedback control. It consists of five essential components. Such a set-up works as follows [37, 48]: The process to be controlled is the depth of anesthesia for a given patient. This process is represented by a measurable pharmacodynamic signal, e.g. median EEG frequency. A model of the process to be controlled serves to compare measured effect with the effect as predicted by the model. Discrepancies between the prediction of the model and measured effect are not only used to correct drug dosing, but are also used to adapt model parameters. Thus, starting with the data of an average patient, the systems learns with time and individualizes the average model parameters to the individual patient under consideration. Finally, a controller is required to transform differences between desired and measured effect into drug administration.

For methohexitone and propofol [21, 23], we used median EEG frequency as pharmacodynamic signal and the interval of 2–3 Hz as desired range of control. A pharmacokinetic dynamic model as described by Eqs. 5 and 6 was used.

Adaptation Algorithm

The entire pharmacokinetic dynamic model used in this study is determined by a set of eight parameters (A, B, β, E_0, E_{max}, c_0 and γ). Relating the drug input function $I(t)$ directly to drug effect E by inserting $c(t)$ of Eq. 5 into the

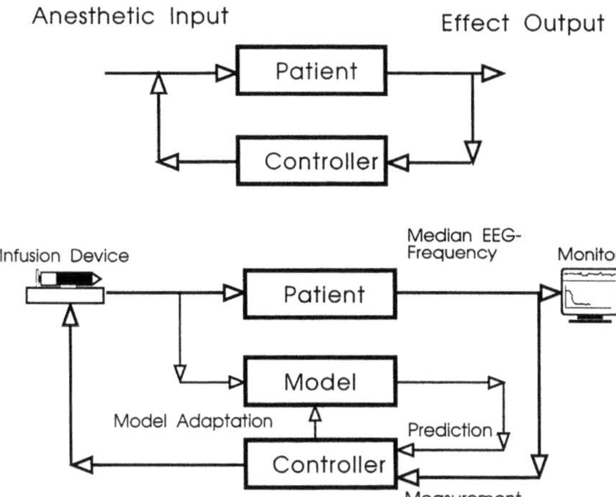

Fig. 5. An adaptive model-based feedback control system (*bottom*) compared with a non-adaptive system (*top*). The process to be controlled, e.g. depth of anesthesia in a given patient, is represented by a measurable pharmacodynamic signal. The model of this process relates drug input to effect output. Such a model can be used in two directions: in the forward direction it can predict the effect on the basis of drug input, and in the inverse direction it can be used to determine the rate of administration of achieve and maintain a given effect. The information content of the difference between measured effect and predicted effect is used to adapt the model to the individual patient. On the basis of the adapted model, the controller uses the difference between measured effect and set-point to determine the rate of administration

pharmacodynamic equation

$$E = E_0 - E_{max}\frac{c^\gamma}{c_0^\gamma + c^\gamma} = E_0 - E_{max}\frac{(c/c_0)^\gamma}{1 + (c/c_0)^\gamma} \tag{16}$$

one observes that this relation depends only on seven parameters $(A/c_0, \alpha, B/c_0, \beta, E_0, E_{max}. \gamma)$. This is due to the scale invariance mentioned above.

A full adaptation algorithm requires an updating of all seven parameters. To get a confident estimation of all parameters, one would have to choose an experimental set-up such that the signal output is sensitive to all parameters, i.e. wide varying concentrations inducing varying effects between baseline and maximum. However, this study has exactly the opposite aim, i.e. to establish a constant effect. It is therefore advisable to update only a minimal set of parameters necessary to adjust for the individual. We chose to estimate and update A and B and to fix all other parameters. This set allows us to make adjustments according to the subject's response to a bolus (short-term adjustment) and to the subject's response to a constant-rate infusion at steady state (long-term adjustment). Immediately after a bolus D, the concentration is given by

$$c_D = D(A + B) \tag{17}$$

whereby during constant-rate infusion I_{ss}, concentration c_{ss} is given by

$$c_{ss} = I_{ss}(A/\alpha + B/\beta) \tag{18}$$

The effect E may be regarded as a function of A, B and the drug input $I(t)$:

$$E = E(A, B, I(t)) \tag{19}$$

By denoting $A + \delta A$ and $B + \delta B$ the true microconstants for an individual subject, the difference ΔE between measured and predicted concentration can be expanded in a Taylor series:

$$\Delta E = E(A + \delta A, B + \delta B, I(t)) - E(A, B, I(t))$$

$$= (\partial E/\partial A)\delta A + (\partial E/\partial B)\delta B + \ldots \tag{20}$$

In conjunction with the condition to minimize the expression $\delta A^2 + \delta B^2$, Eq. 20 was used to solve δA and δB. From the updated values, new microconstants were calculated, which served to correct the drug input function.

Figure 6 depicts the results in 11 volunteers who were submitted to feedback control of methohexital anesthesia [37]. As target range of control, the interval of 2–3 Hz was chosen. Figure 6 depicts the time course of median EEG frequency (mean $\pm$ SD) for the 120-min procedure and the subsequent recovery. To achieve this result, as steady random stimulation of the volunteers by acoustic sensations, verbal commands, cold stimuli, pin picks and testing of eyelid and corneal reflex was necessary.

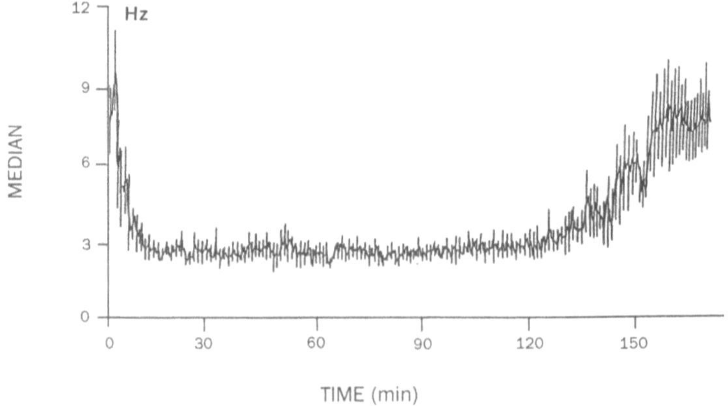

Fig. 6. Mean $\pm$ SD of median electroencephalogram (EEG) frequency in 11 volunteers receiving methohexitone to maintain a median EEG frequency between 2 and 3 Hz for 2 h

Though this figure demonstrates the functioning of feedback controlled methohexitone administration in stimulated volunteers, it does not add anything beyond our expectations. The true added value of feedback control systems is the dose which was required to maintain median EEG frequency at the set-point. The cumulative amount (mean $\pm$ SD) of methohexitone required is depicted in Fig. 7. In this context, a special feature of feedback systems has to be noted: in conventional therapeutic studies, a specified dose is given and the time course of effect emerges. Feedback control systems invert this handling of the dose–effect relationship. Instead of giving a dose and observing the emerging effect, a feedback system allows us to preset an effect and to observe the dose necessary to achieve and maintain the effect. This feature could for instance be of high value in dose-finding studies. Using such a feedback system for propofol administration, we were able to identify similar dose-requirement curves as with methohexitone. Comparing the dose requirement of both, one is now able to exactly define the relative potency of propofol with respect to methohexitone by determining the ratios of the cumulative dose-requirement curves. This ratio as a function of time is depicted in Fig. 8. One recognizes from Fig. 8 that the relative potency of propofol with respect to methohexitone is relatively stable with time and lies in the order of 0.72.

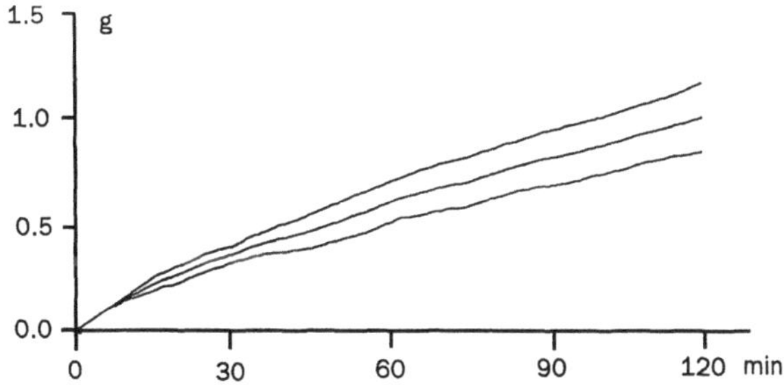

Fig. 7. The cumulative dose requirement (y *axis*) of methohexitone (mean $\pm$ SD) for 120 min (x *axis*) in the 11 volunteers for which median electroencephalogram (EEG) frequency was maintained between 2 and 3 Hz (see Fig. 6)

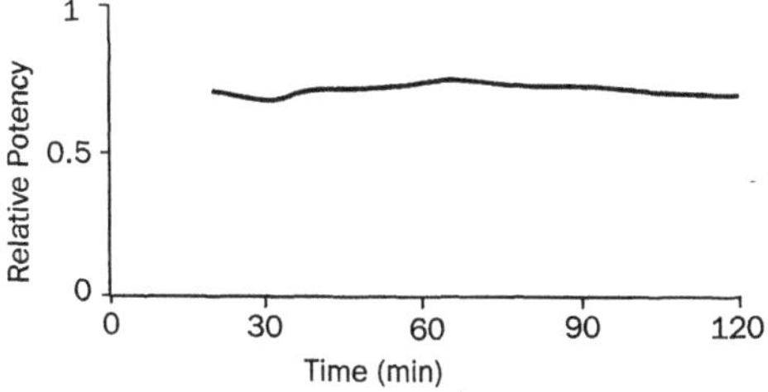

Fig. 8. The relative potency of propofol with respect to methohexitone as determined from the ratio of cumulative dose requirement during feedback-controlled drug administration to maintain median electroencephalogram (EEG) frequency between 2 and 3 Hz

The Concept of Effective Therapeutic Infusions

Figure 9 depicts an idealized cumulative dose-requirement curve together with its asymptote. The asymptote is defined by its intercept and slope, and both may be readily interpreted. The intercept gives the amount the drug in the body at steady state (so-called body load), and the slope gives the rate of infusion to maintain the desired effect. As this rate of infusion is determined by the feedback system as the infusion maintaining a given effect, it is an effective infusion. Where the effect was chosen such that the effect is within the therapeutic range, we named it effective therapeutic infusion (ETI) [40, 41]. Figure 10 depicts the cumulative distribution of the ETI of methohexitone in 11 surgical patients during a steady state fentanyl infusion of 0.22 mg/h for a median EEG frequency between 2–3 Hz [41]. From Fig. 10, one may easily derive the ETI_{50} (by analogy to ED_{50}) or any other quantile. Unlike the determination of MAC [22, 24] or the minimum infusion rate [23], which by definition require that 50% of the investigated patients are not treated in the therapeutic range (movement to skin incision), this approach allows similar and even more conclusions, but each individual is treated within his or her therapeutic optimum. The ETI concept together with the feedback control method is also a suitable tool to study drug–drug interactions. The characterization and quantitation of drug–drug interaction in terms of additivity or non-additivity requires the identification of pairs of doses d_A–d_B of drug A and drug B, leading to the specified effect E at which the drug–drug interaction is studied. The common approach used today is a search among numerous doses d_A with numerous doses d_B until eventually enough dose pairs are identified yielding the

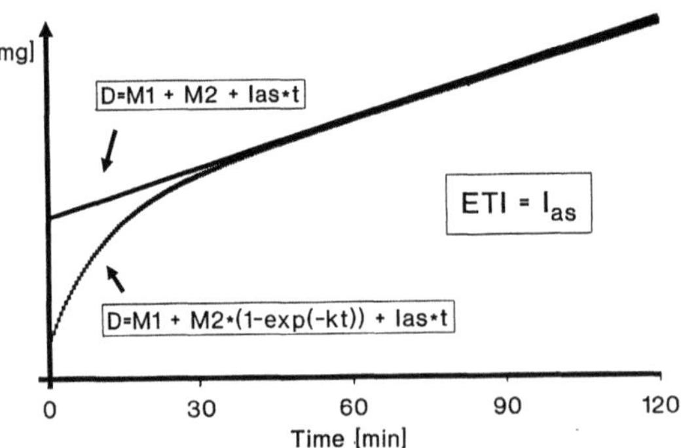

Fig. 9. Idealized dose requirement curve during feedback-controlled drug delivery. The asymptote to the curve is defined by its intercept and its slope. The intercept is the total amount of drug in the body at steady state; the slope I_{ss} gives the rate of infusion to maintain the set-point at steady state. In maintaining a given effect, I_{ss} is thus an effective rate of infusion. If the effect is chosen such that it lies within the therapeutic range, it obviously defines an effective therapeutic infusion (ETI)

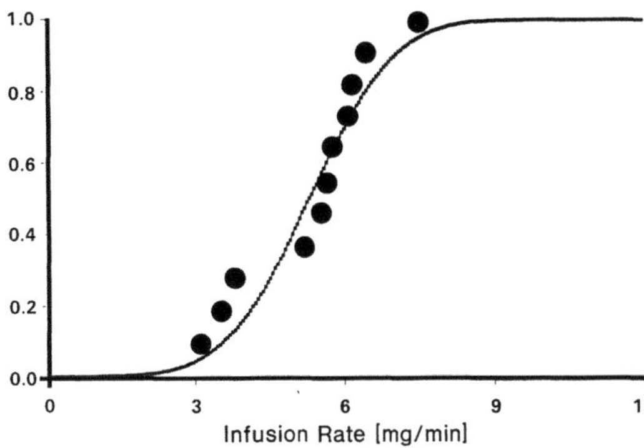

Fig. 10. Cumulative distribution of the effective therapeutic infusion (ETI) of methohexitone in 11 surgical patients who were coadminisitered fentanyl at a steady state infusion of 0.22 mg/h

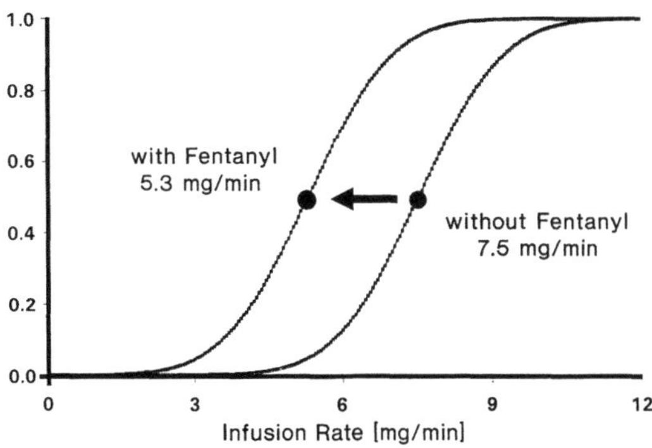

Fig. 11. Comparison of the distribution of the effective therapeutic infusion (ETI) of methohexitone for the 11 surgical patients (see Fig. 10) and the distribution in 11 volunteers receiving no fentanyl. From this data it can be estimated that the coadministration of fentanyl (0.22 mg/h) reduces the methohexitone requirement by 2.2 mg/min at steady state

effect E. Using a feedback system, such a trial and error approach may be streamlined to a few straigthforward studies. The first drug A is given at various doses and the second drug B is administered by the feedback system maintaining effect E, i.e. the feedback system identifies immediately those dose pairs leading to the effect E, and superfluous studies of dose pairs not leading to effect E are avoided.

Figure 11 depicts the interaction of methohexitone with fentanyl. Methohexitone given alone to volunteers required an ETI_{50} of 7.5 mg/min (right curve), and co-administration of fentanyl at a steady state infusion rate of 0.22 mg/h yielded an ETI_{50} of 5.3 mg/min. Hence, the fentanyl administration reduced the methohexitone requirement by approximately 30%. These are not enough data to decide whether this interaction is additive or supra- or infra-additive. Gross estimations, however, lead to the conclusion that an infusion of 0.8 mg/h fentanyl at steady state will yield a median EEG frequency between 2 and 3 Hz. Given this estimation, the index of additivity is computed to 0.22 mg

Table 1. Effective therapeutic infusion (ETI), duration of surgery and recovery time in an investigation of feedback-controlled delivery of alfentanil

Patient	ETI (mg/min)	Duration of surgery (min)	Recovery time (min)
1	0.139	200	19
2	0.217	75	7
3	0.135	105	17
4	0.108	255	6
5	0.113	125	14
6	0.121	120	13
7	0.122	125	20
8	0.165	190	20
9	0.163	130	31
10	0.134	155	22
11	0.121	170	35
Mean	0.140	150.0	18.5
SD	0.032	50.7	8.9

SD, Standard deviation.

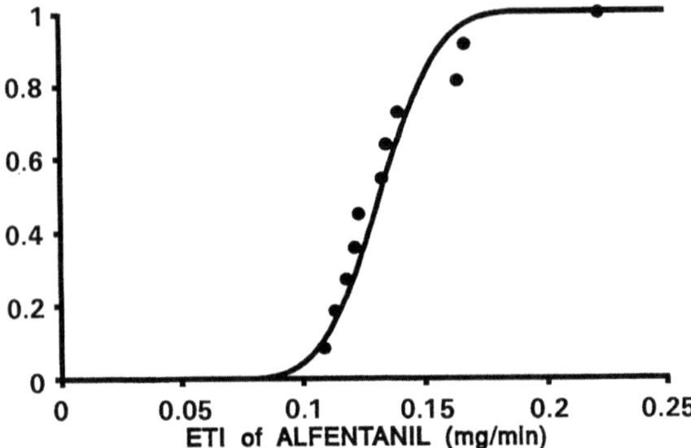

Fig. 12. Cumulative distribution of the effective therapeutic infusion (ETI) of a alfentanil in a surgical patient (major abdominal surgery) during anaesthesia with 60% nitrous oxide and alfentanil. The mean ETI is about 0.140 mg/min, being a reasonable agreement with other estimations of alfentanil requirement in surgical patients (see "Discussion")

per h/0.8 mg per h + 5.3 mg per min/7.5 mg per min = 0.98 ≈ 1. As the sum of both fractional doses of methohexitone and fentanyl add up to 1, it is concluded that, given the available information, there are no indications for a non-additive interaction between both agents. So far only hypnotic compounds have been considered for the administration by EEG-based feedback systems. In a recent study we investigated feedback-controlled delivery of alfentanil during anesthesia with alfentanil–nitrous oxide [42]. Table 1 gives the data for the individual ETI of alfentanil, duration of surgery and recovery times for the 11 surgical patients studied. We found a mean ETI of alfentanil of 0.140 ± 0.032 mg/min. Figure 12 depicts the cumulative distribution of the ETI in the group of 11 patients.

Discussion

In considering the use and usefulness of EEG-controlled feedback systems, two major points have to be addressed: firstly, does the EEG and especially the derived quantity median EEG frequency reflect therapeutic drug action and secondly, does feedback-controlled drug administration lead to "better" anesthesia? It has been argued that different anesthetic agents produce different EEG patterns, and thus it might not be possible to use the EEG for the control of anesthetic drug administration. Indeed, approaches such as the pharmaco-EEG try to identify differences in drug action on the EEG with the aim of anticipating the pattern of behavioural effects a drug and especially a new compound will exhibit. The use of EEG for monitoring anesthesia and anesthetic drug action looks at the problem from an opposite point of view: obviously there are a plethora of anesthetic drugs and their combinations all used with the same goal of inducing and maintaining anesthesia. Given this fact, the natural question is whether this common aim and achievement is reflected in some (not all) characteristics of the EEG. In trying to answer this question, a parametrization of EEG is necessary which is gross enough not to be influenced by the differences of different anesthetics on the EEG, but specific enough to reflect the induced mental state of reduced or lacking perception of stimuli. We believe that median frequency is at least a candidate for such parametrization. As shown in this review, many drugs used for anesthesia such as methohexitone, etomidate, thiopental, propofol, alfentanil–nitrous oxide [42], fentanyl and fentanyl-methohexitone are able to slow EEG frequency. In addition, we have shown that a median EEG frequency between 2 and 4 Hz could be reasonable set-point for clinical uses. Especially for studies using alfentanil–nitrous oxide, there is enough world-wide experience to compare the feedback approach with other approaches of handling this drug combination for the maintenance of anesthesia.

Several years ago, on the basis of pharmacokinetic investigations and clinical experience [27, 31, 47] our group proposed a target concentration for alfentanil of 400–500 ng/ml during anesthesia with 60% nitrous oxide for major surgery. Given the pharmacokinetic data of Schüttler et al. [27], this concentration corresponds to a maintenance infusion rate of approximately 0.14 mg/min for adult patients. Meanwhile, several authors have performed studies of identifying appropriate dose requirements for alfentanil, mostly in addition to 66% nitrous oxide. They usually used blood pressure and heart rate to define additional alfentanil requirements. Ausems et al. [1] reported an average rate of infusion of 1.72 ± 0.15 µg/kg per min during intraabdominal surgery, corres-ponding to 0.123 ± 0.011 mg/min for a patient with 72 kg body weight. This group used 66% nitrous oxide. Lemmens et al. [20] reported for a group of elderly patients (n, 18; 65–86 years) and an average body weight of 67 kg a total requirement of 23.5 ± 10.7 mg alfentanil for an average duration of anesthesia of 162 ± 42 min with the addition of 66% nitrous oxide. Gesink-van der Veer et al. [12] found for patients with Crohn's disease with an average weight of 58.8 kg a mean rate of infusion of 2.1 µg/kg per min and for a control group with an average weight of 64.4 kg a mean rate of 1.05 µg/kg per min, corres-ponding to 0.151 mg/min and 0.076 mg/min for a 72-kg patient in each groups, respectively.

A later study by Ausems et al. [2] found an average rate of infusion of 1.43 ± 0.55 µg/kg per min for lower abdominal surgery and 1.99 ± 0.55 µg/kg per min for upper abdominal surgery, with 66% nitrous oxide necessary to avoid "response", as defined by a set of criteria with respect to haemodynamics and autonomic signs. These data correspond to 0.1 ± 0.04 mg/min and 0.14 ± 0.04 mg/min for a 72-kg patient. In the study by Ausems et al., again 66% nitrous oxide was administered. The comparison of these data with the data presented in this review seems to indicate that median EEG frequency may alternatively serve as a guide in defining appropriate doses of alfentanil during alfentanil – nitrous oxide anesthesia.

In answering the second question as to whether feedback-controlled drug delivery leads to better anesthesia, the key point is "better". It has many aspects, including outcome in terms of morbidity/mortality, patient's and doctors' comfort and convenience as well as economic and ecological aspects. With respect to none of these aspects has it been proven or falsified that feedback-controlled drug delivery is superior to other techniques. There are, however, arguments why this might be so. One may consider feedback-controlled drug administration as a very careful computer-controlled titration of drug dosing within the very narrow limits of the set-point. As such, it mimics the anesthetist, but on the grounds of sound pharmacokinetic–pharmacodynamic models and their handling. On the other hand, the feedback system as used in the presented studies has very limited information (the EEG only) compared to the anesthetist. Given this limited information, it is not surprising that these feedback systems have one major disability, and that is in terms of anticipating action. While the anesthetist can take measures and action before, for example, a skin incision or celiotomy, the feedback system in its present state lacks the ability to foresee such events and therefore cannot take advantage of anticipatory action. We therefore conclude that supervisory feedback control might be a reasonable compromise at this moment in time for the use of feedback-controlled drug delivery in the clinical setting. Another evaluation of the use and usefulness of EEG feedback-controlled anesthetic drug delivery is achieved if one considers it as a research tool in the clinical pharmacology of anesthesia. In this field, it appears to be a useful tool in objectively determining drug requirements at constant and therapeutic effects. It is able to study drug–drug interactions or the interaction of drugs with other circumstances, it eliminates the otherwise existing need to under- or overdose a certain portion of patients or volunteers, it reduces the number of individuals to be studied and allows a more straightforward study design for a number of questions. Thus, it is a more effective and economic research tool compared to trial and error or other search methods.

References

1. Ausems EM, Hug CC, de Lange S (1983) Variable rate infusion of alfentanil to nitrous oxide anesthesia for general surgery. Anesth Analg 62:982–986
2. Ausems EM, Hug CC, Stanski DR, Burm AGL (1986) Plasma concentrations of alfentanil required to supplement nitrous oxide anesthesia for general surgery. Anesthesiology 65:362–373
3. Barrett R, Graham GG, Torda TA (1984) The influence of sampling site upon the distribution phase kinetics of thiopentone. Anaesth Intensive Care 12:5–9
4. Bellville JW, Attura GM (1957) Servo-control of general anesthesia. Science 126:827
5. Bellville JW, Fennel PJ, Murphy T (1960) The relative potencies of methohexital and thiopental. J Pharmacol Exp Ther 129:108
6. Bickford RG (1950) Automatic electroencephalographic control of general anesthesia. EEG Clin Neurophysiol 2:93
7. Bickford RG (1951) Use of frequency discrimination in the automatic electroencephalographic control of anesthesia (servo-anesthesia). EEG Clin Neurophysiol 3:83
8. Brazier MAB, Finesinger JE (1945) Action of barbiturates on cerebral cortex: electroencephalographic studies. Arch Neurol Psychiatrie 53:51
9. Desiderio DP, Thorne AC (1990) Awareness and general anesthesia. Acta Anaesthesiol Scand 34 (S92):45–50
10. Eger EI II, Saidman LJ, Brandstater B (1965) Minimum alveolar concentration: a standard of anesthetic potency. Anesthesiology 26:756–763
11. Fink M (1977) Quantitative EEG analysis and psychopharmacology. In: Remond S (ed) EEG informatics. A didatic review of methods and application of EEG data processing. Elsevier/North Holland, Amsterdam, p 301
12. Gesink-van der Veer BJ, Burm AGL, Hennis PJ, Bovill JG (1989) Alfentanil requirement in Crohn's disease. Anaesthesia 44:209–211
13. Gibbs FA, Gibbs EL, Lennox WG (1937) Effect on the electro-encephalogram of certain drugs which influence nervous activity. Arch Intern Med 60:154
14. Grassberger P, Procaccia I (1983) Measuring the strangeness of strange attractors. Physica 9D:189–209
15. Herrmann WM (1982) Development and critical evaluation of an objektive procedure for the electroencephalographic classification of psychotropic drugs. In: Herrmann WM (ed) Electroencephalography in drug research. Fischer, Stuttgart, p 249
16. Holford NHG, Sheiner LB (1981) Understanding the dose-effect relationship. Clin Pharmacokinet 6:429–453

17. Hull CJ, van Beem HBH, McLeod K (1978) A pharmacodynamic model of pancuronium. Br J Anaesth 50:1113–1123
18. Kaplan DT, Glass L (1992) Direct test for determinism in a time series. Phys Rev Lett 68:427–430
19. Keeri-Szanto M (1961) Anesthetic time/dose curves. The limiting factor in the utilization of intravenous anesthetics during surgical operations. Clin Pharmacol Ther 2:45–51
20. Lemmens HJM, Burm AGL, Bovill JG, Hennis PJ (1988) Pharmacodynamics of alfentanil as a supplement to nitrous oxide anaesthesia in the elderly. Br J Anaesth 61:173–179
21. Levy WJ, Shapiro HM, Maruchak G (1980) Automated EEG processing for intraoperative monitoring – a comparison of techniques. Anesthesiology 53:223–236
22. Merkel G, Eger EI II (1963) A comparative study of halothane and halopropane anesthesia including a method for determining equipotency. Anesthesiology 24:346–352
23. Prys-Roberts C, Davis JR, Calverley RK, Goodman NW (1983) Haemodynamic effects on infusions of diisopropyl phenol (ICI 35868) during nitrous oxide anaesthesia in man. Br J Anaesth 55:105–111
24. Quasha AL, Eger EI II, Tinker JH (1980) Determination and applications of MAC. Anesthesiology 53:315–334
25. Rampil IJ, Sasse FJ, Smith NT (1980) Spectral edge frequency – a new correlate of anesthetic depth. Anesthesiology 53:S12
26. Richard W, Bromley D, Pickett D, Passamante A (1993) Recognizing determinism in a time series. Phys Rev Lett 70:580–583
27. Schüttler J, Stoeckel H (1982) Alfentanil (R 39209). Ein neues kurzwirkendes Opioid. Pharmakokinetik und erste klinische Erfahrungen. Anaesthesist 31:10–14
28. Schüttler J, Schwilden H, Stoeckel H (1983) Pharmacokinetics as applied to total intravenous anaesthesia. Practical implications. Anaesthesia [Suppl] 38:53–56
29. Schüttler J, Schwilden H, Stoeckel H (1985) Infusion strategies to investigate the pharmacokinetics and pharmacodynamics of hypnotic drugs: etomidate as an example. Eur J Anaesthesiol 2:133–142
30. Schüttler J, Schwilden H, Stoeckel H (1985) Pharmacokinetic and pharmacodynamic modelling of propofol ('diprivan') in volunteers and surgical patient. Postgrad Med Journal 61 [Suppl 3):53–54
31. Schüttler J, Stoeckel H, Schwilden H, Lauven PM (1986) Pharmakokinetisch begründete Infusionsmodelle für die Narkoseführung mit Alfentanil. In: Doenicke A (ed) Alfentanil. Springer, Berlin Heidelberg New York, pp 42–51
32. Schwilden H, Stoeckel H (1980) Untersuchungen über verschiedene EEG-Parameter als Indikatoren des Narkosezustandes, Der Median alo quantitative Maß der Narkosetiefe. Anaesth Intensivther Notfallmed 15:279–286
33. Schwilden H, Schüttler J, Stoeckel H (1983) Pharmacokinetics as applied to total intravenous anaesthesia. Theoretical considerations. Anaesthesia [Suppl] 38:51–52
34. Schwilden H, Schüttler J, Stoeckel H (1985) Quantitation of the EEG and pharmacodynamic modelling of hypnotic drugs: etomidate as an example. Eur J Anaesthesiol 2:121–131
35. Schwilden H, Stoeckel H, Schüttler J, Lauven PM (1986) Pharmacological models and their use in clinical anaesthesia. Eur J Anaesthesiol 3:175–208

36. Schwilden H, Stoeckel H (1987) Quantitative EEG analysis during anaesthesia with isoflurane in nitrous oxide at 1.3 and 1.5 MAC. Br J Anaesth 59:738–745

37. Schwilden H, Schüttler J, Stoeckel H (1987) Closed-loop feedback control of methohexitone anesthesia by quantitative EEG-analysis in humans. Anesthesiology 67:53–59

38. Schwilden H, Stoeckel H, Schüttler J (1989) Closed-loop feedback control of propofol anaesthesia by quantitative EEG analysis in humans. Br J Anaesth 62:290–296

39. Schwilden H (1989) Use of the median EEG frequency and pharmacokinetics in determining depth of anaesthesia. In: Jones JG (ed) Baillièrs clinical anaesthesiology. Bailllier, London, pp 603–622

40. Schwilden H, Stoeckel H (1990) Effective therapeutic infusions produced by closed-loop feedback control of methohexital administration during total intravenous anesthesia with fentanyl. Anesthesiology 74:225–229

41. Schwilden H, Schüttler J (1990) Bestimmung effektiver therapeutischer Infusionsraten (ETI) für intravenöse Anaesthetika durch feedback-geregelte Dosierung. Anaesthesist 39:603–606

42. Schwilden H, Stoeckel H (1993) Closed-loop feedback controlled administration of alfentanil during alfentanil-nitrous oxide anaesthesia. Br J Anaesth 70:389–393

43. Scott JC, Ponganis KV, Stanski DR (1985) EEG quantitation of narcotic effect. The comparative pharmacodynamics of fentanyl and alfentanil. Anesthesiology 62:234–241

44. Sheiner LB, Stanski DR, Vozeh S (1979) Simultaneous modeling of pharmacokinetics and pharmacodynamics: application to D-tubocurarine. Clin Pharmacol Ther 25:358

45. Stoeckel H, Schwilden H, Lauven PM, Schüttler J (1981) EEG parameters for evaluation of depth of anaesthesia. The median of frequency distribution. In: Vickers MD, Crul J (eds) Proceedings of the European Academy of Anaesthesiology 1980. Springer. Berlin Heidelberg New York, pp 73–78

46. Stoeckel H, Schwilden H (1984) Quantitative EEG-analysis and monitoring depth of anaesthesia. In: Gomez QJ, Egay LM, de la Cruz-Odi MF (eds) Anesthesia–safety for all. Elsevier, Amsterdam, p 151

47. Stoeckel H, Schüttler J, Schwilden H (1985) Grundlagen der Infusionsnarkose mit Alfentanil. In: Zindler M, Hartung E (eds) Alfentanil. Ein neues, ultrakurzwirkendes Opioid. Urban and Schwarzenberg, Munich, pp 141–150

48. Stoeckel H, Schwilden H (1986) Methoden der automatischen Feedback-Regelung für die Narkose, Konzepte und klinische Anwendung. Anaesth Intensivther Notfallmed 21:60–67

49. Ulrych TJ, Bishop TN (1975) Maximum entropy spectral analysis and autoregressive decomposition. Rev. Geophys Space Phys 13:183

The Use of Processed EEG in the Operating Room

M. J. Bloom

Even in an age of contracting medical resources, intraoperative EEG monitoring is gaining increasing acceptance as a tool to identify the need for intervention. However, central nervous activity is characterized by very rapid fluctuations. Neurologic injury can happen so quickly that most tests of cerebral well-being do not give useful information in time to intervene. Offline processing techniques for sophisticated derivation of EEG parameters are too slow to provide effective monitoring. To be effective, monitors must be able to detect a change soon enough for something to be done immediately; even 2 h later is too late.

Whether therapy can be effectively modified based on changes in the EEG is still a controversial issue. Some encouraging initial reports show that intervention based on EEG changes can significantly reduce the incidence of neuropsychiatric disorders after cardiac surgery (Arom et al. 1989; Edmonds et al. 1992). Other reports indicated that monitoring can be used both for intraoperative assessment, and to a certain extent in the intensive care unit (ICU), for prediction of outcome. (Bricolo et al. 1978). The latter application is particularly useful at times when a lot of effort, materials, and resources are expended in the care of a critically ill patient who may not be getting any better.

Indications

The applications of processed EEG monitoring that are most clear-cut include: (1) The detection of a cerebral insult particularly during carotid endarterectomy – statistically significant differences in outcome from carotid endarterectomy have been found in response to selective shunting based on EEG (Bloom et al. 1990). (2) Monitoring cerebral perfusion during temporary vessel occlusion (Cloughesy et al. 1993) – surgeons now frequently use a technique of proximal vessel control, that is, they deliberately clip the feeder vessels to the aneurysm sack temporarily before applying the aneurysm clip. Doing this may decrease perfusion rather profoundly. The hope is that collateral perfusion is adequate or that temporary clips can be removed in time to avoid permanent injury. However, on several occasions, EEG changes have suggested that the surgeon had to hasten to remove the temporary clips that were applied for the aneurysm clipping. (3) Monitoring of cerebral oxygenation during cardiopulmonary by-pass – many subtle neurologic deficits are generated during cardiopulmonary bypass (Nussmeier 1986). Initially the thought was that these changes are embolic, and that once they have happened, there is little that can be done. But a recent study from Louisville (Edmonds et al. 1992) showed that the incidence of neurologic deficits after cardiopulmonary bypass could be decreased dramatically from 29% down to as little as 4% by intervening with increased perfusion measures in response to EEG changes during cardiopulmonary bypass. (4) Monitoring of the depth of anesthesia (Sebel et al. 1991; Vernon et al. 1992). This new challenge was the focus of a recent conference in Israel. EEG and anesthetic effects are weak in their association, but, with the advent of more advanced processing, in particular the use of discriminant analysis, it may be possible to derive more quantitative measures of the anesthetic effects on EEG. (5) Achieving an effective barbiturate coma – the dose of barbiturate necessary for cerebral protection in humans is between 7 and 45 mg/kg per hour. (Zaidan 1991). With such a wide dosage range, the proper dose cannot be chosen arbitrarily and then be certain to be effective. (6) Evaluation of hyperventilation for the treatment of elevated intracranial pressure – hyperventilation is not uniformly beneficial and, in certain situations may exacerbate the ischemic insult (Wolf et al. 1993) EEG may be a good way to identify cases of inappropriate hyperventilation in the treatment of intracranial pressure. (7) Identification of seizure activity during muscle relaxation in patients with head trauma – in patients subjected to neuromuscular blockade, EEG monitoring provides the only way to detect seizure activity, which generates very high cerebral metabolic demands.

In summary, the most common indications for EEG monitoring in the operating room are carotid endarterectomy, cardiac surgery, cerebrovascular surgery, and barbiturate coma.

Other Uses

EEG monitoring may also be indicated in less well-documented situations. (1) The use of EEG to detect brain death is quite clear, but unproved is the use of continuous EEG monitoring for early detection of brain death or for the prognostication of brain death before a neurologist has documented death by the traditional EEG. (2) Bricolo et al. (1978) looked at the outcome of coma as a function of variability and change in the EEG in ICU patients and found very good results in terms of prognostic value or even therapeutic value. Particular patterns in EEG have very good prognostic value. A very depressed spectral pattern that is monotonous or has very few variations over the course of the day is a very poor prognostic sign. (3) Another use is the potential detection of anxiety and pain in patients under muscle relaxants who cannot indicate when they are in severe pain. Patients sedated with doses of midazolam to tolerate mechanical ventilation may also be in severe pain but unable to communicate.

EEG arousal responses similar to the EEG pattern during arousal from anesthesia may be seen in patients given relaxants but were inadequately sedated in the ICU (Veselis et al. 1989).

Why Process the EEG?

Traditional, raw EEG monitoring could be brought into the operating room. However, an anesthesiologist trying both to deliver an anesthetic and to monitor EEG at the same time is drawn away from the care of the patient. It is difficult to remain observant of the anesthesia delivery system, to monitor all other physiologic parameters, and still try to make annotations and control all of the equipment needed for monitoring raw EEG. An even greater problem is attempting to look for and compare EEG changes over time in the mountain of paper that is the record of prior electrical activity. To accomplish such a task would force the anesthesiologist to neglect the care of the rest of the patient. The resultant mass of data is so large as to preclude any way to process it and make sense of it in an appropriate time frame to allow intervention in the operating room.

The shortcomings of raw EEG need to be addressed in order to make processed EEG successful. It is difficult to quantify the EEG because it is difficult to measure parameters of EEG on paper. The EEG must be quantified in a way that can be rapidly interpreted by the anesthesiologist or the monitoring physiologist in the operating room. Gradual changes are very difficult to detect in EEG, particularly in the raw EEG, so we need ways that are not ambiguous to compress time and express gradual trends and changes in EEG. Long-term data storage is another very big problem, particularly in the USA, where there is a high liability risk. If something adverse goes undetected until a patient wakes up in the recovery room and then shows some sort of deficit, the questions arise: "When did this happen? How did this happen? Why did this happen?" If raw EEG has only been recorded on paper and not stored for further review and processing, these questions become almost impossible to answer.

A final problem is that EEG monitoring equipment is costly, complex, and difficult to use. Processing EEG may make EEG affordable and simple enough to be manipulated and used in the operating room, with relatively little attention to dials and switches.

Procedures to Optimize EEG Monitoring

To interpret changes in the EEG, a number of important principles must be followed. A baseline in the EEG must be established. Each patient's individual EEG varies so much from that of the population that, if only population numbers are used to measure EEG changes, many deviations from the individual patient's own baseline will be missed. Further, a steady-state value must be established. That is, the state of the brain during anesthesia differs from the awake state. The EEG changes associated with anesthetic drugs may be interesting, but to separate the physiologic changes from the pharmacologic changes, a steady pharmacologic history and a steady physiologic history must be generated so that unexpected changes can be identified. The use of pharmacokinetics and pharmacodynamics in neurophysiologic monitoring is a significant advance to aid in tracking the pharmacologic history (Schwilden et al. 1989) It is important to keep track of the expected cumulative effects of the administered drugs in the interpretation of the EEG. For example, something as simple as a mild benzodiazepine premedication before surgery will increase the high frequency activity in the EEG which might otherwise be interpreted as an arousal response. There is often a clear and important temporal association between the actions taken by the surgeon and the changes observed in the EEG. Surgical clips may be applied to the patient before incision, accidentally or intentionally. The same is true of changes generated by the anesthesiologist and other significant external influences. Thermal changes and electrical noise in the room may be associated with EEG changes. Painful stimulation from the electrical stimulus applied for somatosensory-evoked potentials (SEP) may affect the EEG. If SEP responses are weak, a 120- or 130-V stimulus may be applied directly onto a peripheral nerve. (Perhaps this becomes a pain-evoked potential rather than a somatosensory evoked potential; Kochs et al. 1990)

Challenges in Processed EEG

Many approaches to EEG monitoring have failed in the past because they attempted to derive all the necessary information from a single channel. A frequently asked question is, "How many leads do you need?" The answer lies in the degree of the regional selectivity of the EEG needed to detect the changes during intraoperative physiologic monitoring. No regional selectivity is possible with only one channel. Left–right brain asymmetries must be examined. Although two channels may be enough to detect changes in anesthetic effect, in situations in which the anterior or carotid-supplied circulation may be affected differently from the vertebrobasilar circulation, EEG activity supported by the anterior and posterior cerebral circulation must be examined separately. Therefore, four channels are preferred for intraoperative monitoring. Some centers are now looking at the number of channels necessary for adequate topographic mapping; here, four channels are not enough. Ashida et al. (1984) derived an unbiased polynomial interpolation map of EEG under various anesthetic conditions and calculated the number of channels needed to preserve the basic features of the map. Eight to ten channels appear to be enough to provide adequate regional sensitivity, but some neurologists would argue that the use of less than 16 leads is suboptimal.

An interpolated map of EEG provides considerable detail. Nevertheless, another important point, brought out initially by Duffy et al. (1981) is being overlooked in present monitoring trends. For any parameter being mapped, in addition to changes in a particular patient, the expected distribution for the population must be examined and used to determine a probability density map describing the likelihood that any observed change is simply a normal variant rather than reflection of a cerebral insult. It is important to use normative databases and to compare with a statistical probability map of this kind in order to get accurate interpretation out of mapping-type monitors (John 1980).

Another challenge in EEG monitoring is time compression. Most often, compression and averaging will improve the readability of an EEG that may contain interesting features. Figure 1 shows a density spectral array of four channels, with frequency on the vertical axis and time going from left to right on the horizontal axis. A gradual onset of ischemic change can be seen. Three quarters of the way through the recording, the patient had a brief seizure as well. This activity is even easier to see when time is compressed and the variability in the EEG is removed by averaging techniques. These procedures preserve the information in the original display while enhancing its readability. One advantage of processed EEG is its ability to smooth out variability and reveal the "big picture" in a compressed format, which allows more information to be visible on one screen. A second advantage relates to the fact that the EEG changes of interest to an anesthesiologist, or other clinician in the operating room, are not transient. Neurophysiologic changes that persist for less than 10–20 s are of little

COMPRESSION & EPOCH AVERAGING IMPROVES READABILITY

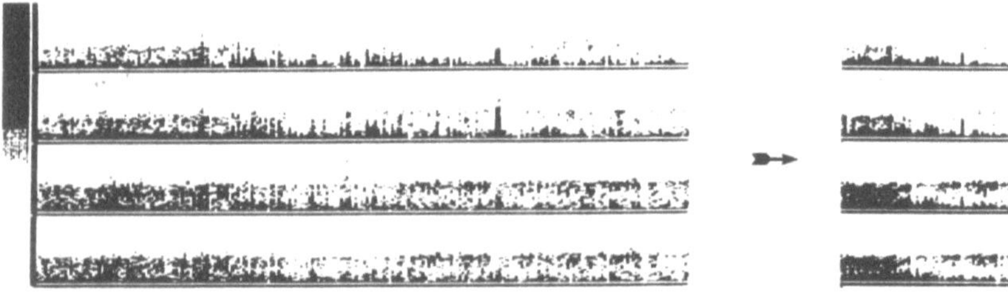

Fig. 1. Four-channel density spectral array of EEG in a patient suffering progressive cerebral ischemia. *Upper channel* is left frontal area; *second channel* is right frontal area; *third channel* is left parietal occipital area; *bottom channel* is right parietal occipital area. Brightness indicating increasing power at each frequency is shown in the *left hand margin*. For each channel, the *vertical scale* represents 0–30 hz. The portion of the display *left of the arrow* represents approximately 1 h of time. The display to the *right of the arrow* represents the identical period of time using 4:1 averaging and compression

interest intraoperatively. More important are changes that show a significant trend and persist long enough to have some lasting effect. Smoothing across epochs removes second-by-second variation in the EEG (Moberg 1987). Smoothing is useful in two dimensions: time and frequency. Frequency smoothing is done across individual bins, since separation of EEG frequencies into 1/4-Hz bins is probably nonphysiologic. Smoothing across the frequencies reveals subtle shifts in the range of a band that might be inapparent in the raw EEG. The resultant frequency-smoothed and time-smoothed EEG preserves the most important features and may remove localized variation present in the raw EEG.

Because it preserves all the information in the original signal, the fundamental processing most commonly used is Fourier analysis. However, the phase information of the Fourier transform is lost when power spectra are derived. There is renewed interest in looking more carefully at techniques such as bispectral analysis, which includes the phase information. Phase information is also included in the calculation of generalized coherence, which is a calculation of the coherence encompassing all the channels (Gish and Cochran 1988). Another calculation under investigation is instantaneous frequency derived from the Hilbert transform (Sclabassi et al. 1992). Moberg (1987) has reported a calculation that separates the power spectrum into percentage steps at equal intervals and presents each range as a density. This spectral percent power array (or relative power distribution) can be displayed just like the power spectrum itself. Particularly in low-power situations, this method emphasizes the redistribution of power across the spectrum.

Since power spectra of the EEG are multimodal (Levy 1986), more than one parameter clearly must be necessary to describe this multimodal distribution statistically. Any simple parameter derived to characterize all EEG features is an oversimplification (Bashein et al. 1992). One parameter may correlate in particular situations, for a particular drug, under certain controlled conditions, but for clinically useful applications, multiparametric measures are the only ones that achieve high enough specificity (Pronk and Simons 1984). Because of the high variance in derived parameters, smoothing is an important function.

Therapeutic urgency, the importance of responding to a particular change, is not linear. It is particularly important for a sudden change to be caught very quickly, and yet gradual changes with a subtle trend are best observed on a compressed time scale. Logarithmic time scales allow the observer to look quickly with a high resolution at short-latency changes close to the current time in monitoring and still compress time progressively backward so that an hour or more of EEG can be examined at the same time.

EEG challenges have been addressed using a number of new techniques. Most of the new monitors will be available with four or eight channels. In the past, asymmetry measures, i.e., differences between the two sides of the brain, have been ignored to a great extent. Either multiparametric asymmetry scores (Kopruner and Pfurtscheler 1984) or coherence (Gish and Cochran 1988) and other types of measures of phase difference between channels (Watt and Hameroff 1988) will become increasingly important to examine these differences.

Most important, studies are now beginning which submit all of these parameters to discriminant analysis to create descriptive factors (Thomsen et al. 1991). These discriminant factors can be plotted to examine the clustering of EEG parameters and then identify regions in the parameter space that are believed to represent danger. The parameters used in the plots defy simple description because they are multiparametric descriptors from a discriminant function optimized to detect a particular change in EEG caused by ischemia or anesthetic effect (Kearse et al. 1991). These discriminant functions are difficult to describe in terms that would be apparent or intuitive.

Currently under investigation is the application of cluster analysis and discriminant analysis to EEG (Thomsen et al. 1991) in an effort to draw out information which is germaine to clinical care. For example, there are particular shifts in cerebral state associated with hypoxia. The EEG undergoes not only slow gradual changes, but cluster analysis of EEG during progressive hypoxia reveals sudden shifts from one cluster of EEG to another, and these shifts are asymmetric. That is, the change during the onset of hypoxia is not the inverse of the change during the recovery period. The cerebral activity shifts more gradually during the recovery, and there appears to be a nonhomogenous sensitivity to hypoxic insult (Bloom 1993b). Various areas of the brain react more adversely or recover more quickly, reflecting the metabolic heterogeneity of the brain (Rosner 1986).

When discriminant analysis is applied to hypoxic EEG, the classical band descriptors (alpha, beta, delta, theta) do not appear in the results. These bands have been derived in the neurology sleep laboratory. The drugs used and the conditions generated in the operating room severely change the conditions that determine the boundaries for these classic band descriptors. In particular, new boundary conditions have been discovered during cardiopulmonary bypass under high-dose narcotic anesthesia with hypothermia (Bloom 1993a).

Practical Challenges

There are practical as well as technical challenges to the monitoring of the EEG. Monitoring services often need to be supplied in many locations simultaneously. A problem in the USA is that payment for expert interpretation is possible only if a neurophysiologist is monitoring the patient without providing other clinical care. If a separate neurophysiologist or Scientist, or even a clinician, must be present for each monitored patient, it is very inefficient to have one in every operating room or next to every ICU bed or in every location where invasive radiology is done. This is a particular problem given the temporally and spacially distributed needs for expertise. This expert manpower may be needed only intermittently. The time when a physiologist is particularly needed is when a particular problem, insult, or procedure occurs. The physiologist is not needed for expert interpretation throughout every procedure. The challenge is to have this expert available at just the right time. A surgeon may expect to be compromising an artery shortly, only to find several hours later that he is still searching for the artery. It is, thus, difficult to assume that an expert will be available at the particular moment of need.

Another challenge is that the technical requirements for any particular modality of monitoring may vary significantly. The number of channels required will vary among applications, depending upon the expected risks. The required sampling rates vary, as do other technical aspects such as filter settings. It is difficult to design a single monitor that can handle all the various needs of these situations and still be understood and operated by a technician with a small amount of training.

During EEG monitoring, a massive volume of data is acquired and needs to be stored. Optical disk storage is quickly becoming a very practical matter without large cost. Presentation preferences vary with the personnel doing the monitoring. Trying to build the monitor for all people results in a monitor with so many variations and complexities that it satisfies no one in particular. Analytic techniques change, too. What is in vogue today is not what will be done in processing EEG even 3 years from now. Beware of a monitor that has been locked into a particular kind of analysis and can do no more.

The Center for Clinical Neurophysiology (CCN) in Pittsburgh has addressed a number of these challenges with some unique solutions (Krieger et al. (1991). A network has been installed to all of the locations in which monitoring is done or observed. There is a complete network throughout the hospital–next to every ICU bed and in every operating room. Physicians can even dial-in from home or elsewhere and observe the EEG of a patient in an operating room. This network allows one expert to oversee many locations simultaneously. On one workstation, as many as 16 different locations may be overseen at once. Flexible protocols are stored in advance. These may be procedure-based (a carotid endarterectomy protocol, bypass protocol), or personnel based, according to individual preferences. The user can call up an individual protocol in which the entire configuration of both acquisition and display can be tailored to individual needs.

The need for backup is addressed as well. The data must not only be stored, but stored in a form that can be recovered, should the system crash. CCN stores everything to write-once-read-many (WORM) optical disks across the same network. A standardized data storage format is used, so that when a promising new way to process EEG emerges, the data can be retrieved from the standardized database and submitted to new analytic techniques.

Current Workstation

Figure 2 shows the current version of the CCN workstation. It a is homemade configuration with a UNIX workstation on the bottom, standard Grass EEG amplifiers at the top, a standard oscilloscope to assure an acceptable raw EEG is into the computer, and a video screen on which the preferred output can be configured.

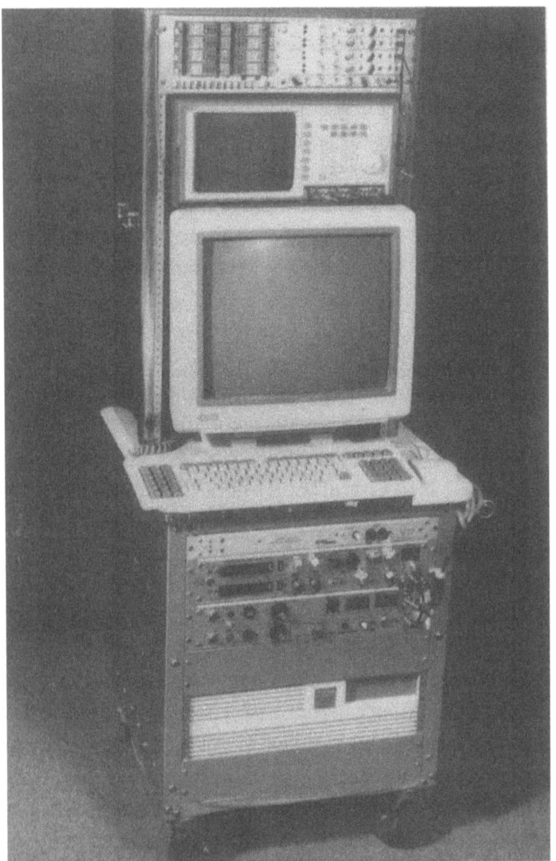

Fig. 2. Custom neurophysiologic workstation. Components shown from top to bottom are: (1) Grass model 10 EEG amplifiers, (2) oscilloscope for raw signal observation, (3) 19-in high-resolution video display, (4) keyboard and mouse or trackball for program control and interaction, (5) Grass stimulator modules, and (6) Apollo Domain series computer. This system is flexible, portable, expandable, and relatively compact. It has been organized into a commercial product by Computational Diagnostics, Inc

The Application of Processed EEG in Cardiac Surgery

An initial reluctance to monitor EEG in cardiac surgery came from the belief that most of the insults from cardiopulmonary bypass were of embolic phenomena. Indeed, a large proportion of these phenomena are embolic. As investigators have begun to use transcranial Doppler sonography to look for emboli in the middle cerebral artery, surgeons have been horrified to find how often they occur. Two things are clear, however. First, all of the damage is not done by the time changes are seen in the EEG. The argument was that once an embolism has occurred, the damage is done and cannot be treated. But, in fact, dramatic responses have been seen to elevations of the mean perfusion pressure during cardiopulmonary bypass (Edmonds et al. 1992) or increased perfusion by deliberate hemodilution in the cases where the hematocrit may have been slightly elevated. There may also be situations in which (although the PaO_2 is adequate) the FiO_2 can be increased, which may provide satisfactory oxygenation to those marginally ischemic areas. Second, even if damage has been done and cannot be undone in a particular patient, a surgeon may find that a particular way of unclamping the aorta consistently produces a shower of emboli. Appropriate monitoring may cause him to change his technique so that even though therapy may not be completely effective for a particular patient, in a series of hundreds of patients, as the surgeon modifies his technique and begins to identify those situations in which particular ischemic or embolic risks are present, improvements are made in the risks to all patients who follow.

References

Arom KV, Cohen DE, Strobl FT (1989) Effect of intraoperative intervention on neurological outcome based on electroencephalographic monitoring during cardiopulmonary bypass. Ann Thorac Surg 48:476–483

Ashida H, Tatsuno J, Okamoto J, Maru E (1984) Field mapping of EEG by unbiased polynomial interpolation. Comput Biomed Res 17:267–276

Barlow JS (1986) Artifact processing (rejection and minimization) in EEG data processing. In: Lopes da Silva FH, Storm van Leeuwen W, Remond A (eds) Clinical applications of computer analysis of EEG and other neurophysiological variables, vol 2. Elsevier, Amsterdam, pp 15–62. (Handbook of electroencephalography and clinical neurophysiology)

Bashein G, Nessly ML, Bledsoe SW, et al. (1992) Electroencephalography during surgery with cardiopulmonary bypass and hypothermia. Anesthesiology 76:878–891

Bloom MJ (1993a) EEG changes associated with the start of cardiopulmonary bypass. Anesth Analg 76:S21–22

Bloom MJ (1993b) Techniques to identify clinical contexts during automated data analysis. Int. J Clin Monit Comput 10:17–22

Bloom MJ, Schwartz DM, Berkowitz HD, Pratt RE Jr (1990) DSA processing of EEG is an effective in CEA. Presented at the Society of Neurosurgical Anesthesia and Critical Care Meeting, Waikiki, Hawaii, 14–15 March

Bricolo A, Turazzi S, Faccioli F, Oroizzi F, Sciarretta G, Erculiani P (1978) Clinical application of compressed spectral array in long-term EEG monitoring of comatose patients. Electroenceph Clin Neurophysiol 45:211–225

Cloughesy TF, Nuwer MR, Hoch D, Vinuela F, Duckwiler G, Martin N (1993) Monitoring carotid test occlusions with continuous EEG and clinical examination. J Clin Neurophysiol 10(3):363–369

Demetrescu M, Kavan E, Smith N (1981) Monitoring the brain condition by advanced EEG. Anesthesiology 55:A130

Duffy FH, Bartels PH, Burchfield JL (1981) Significant probability maping: an aid in the topographic analysis of brain electrical activity. Electroenceph Clin Neurophysiol 51:455–462

Edmonds HL Jr, Griffiths LK, Shields CB Online statistical analysis of the EEG predicts postoperative neurologic dysfunction during cardiopulmonary bypass (CPB) surgery. 12th annual meeting of the Society of Cardiovascular Anesthesiologists, Orlando, FL, 13–16 May

Edmonds HL Jr, Griffiths LK, van der Laken J, Slater AD (1992) Quantitative electroencephalographic monitoring during myocardial revascularization predicts postoperative disorientation and improves outcome. J Thorac Cardiovasc Surg 103:555–563

El-Fiki M, Fish KJ (1987) Is the EEG a useful monitor during cardiac surgery? Case Report. Anesthesiology 67:575–578

Gish H, Cochran D (1988) Generalized coherence. Proc IEEE ICASSP, New York, pp 2745–2747

Gregory TK, Pettus DC (1986) An electroencephalographic processing algorithm specifically intended for analysis of cerebral electrical activity. J Clin Monit 2:190–197

Hanowell LH, Soriano S, Bennett HL (1992) EEG power changes are more sensitive than spectral edge frequency variation for detection of cerebral ischemia during carotid artery surgery: a prospective assessment of processed EEG monitoring. J Cardiothorac Vasc Anesth 6:292–294

Jansen BH (1986) Quantitative EEG analysis in renal disease. In: Lopes da Silva FH, Storm van Leeuwen W, Remond A (eds) Clinical applications of computer analysis of EEG and other neurophysiological variables, vol 2. pp 239–256, Elsevier, Amsterdam (Hand book of Electroencephalography and Clinical Neurophysiology, Chap. 8)

John ER (1980) Developmental equations for the Electroencephalogram. Science 210:1255–1258

Kalkman CJ, Boezeman EH, Ribberink AA, Oosting J, Deen L, Bovil JG (1991) Influence of changes in arterial carbon dioxide tension on the electroencephalogram and posterior tibial nerve somatosensory cortical evoked potentials during Alfentanil/nitrous oxide anesthesia.Anesthesiology 75:68–74

Kearse L, Saini V, deBros F, Chamoun N (1991) Bispectral analysis may predict anesthetic depth during narcotic induction. Anesthesiology 75:A175

Kochs E et al. (1990) Modulation of pain-related somatosensory evoked potentials by general anesthesia. Anesth Analg 71:225

Kopruner V, Pfurtscheler G (1984) Multiparametric Asymmetry Score (MAS) distinction between normal and ischemic brains. EEG Clin Neurophysiol 57:343

Krieger DN, Burk G, Sclabassi RJ (1991) NeuroNet: a distributed real-time system for monitoring neurophysiological function in the medical environment. Computer 24:45–55

Levy WJ (1984) Intraoperative EEG patterns: implications for EEG monitoring. Anesthesiology 60:430–434

Levy WJ (1986) Power spectrum correlates of changes in consciousness during anesthetic induction with enflurane. Anesthesiology 64:688–693

Levy WJ, Parcella P (1987) Electroencephalographic evidence of cerebral ischemia during acute extracorporeal hypoperfusion. Cardiothorac Anesth 1:300–304

Levy WJ, Shapiro HM, Maruchak G, Meathe E (1980) Automated EEG processing for intraoperative monitoring: a comparison of techniques. Anesthesiology 53:223–236

Leib JP, Sperling MR, Mendius JR, Smoker CE, Englel J Jr (1986) Visual versus computer evaluation of thiopental-induced EEG changes in temporal lobe epilepsy. Electroenceph Clin Neurophysiol 63:395–407

Long CW, Shah NK, Loughlin C, Spydell J, Bedford RF (1989) A comparison of EEG determinants of near-awakening from isoflurane and fentanyl anesthesia. Anesth Analg 69:169–173

Lopes de Silva FH (1981) Pattern recognition and automatic EEG analysis. Trends Newosci 297

Matthis P, Scheffner D, Benninger C (1981) Spectral analysis of the EEG: comparison of various spectral parameters. Electroenceph Clin Neurophysiol 52:218–221

Maynard DE, Jenkinson JL (1984) The cerebral function analysing monitor. Anesthesia 39:678–690

Moberg (1987) Electroencephalographic and evoked potential processing for continuous monitoring in the intensive care unit and operating room. J Clin Monit 3:332

Myers RR, Stockard JJ, Saidman LJ (1977) Monitoring of cerebral perfusion during anesthesia by time-compressed Fourier analysis of the electroencephalogram. Stroke 8:331–337

Nagata K, Mizukami M, Araki G, Kawase T, Hirano M (1982) Topographic electroencephalographic study of cerebral infarction using computer mapping of the EEG. J Cereb Blood Flow Metab 2:79–88

Nussmeier N, Arlund C, Slogoff S (1986) Neuropsychiatric complications after cardiopulmonary bypass: cerebral protection by a barbiturate. Anesthesiology 64:165–170

Prichep LS, John ER (1986) Neurometrics: clinical applications. In Lopes da Silva FH, Storm van Leeuwen W, Remond A (eds) Clinical applications of computer analysis of EEG and other neurophysiological variables, vol 2. Elsevier, Amsterdam, pp 153–170 (Handbook of Electroencephalography and Clinical Neurophysiology, Chap. 5)

Pronk RAF, Simons AJR (1984) Processing of the electroencephalogram in cardiac surgery. Comp methods Programs Biomed 18:181–190

Rosner G, Graf R, Kataoka K, Heiss WD (1986) Selective functional vulnerability of cortical neurons following transient MCA-occlusion in the cat. Stroke 17:1

Sainio K, Stenberg D, Keskimaki I, Muuronen A, Kaste M (1983) Visual and spectral EEG analysis in the evaluation of the outcome in patients with ischemic brain infarction. Electroenceph Clin Neurophysiol 56:117–124

Salerno TA, Lince DP, White DN, et al. (1978) Monitoring of electroencephalogram during open-heart surgery. J Thorac Cardiovasc Surg 76:97–110

Schuttler J, Schwilden H, Stoeckel H (1983) Pharmacokinetics as applied to total intravenous anaesthesia. Practical implications. Anaesthesia 38:53–56

Schuttler J, Schwilden H, Stoeckel H (1985) Pharmacokinetic and pharmacodynamic modeling of propofol (Diprivan) in volunteers and surgical patients. Postgrad Med J 61:53–54

Schwilden H, Stoeckel H, Schuttler (1989) Closed-loop feedback control of propofol anesthesia by quantitative EEG analysis in humans. Br J Anaesth 2:290–296

Sclabassi RJ, Sun M, Krieger SN, Scher MS (1990) Time-frequency analysis of the EEG signal. Proc ISSPA 935–938

Sclabassi RJ, Sun M, Krieger DN, et al. (1992) Time-frequency domain problems in the neurosciences. In: Boashash B (ed) Time-frequency signal analysis. Longman Chesire, Australia, pp 498–519

Scott JC, Stanski DR, Ponganis KV (1983) Quantitation of fentanyl's effect on the brain using the EEG. Anesthesiology 59:A370

Sebel PS, Bowles S, Saini V, Chamoun N (1991) Accuracy of EEG in predicting movement at incision during isoflurane anesthesia. Anesthesiology 75:A446

Snyder MM, Core RC, Watt RC (1988) A comparison of derived parameters used in electroencephalography. Anesth Analg 67:S214

Sotaniemi KA, Sulg IA, Hokkanen TE (1980) Quantitative EEG as a measure of cerebral dysfunction before and after open-heart surgery. Electroenceph Clin Neurophysiol 50:81–95

Spackman TN, Faust RJ, Cucchiara RF, Sharbrough FW (1987) A comparison of aperiodic analysis of the EEG with standard EEG and cerebral blood flow for detection of ischemia. Anesthesiology 66:229–231

Thomsen CE, Rosenfalck A, Christensen KN (1991) Assessment of anaesthetic depth by clustering analysis and autoregressive modelling of electroencephalograms. Comput Methods Programs Biomed 34:125–138

Tolonen U, Sulg IA (1981) Comparison of quantitative EEG parameters from four different analysis techniques in evaluation of relationships between EEG and CBF in brain infarction. Electroenceph Clin Neurophysiol 51:177–185

Vernon J, Bowles S, Sebel PS, Chamoun N (1992) EEG bispectrum predicts movement at incision during isoflurane or propofol anesthesia. Anesthesiology 77:A502

Veselis RA, Carlon GC, Bedford RF (1989) Spectral edge frequency correlates with sedation level in icu patients receiving continuous IV midazolam. Anesthesiology 17:157

Watt RC, Hameroff SR (1988) Phase space electroencephalography (EEG): a new mode of intraoperative EEG analysis. Int J Clin Monit Comput 5:3–13

William GW, Luders HO, Brickner A, Goormastic M, Klass DW (1985) Interobserver variability in EEG interpretation. Neurology 35:1714–1719

Wolf AL, Levi L, Marmarou A, Ward JD, Muizelaar PJ, Choi S, Young H, Rigamonti D, Robinson WL (1993) Effect of THAM upon outcome in severe head injury: a randomized prospective clinical trial. J Neurosurg 78 (1):54–59

Zaidan JR, Klochany A, William MM, Ziegler JS, Harless DM, Andrews RB (1991) Effect of thiopental on neurologic outcome following coronary artery bypass grafting. Anesthesiology 74:406–411

Bispectral Electroencephalogram Analysis for Monitoring Anesthetic Adequacy

J. M. Vernon, P. S. Sebel, S. M. Bowles, N. Chamoun, and V. Saini

Introduction

One avenue of research for a reliable monitor of anesthetic efficacy has been the computer-processed electroencephalogram (EEG), investigation being centered around derivatives of the power spectrum. Schüttler [1] has used the median frequency to control drug administration for closed loop anesthesia. However, the ability of power spectral derivatives to act as precise indices of anesthetic efficacy has not been consistently demonstrated. Rampil [2] showed that the 95% spectral edge may predict a hypertensive response to laryngoscopy, whereas Mills [3] was unable to correlate EEG and hemodynamic changes during induction, intubation, and skin incision and Dwyer [4] showed no correlation between movement at skin incision and several EEG power spectral derivatives during 1.0 MAC (minimum alveolar concentration to prevent movement at incision in 50% of patients) isoflurane anesthesia.

Conventional EEG analysis with Fourier transformation gives the EEG frequency, the EEG power in terms of voltage, and the phase of the signal. In conventional power spectral analysis, the phase information is discarded. It is possible that important information about anesthetic effects may be obtained by analyzing these phase relationships. Bispectral analysis is one such analytical technique which allows us to examine interfrequency phase relationships in the EEG.

The principles underlying bispectral analysis may be clarified by considering the two simulated EEG tracings in Fig. 1. The top tracing contains a simulated EEG with 2-, 3-, and 5-Hz components which have no phase relationship. In the lower tracing, the 5-Hz signal is a harmonic of the 2- and 3-Hz signals. Note that, although the raw signals look different, the power spectra (middle panels) are identical. In bispectral analysis (right panels), the peak at the intersection of the 2- and 3-Hz lines is clearly seen. Thus, the bispectrum provides us with useful information that has been lacking in all previous forms of EEG data compression, i.e., information about interfrequency phase relationships.

The bispectral index is a probability function for a univariate index generated by a linear combination of multiple bispectral variables, as well as the burst suppression ratio. Figure 2 is a graph of probability of no response (movement

                                        J. M. Vernon et al.

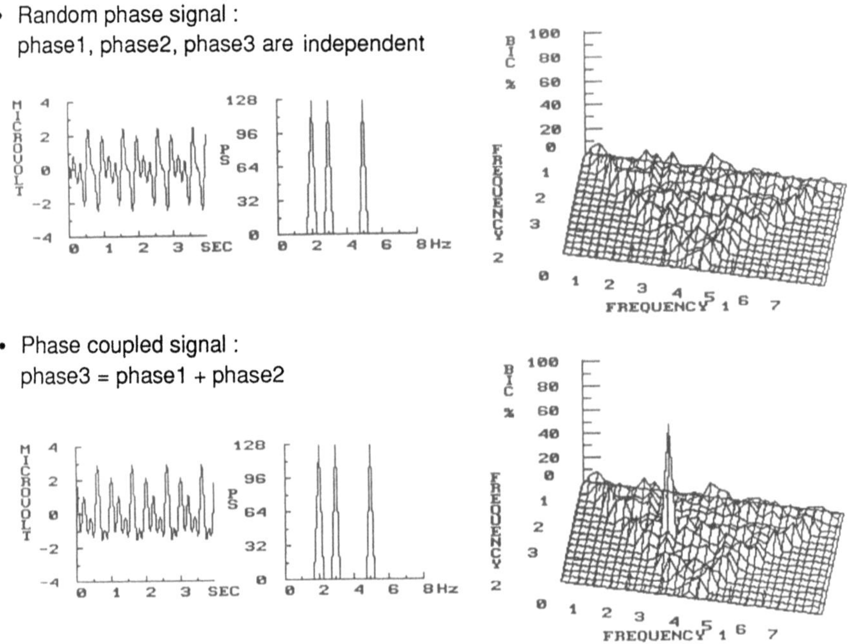

Fig. 1. Simulated electroencephalogram tracing: schematic representation of bispectral analysis. In the *upper* portion, the 2-, 3-, and 5-Hz waveforms are phase independent. In the *lower* portion, the 5-Hz component is the result of the phase relationship between the 2- and 3-Hz components

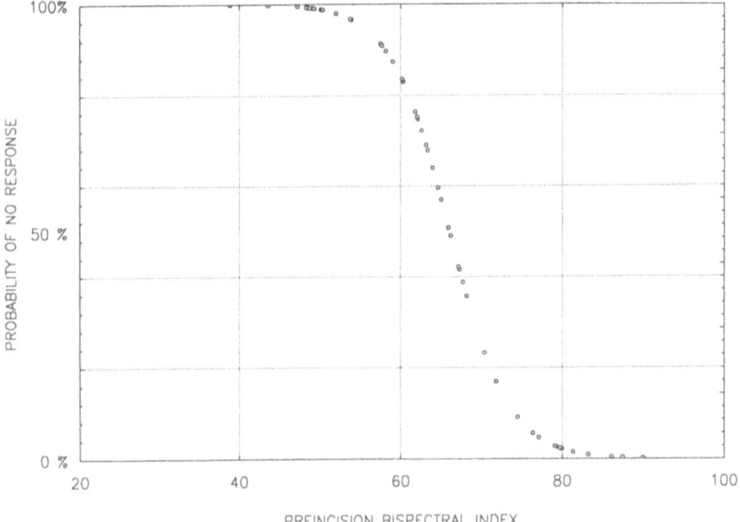

to skin incision) against bispectral index. The curve is sigmoid in shape, with 50% probability of response at a bispectral index of 65; these features are due to stretching and molding of the scale to magnify the clinically significant regions of the index. Further discussion of bispectral analysis may be found in the work of Dumermuth [5] and Barnett [6].

The first of our two studies was the investigation of the ability of the bispectral index to predict movement at skin incision during isoflurane/oxygen general anesthesia.

Study 1

Method

Forty-two surgical patients participated in this study, which was approved by the local Human Investigations Committee. Informed consent was obtained from patients of American Society of Anesthiologists' (ASA) status 1–3, aged 17–70, undergoing surgery involving an incision of greater than 8 cm in length.

Sedative premedication was omitted. Patients were randomly assigned to three isoflurane dose groups, of 0.75, 1.0, and 1.25 MAC (age adjusted). Gold-cup electrodes were placed in positions P3, P4, F3, and F4, referenced to Cz, and, following skin preparation with Omni Prep, (Weaver and Co., Aurora, CO, USA) ground to mastoid and attached with conductant jelly and collodion. The EEG signal was preamplified, and four channels were displayed and stored on a Toshiba lap-top computer. EEG, vital signs, and manually entered data were transferred to magnetic tape for off-line analysis at the laboratories of Aspect Medical Co., Framingham, MA, analysis taking place blind of the patient's response to skin incision. Noninvasive blood pressure (Dinamap), pulse oximeter (Nelcor 2000), and anesthetic gas (Datex) data were recorded on the computer, and electrocardiogram (ECG), esophageal temperature, and neuromuscular function were monitored without automatic recording.

Following preoxygenation, anesthesia was induced with thiopental (up to 5 mg/kg) and intubation accomplished following 100 mg succinylcholine. Isoflurane was maintained at the predetermined concentration prior to skin incision, which occurred after the return of neuromuscular function was demonstrated. If any movement occurred within the first 60 s following skin incision, this was regarded as a positive response.

The bispectral index used for analysis was constructed retrospectively from a data base of 160 patients and consisted of 33 bispectral variables and the burst suppression ratio.

The EEG indices used for analysis of the 60 s of EEG prior to skin incision included the power spectral index (PSI), created from eight power spectral variables plus the burst suppression ratio.

Fig. 2. Probability of no response to skin incision against bispectral index

Statistical analysis was performed using appropriate t-tests and analysis of variance (ANOVA). Sensitivity, specificity, and accuracy were calculated as follows:

$$\text{Sensitivity} = \frac{\text{Number of correctly predicted movers}}{\text{Total number of movers}}$$

$$\text{Specificity} = \frac{\text{Number of correctly predicted nonmovers}}{\text{Total number of nonmovers}}$$

$$\text{Accuracy} = \frac{\text{Total number of correctly predicted patients}}{\text{Total number of patients}}$$

Results

No recall of operative events occurred. Prior to incision, three patients in the 0.75-MAC group required neuromuscular blockade due to movement. Steady state end-tidal isoflurane concentrations had been achieved, so these patients were classified as movers.

Bispectral index, power spectral index, and traditional power spectral variables were evaluated as predictors of movement. There was a statistically significant difference between the move and no-move groups for the bispectral index and for relative delta (Table 1). The sensitivity, specificity, and accuracy values are most favorable for the bispectral index.

A further evaluation of the bispectral index was made using two different anesthetic techniques: alfentanil/isoflurane and alfentanil/propofol anesthesia, again assessing the ability to predict movement at skin incision.

Table 1. Diagnostic accuracy of electroencephalogram predictors. Isoflurane/oxygen anesthesia

	Move	No move	Sensitivity	Specificity	Accuracy
BIS	0.65 ± 0.03	0.46 ± 0.04^a	96	63	83
PSI	0.56 ± 0.01	0.52 ± 0.02	96	25	69
Relative delta (%)	-3.47 ± 0.23	-2.64 ± 0.2^a	73	75	74
Median frequency (Hz)	4.5 ± 0.24	3.87 ± 0.39	73	75	74
95% SEF (Hz)	15.4 ± 0.72	14.07 ± 0.73	62	81	69

Data are mean $\pm$ SE.
BIS, bispectral index; PSI, power spectral index; 95% SEF, 95% power spectral edge.
[a] $P < 0.05$.

Study 2

Method

The method of this study was similar to that of the previous study with the following exceptions: 50 surgical patients aged 18–65 years participated in this study, exclusion criteria being a known neurological disorder or recent use of anticonvulsant medication.

The EEG electrode positions were FP1, FP2, P3, and P4, ground to mastoid, and reference to Cz. Patients were randomly assigned to one of two groups, propofol/alfentanil (propofol group) or isoflurane/alfentanil (isoflurane group) anesthesia. Induction for both groups was identical. Following pre-oxygenation, propofol and alfentanil were given to achieve predicted plasma concentrations of 10 μg/ml and 125 ng/ml, respectively. This was achieved by use of an Ohmeda syringe pump, driven by a Psion microcomputer using software provided by Dr. Gavin Kenny of the University of Glasgow, U.K. A dose of succinylcholine 1.5 mg/kg was given to facilitate intubation.

Following loss of consciousness, the propofol infusion was set to achieve a maintenance-predicted propofol plasma concentration (propofol group), or terminated and isoflurane/oxygen was administered to achieve the desired maintenance end-tidal concentration, (isoflurane group). The alfentanil infusion continued unchanged in both groups.

The maintenance concentrations of propofol and isoflurane were determined by the response to skin incision of the last patient in that group. If movement occurred within 2 min of skin incision, the maintenance concentration would be increased for the next patient; if no movement occurred, then it would be decreased. The increments or decrements and maintenance concentration for the first patient in the propofol and isoflurane groups were 0.5 μg/ml starting at 6.5 μg/ml and 0.1% starting at 0.5%, respectively.

Unless clinically indicated, the target maintenance concentration was maintained unchanged for 10 min before, and 3 min after, the initial skin incision. Any movement in the following 2 min was regarded as a positive response. For the remainder of the operation, the use of nitrous oxide, neuromuscular blocking agents, and adjustment of anesthetic dose was left to the discretion of the anesthesiologist. EEG data for the 1 min before incision was analyzed.

Results

There were no significant differences in age, sex, weight, or ASA status between the two study groups.

The graph of end-tidal isoflurane concentration at incision against bispectral index (Fig. 3) shows a reasonable separation of the move and no-move

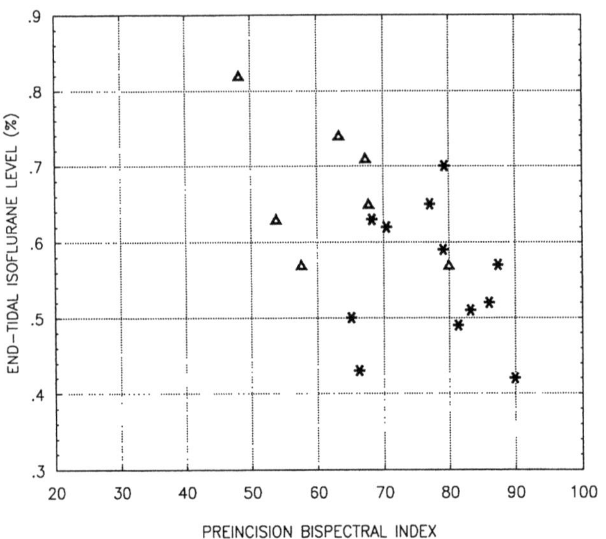

Fig. 3. Preincision end-tidal isoflurane concentration against bispectral index. Patients who moved at skin incision are designated by a *star*. Patients who did not move are designated by a *triangle*

Table 2. Diagnostic accuracy of electroencephalogram descriptors. Isoflurane/alfentanil anesthesia

	Move (n = 12)	No move (n = 7)	Sensitivity	Specificity	Accuracy
BIS	77.8 ± 8.5	62.6 ± 10.5[a]	83	86	84
Relative delta (%)	43.9 ± 17.2	53.4 ± 14.5	67	71	68
Median frequency (Hz)	6.8 ± 3.9	3.7 ± 2.9	100	79	77
95% SEF (Hz)	15.3 ± 4.1	13.0 ± 2.5	92	29	68

Data are mean ± SD.
BIS, bispectral index; 95% SEF, 95% power spectral edge.
[a] $P < 0.05$.

groups. There are no movers below a value of 65 and only one mover above a value of 68. There is a trend for the bispectral index to increase as isoflurane concentration decreases.

This data is tabulated in Table 2 and shows a significant difference between the move and no-move groups for bispectral analysis alone, which also has higher values for sensitivity, specificity, and accuracy.

A graph of predicted propofol plasma concentration against bispectral index (Fig. 4) shows no clear relationship between these variables and, again, a reasonable separation of the move and no-move groups. The overlap occurring between the move and no-move groups is to be expected, as the bispectral index is a probability function. The table of this data (Table 3) again shows a statistically significant difference between the move and no-move groups for

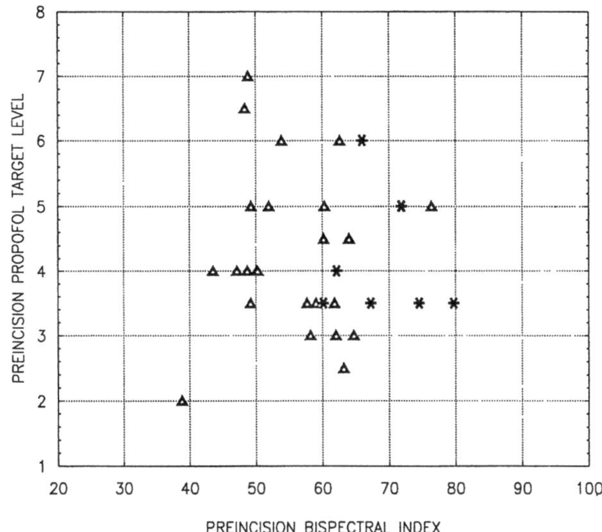

Fig. 4. Preincision predicted propofol concentration against bispectral index. Patients who moved at skin incision are designated by a *star*. Patients who did not move are designated by a *triangle*

Table 3. Diagnostic accuracy of electroencephalogram descriptors. Propofol/alfentanil ancsthcsia

	Move (n = 7)	No move (n = 24)	Sensitivity	Specificity	Accuracy
BIS	68.8 ± 7.0	55.4 ± 8.4[a]	71	96	90
Relative delta (%)	61.2 ± 11.3	70.6 ± 15.2	71	79	77
Median frequency (Hz)	2.4 ± 1.3	2.0 ± 2.0	71	79	77
95% SEF (Hz)	13.1 ± 0.9	10.7 ± 3.0	71	83	80

Data are mean ± SD.
BIS, bispectral index; 95% SEF, 95% power spectral edge.
[a] $P < 0.05$.

Table 4. Diagnostic accuracy of electroencephalogram descriptors. Combined groups

	Move (*n* = 12)	No move (*n* = 7)	Sensitivity	Specificity	Accuracy
BIS	74.5 ± 8.9	57.1 ± 9.3	89	87	88
Relative delta (%)	50.3 ± 17.2	66.7 ± 16.5	79	68	72
Median frequency (Hz)	5.2 ± 3.8	2.4 ± 2.3	84	68	73
95% SEF (Hz)	14.5 ± 3.4	11.3 ± 3.0	47	94	78

Data are mean ± SD.
BIS, bispectral index; 95% SEF, 95% power spectral edge.
[a] $P < 0.05$.

bispectral analysis alone, as well as better sensitivity, specificity, and accuracy values overall.

It is not statistically correct to combine the results from the two treatment groups. However, as the bispectral index may be useful in measuring anesthetic effect, it is an interesting exercise (Table 4). All four methods of EEG data compression show a statistically significant difference between the move and no-move groups, with the bispectral index again having superior discriminatory ability.

ANOVA showed an effect of treatment group, i.e., drugs used. This is related to the fact that the propofol movers and nonmovers had lower bispectral indices than the respective isoflurane groups.

Discussion

The results of these two studies suggest that the bispectral index may be a far better indicator of anesthetic efficacy as defined by movement at incision than more conventional EEG indices. This is supported by a study by Kearse [7] showing that the bispectral index, but not the 95% spectral edge values, were significantly different for patients with a hypertensive response to laryngoscopy during narcotic induction.

It is possible that the index is not entirely independent of anesthetic agent; however, small numbers in studies to date prevent any firm conclusion. Future studies will investigate the effects of different anesthetic drugs, using the index to guide dosing in order to achieve a high or low index, with a high or low probability of movement at incision.

References

1. Schüttler J, Kloos S, Ihmsem H, Schwilden H (1992) Clinical evaluation of a closed-loop dosing device for intravenous anesthesia based on EEG depth of anesthesia monitoring. Anesthesiology 77:A501
2. Rampil IJ, Matteo RS (1987) Changes in EEG spectral edge frequency correlate with the hemodynamic response to laryngoscopy and intubation. Anesthesiology 67:139–142
3. Mills AK, Ghouri A, Monk TG, White PF (1990) Lack of correlation between electroencephalographic and hemodynamic changes during general anesthesia. Anesth Analg 70:S268
4. Dwyer R, Rampil I, Eger II EI, Bennett HL (1991) The EEG does not predict movement in response to surgical incision at 1.0 MAC isoflurane. Anesthesiology 75:A1025
5. Dumermuth G, Huber PJ, Kleiner B, Gasser TH (1971) Analysis of the interrelations between frequency bands of the EEG by means of the bispectrum. A preliminary study. Electroencephalogr Clin Neurophysiol 31:137–148
6. Barnett TP, Johnson LC, Hicks N, Nute C (1971) Bispectral analysis of electroencephalogram signals during waking and sleeping. Science 172:401–402
7. Kearse L, Saini V, deBros F, Chamoun N (1991) Bispectral analysis of EEG may predict anesthetic depth during narcotic induction. Anesthesiology 75:A715

Does Spectral Edge Frequency Assess Depth of Anesthesia?

G. M. Gurman

Introduction

The need for monitoring depth of anesthesia was emphasized by Hamerhoff and Grantham in their recent review of this topic [25]: it results in prevention of pain perception, awareness, and recall; reduction of untoward autonomic effects of excessively light or deep anesthesia; minimization of stress and its manifestations; and facilitation of prompt recovery.

The precurare period of modern anesthesiology solved most of the above problems by monitoring the most important clinical sign related to depth of anesthesia, i.e., patient's movements.

Early studies on depth of anesthesia such as those published by Guedel [21], Artusio [2], and Woodbridge [56], included patient's motor activity in the lists of clinical parameters to be observed during general anesthesia.

The introduction of neuromuscular blockade in 1942 [20] completely changed the picture. Until neuromuscular blocking agents became part of routine anesthesia technique, the patient motor response was an adequate parameter of depth of anesthesia. Since then, anecdotal reports have described the traumatic experience of unwanted episodes of awareness during general anesthesia [19, 54]. In all cases, neuromuscular paralysis was achieved and the patients were able to reproduce events which occurred during the surgical procedure. Later, Meyer and Blacher [32] described a series of patients who were awake during cardiac surgery and developed signs of anxiety and repetitive nightmares long after the experience.

Finally, Levinson, in his classic paper [29] published in 1965, reported the results of staging a bogus crisis during anesthesia and surgery of ten patients with no immediate postoperative recall, but with significant recollection of events under hypnosis 1 month later.

In the absence of motor activity monitoring, the classic approach was to rely upon the indirect signs of anesthesia, e.g., the measure of sympathetic activity expressed by blood pressure (BP), heart rate (HR), or lacrimation and sweating. However, Cullen et al. [10] found that BP varied predictably with the dose of volatile anesthetics in combination or not with N_2O only during the first hour of anesthesia. After that, BP remained constant, despite increases in the anesthestic concentration.

Evans's score based on BP, HR, sweating, and tear formation [12] was found not to be consistently related to hand movements when using the isolated arm technique [40].

Hypovolemia, decrease in myocardial contractility, overloading, hypoxia, use of atropine, or isoproterenol, and hypercarbia are situations which may directly influence hemodynamic parameters and affect their relationship to depth of anesthesia.

Since all the traditional signs were rendered useless, anesthesiologists have been looking for an instrumental alternative for subjective assessment of depth of anesthesia. Methods proposed include monitoring of esophageal contractility, facial electromyography (FEMG), skin conductance, digital plethysmography, evoked potentials, and electroencephalography (EEG).

Esophageal contractility monitoring [13] was found to be influenced by sodium nitroprusside and anticholinergic drugs. FEMG did not significantly improve the administration of methohexital during brief outpatient procedures [6]. Anticholinergic drugs and autonomic neuropathy are likely to substantially reduce the accuracy of measurement of skin conductance, proposed by Goddard [17] as another tool of monitoring depth of anesthesia. Finally, digital pletismography [28], in spite of being an instrumental method of monitoring anesthesia, remains an expression of sympathetic activity, as unreliable as any other sign of this indirect aspect of anesthetic level,

A description of the use of evoked potentials for measuring the depth of anesthesia is beyond the scope of this paper. The reader is referred to the excellent recent review by Thornton and Newton [50]. They emphasize that using evoked potentials for monitoring depth of anesthesia is still in its early stages and that technical obstacles still prevent daily use of this method in the operating theater for guiding general anesthesia.

Role of Electroencephalography in Measuring Depth of Anesthesia

The spontaneous cerebral biopotentials obtained using scalp electrodes are characterized by a desynchronized pattern, due to the differential activation of parallel cortical columns underlying the surface which are collected by each electrode. Low amplitudes and high frequencies (up to 32 Hz) represent the normal EEG pattern during awareness. In contrast, high amplitudes, and lower frequencies (toward the delta band) are seen during general anesthesia. This pattern reflects a synchronized depolarization of the apical dentrites, as happens when central nervous system (CNS) depressants such as halothane or barbiturates are used [34]. A different type of rhythmic waves is obtained when a CNS excitant such as ether or ketamine is administered to the patient [33].

It was Gibbs in 1937 [16] who showed graded changes in EEG with increasing concentrations of volatile anesthetics. Theoretically , raw EEG could be considered a method for measuring the depth of anesthesia; its pattern is influenced by all anesthetic drugs, while neuromuscular agents do not influence the EEG signal; the method is also noninvasive and can be used continuously without causing any harm to the patient. However, as already mentioned, different anesthetics produce different changes in EEG, a fact which was clinically reported as early as 1973 by Clark and Rosner [9].

From the practical standpoint, the interpretation of raw EEG is often difficult. Standard EEG analysis has to be based on visual identification of individual components. This has often proved to be very difficult, requiring trained professionals and continuous vigilance. Not surprisingly, Volavka et al. [52] found that the highest average correlation in interpretation of EEG traces by seven experienced evaluators was only 56%.

Since the alterations of the EEG follow within seconds those which occur as a result of alterations of cerebral metabolism, the raw recording is still used in some centers for detection of global or focal ischemic changes during carotid and open heart procedures. This kind of surgery is accompanied in a certain percentage of cases by irreversible cerebral alterations which are responsible for postoperative neurologic and psychopathological disturbances. Based on the fact that raw EEG changes are closely correlated with cerebral blood flow and oxygen uptake, several groups have reported its reliability during carotid endarterectomy [7, 49]. Although the literature is divided on the usefulness of EEG monitoring during cardiopulmonary bypass [4], global EEG is still recommended, at least for guiding the administration of thiopental for CNS protection against incomplete ischemia [27]. This rather limited role of raw EEG during specific surgery encouraged research with the aim of finding an EEG system which offers complex information in a simpler form that is easy to understand, convenient, and inexpensive.

Processed Electroencephalography

Early reports describing the relationship between EEG and depth of anesthesia suggested that in order to identify small changes in EEG, new techniques had to replace the routine analytic methods, which were mainly using continuous visual inspection.

Power spectral analysis and fast Fourier transformation (FFT) are the two processes used today for automatic EEG monitoring. Power spectral analysis retains all the information offered by raw EEG and digitizes it at frequent intervals (epochs). The data obtained during an epoch (with a duration of 2–8 s) is analyzed by using the FFT. This separates EEG into a number of component sine waves of different amplitudes, whose sum is the original waveform [30].

After the calculation, the power spectrum is graphically displayed in a continuous manner (even though the mathematical analysis is a discontinuous process).

This much more simplified way of presenting EEG still generates thousands of data points per minute for each separate channel and led to the development of two main display methods for spectral analysis data:

1. Compressed spectral array (CSA), a term first used by Bickford [5], displays a graph of power versus frequency for each epoch of analysis. The relationship between this kind of display and clinical signs of depth of anesthesia [14] or anesthetic concentrations of halothane and enflurane [44] has been assessed. Technical and logistic difficulties causing misreading and some loss of information required the development of another method of display.
2. Density-modulated spectral array (DSA), which was first described by Fleming and Smith as "density modulation" [14], prevents loss of data. It displays data as a line of varying densities or a series of dots of various sizes. The maximum intensity or the largest dots correspond to the most common frequencies on the display. This method is easier for the user and remains legible even when read from a distance [30].

Power spectrum analysis and FFT offer a relatively large number of parameters which can be monitored during anesthesia: relative (or percentage) band power, absolute band power, power ratios, mean frequency, median frequency, and spectral edge frequency (SEF). All these quantitative descriptors reduce the EEG signal to a single number with obvious loss of some of the information contained.

Two spectral parameters have been frequently mentioned as reliable descriptors of electrical activity of the cortex during general anesthesia: median frequency and SEF. Since 1980, Schwilden and Stoeckel have used median frequency for monitoring anesthesia. They found a strong correlation between various stages of anesthesia (induction, maintenance, and recovery) and median frequency when using isoflurane–N_2O [42]. They also found a good relationship between observed clinical signs (BP, HR, etc.) and subdelta activity, median frequency, and spectral edge frequency [43] in volunteers treated either with etomidate, methohexital or propofol. Long et al. [31] also studied 14 patients, who were monitored while emerging from either isoflurane or fentanyl anesthesia at the termination of major surgical procedures. Their results regarding delta ratio, median frequency, and SEF showed that all three parameters significantly shifted at the end of anesthesia, indicating imminent awakening from both isoflurane and fentanyl.

In 1985 we developed a conceptual framework for selection of monitored parameters during anesthesia [35]. As a result of increasing concern about adequate patient anesthetic level and preservation of main hemodynamic parameters, we proposed a cortical activity monitor which uses DSA analysis and measures, among other factors, on-line SEF.

Spectral Edge Frequency and Depth of Anesthesia

Data from the Literature

SEF is conceptually based on a simple model of the frequency spectrum of EEG. It was first described by Rampil et al. [36] in dogs and defined as the highest frequency present in a significant quantity (90%–97%) in the spontaneous EEG on an epoch by epoch basis. A change in the end-tidal concentration of the anesthetic agent (in this case, halothane or enflurane) produced a rapid, reproducible change in SEF. The data were reproducible in individual dogs, but variability between dogs in the sensitivity of SEF to anesthetic concentration was also described.

Hudson and coworkers [26] have correlated SEF with serum concentration of thiopental in eight volunteers. They observed an average baseline SEF of 24.5 Hz in the awake state and 12.7 Hz during maximum thiopental effect.

Rampil and Matteo [37] and Sidi et al. [47] found a correlation between SEF and the pressor response to laryngoscopy and intubation.

The use of SEF for measuring depth of anesthesia has also produced contrasting data. Beside results which showed a good correlation with clinical signs of recovery from N_2O–isoflurane in children [18] and when using various doses of sufentanil in adults [39], others were unable to confirm those findings.

No correlation between SEF and methohexital blood concentration was found by Withington et al. in a small group of patients [55]. Arden et al. [1] described a poor relationship between SEF and blood level of etomidate.

These divergent results, which can be partially explained by the fact that each time a single, but different anesthetic was used, led to the conclusion that intraoperatively EEG cannot be interpreted in isolation; it must include not only the anesthesiologist's knowledge of the ongoing clinical milieu [11], but also supplementary parameters. Schwilden and coworkers [44] controlled depth of anesthesia by using a closed loop feedback system based on propofol blood concentration, with median frequency serving as a control variable.

Stanski developed the hypothesis that depth of anesthesia is a pharmacodynamic measurement [48]. Since much earlier studies showed that there was no correlation between clinical assessment during anesthesia and EEG [15], he suggested the examination of blood drug concentration in connection with certain EEG parameters and some clinical measures of depth of anesthesia.

The Clinical Approach

For technical reasons, we do not yet possess a simple and reliable method for measuring the serum level of an anesthetic drug. Therefore, we concentrated our efforts in the direction of establishing a series of guidelines during general anesthesia using only two out of three kinds of parameters proposed: the hemodynamic variables as clinical signs and SEF as a measure of cortical electrical activity.

The Decision Matrix

The proposed matrix [41] is based on a combination between BP variations and SEF numbers (Table 1). A BP deviation of more than 20% from the baseline value obtained before induction is considered abnormal and demands management [8, 38].

SEF varying between 8 and 12 Hz was interpreted by Archibald and Drazkowski [3] as compatible with normal level of a balanced anesthesia technique during maintenance. They also established that light anesthesia is accompanied by an SEF higher than 15 Hz, deep anesthesia being defined as SEF below 7 Hz.

The proposed matrix is based also on the assumption that oxygen saturation (SpO_2), end-tidal CO_2, and temperature are continuously monitored and maintained within the normal range.

The decision matrix has three BP situations (normal, at least 20% less than normal, and at least 20% higher than normal) and three levels of electrical cortical activity, as expressed by SEF ("normal" between 8 and 12 Hz, less than 8 Hz, and higher than 12 Hz); thus, the decision matrix contains nine possible combinations.

Any combination (except the ideal one in which BP and SEF both stay within the normal range) represents a special condition which is explained and for which a proposal for management is made. For instance, an SEF of 7 Hz in the presence of significant hypotension (situation F) is a sign of too deep anesthesia. The treatment includes decrease of the inspiratory concentration of the volatile drug.

On the contrary, the same significant drop in BP (more than 20% from the preinduction value) in the presence of an SEF of 11 Hz (situation H) most probably indicates hypovolemia. The proposed treatment is fluid administration, and it should not include an adjustment of the anesthetic concentration, which could lead to an episode of unwanted awareness.

The matrix also recognizes incipient situations (such as B and E) during which BP is still within normal limits, but SEF has already deviated from the

Table 1. The proposed decision matrix using spectral edge frequency (SEF) and blood pressure (BP) monitoring

SEF \ BP	> 20% higher than normal	Normal	> 20% lower than normal
> 12 Hz	A	B	C
8–12 Hz	G	Ideal	H
< 8 Hz	D	E	F

8- to 12-Hz range. In both cases, the working hypothesis demands a correction of anesthesia level which would prevent the consequent significant BP change.

Equipment

For continuous monitoring of SEF, we use CEREBRO TRAC 2500 Plus (TM), a commercially available EEG monitor [35] produced by SRD Medical, Shorashim, Israel. This monitor receives the EEG signal from five silver/silver chloride electrodes placed on the forehead with an impedance of less than 5 kΩ. The grounded lead is positioned over the midfrontal region. The other four are symmetrically placed in order to obtain two frontoparietal channels.

The monitor is a dual channel bipolar device which utilizes FFT to convert EEG waveforms from the time domain to the frequency domain. It uses epochs of 2 s each, all information in that epoch (frequency and amplitude) being processed and then displayed in color. For display purposes, the frequency data uses the DSA system. A solid white line over the DSA indicates the SEF. Complementary to SEF (obtained simultaneously for both channels), the monitor offers the following data: amplitude, percentage of bands power, and spontaneous frontal muscle activity.

The monitor can be interfaced with other monitoring devices to measure, for example, HR, BP, SpO_2 (from a pulse oximeter), or end-tidal CO_2 from a capnograph.

Clinical Use of Spectral Edge Frequency

In order to validate the efficiency of continuous monitoring SEF during maintenance of general anesthesia, we initiated a multicenter study in which therapeutical decisions were based on the proposed decision matrix.

The preliminary results, recently presented [41], seem to validate the working hypothesis. In those groups of patients in which the EEG monitor screen was visible to the anesthesiologist in charge of the case, SEF was largely kept within the pre-established normal range (8–12 Hz). This result was statistically different in the groups in which the EEG screen was not visible for the purpose of taking decisions concerning management of anesthesia. For these patients, SEF remained within normal limits for only a smaller proportion of the maintenance time.

These partial results also showed that only a proportion of the "hemodynamic events" (defined as deviations of more than 20% of the preanesthetic values of BP and HR) could be related to the level of anesthesia. When a hemodynamic event occurred and was treated according to the matrix, it was shorter than one which occured in the "nonvisible" groups.

We were also able to demonstrate that keeping SEF in the range of 8–12 Hz for the duration of maintenance of anesthesia (in the visible groups) did not significantly delay recovery time.

The studied intervals (the elapsed time from cessation of N_2O administration to extubation or to correct responses to questions such as "what is your name?" and instructions such as "open your eyes") were not longer in the visible groups than in those patients for whom SEF data were not available to the anesthesiologist responsible for the case.

These results have since been confirmed in other studies which used the same technology and methodology, but on smaller cohorts of patients.

In a clinical study [22], we used either propofol in continuous infusion (3–6 mg/kg/h^{-1}) or isoflurane (0.75%–1.5%) in maintaining general anesthesia for abdominal or orthopedic surgery. The anesthesiologist could see the SEF display for the entire duration of anesthesia. In a third group of parturients having an elective or semielective cesarean section, the anesthesiologist was denied access to SEF data. He delivered general anesthesia which included isoflurane–N_2O according to clinical signs such as HR and BP.

The results showed that for the first two groups, in which the EEG screen was open to view, SEF was kept within the desired range more than 80% of the maintenance time. The same variable in the cesarean section group (in which the SEF data was not available during anesthesia) was only 45% of the maintenance time (Table 2).

Analysis was then performed case per case for each patient in all the above three groups. It showed that the periods of time in which SEF was kept or stayed in the normal range were accompanied by a higher hemodynamic stability than in those moments in which SEF was lower than 8 Hz of higher than 12 Hz.

The initial group of 55 parturients already reported [23] was subsequently enlarged to a total number of 88, and the data was analyzed regarding differences in maternal and fetal behaviour during and after anesthesia. Once again, the EEG screen in all cases could not be seen by the anesthesiologist in charge. The analysis performed included total fentanyl dose administered (in addition to

Table 2. The spectral edge frequency (SEF) distribution in time during maintenance of general anesthesia

Group	EEG screen	Total anesthesia time (min)	SEF distribution in time (min)	
			8–12 Hz	< 8 or > 12 Hz
Propofol (n = 20)	Visible	1782	1594 (89.5%)	188 (10.5%)
Isoflurane (n = 23)	Visible	1927	1614 (83.7%)	313 (16.3%)
Cesarian (n = 55)	Hidden	2727	1229 (45%)	1498 (55%)

EEG, electroencephalogram

Table 3. The level of pain and need for analgetics in the recovery room (RR) (cesarain section group, 88 parturients)

Parameter	In-range group ($n = 29$)	Out-of-range group ($n = 59$)	p
RR pain (on scale 1–10)	5.4	6.7	0.05
RR pain score > 7 (no. patients)	8	30	0.03
Need for analgetics (no. patients)	5	24	0.02

isoflurane–N_2O–vecuronium), rate of spontaneous movements toward the end of anesthesia, the 1- and 5- min Apgar score of the newborn child, and the follow-up of the parturient during the first 24 h postoperatively. The patients were found to belong to one of the following two groups: one in which SEF stayed within the desired range of 8- to 12–Hz for more than 50% of the duration of anesthesia (29 parturients, "in range" group) and a second one for which SEF stayed in the range less than 50% of the time (the remaining 59 parturients, "out of range" group).

A more stable SEF was accompanied in the patients in the first group by a lower level of immediate postoperative pain and, as a consequence, they required less analgetics during their stay in the recovery room (Table 3).

Recently, in a pilot study [24], we investigated the usefulness of continuous SEF monitoring during propofol sedation as a supplement to epidural anesthesia. This time, we randomized the patients into two groups, differentiated only by availability of SEF data to the anesthesiologist in charge. This study showed that in the visible group, in which direct and continuous observation of the EEG screen was available, the mean duration of a BP "event" and the number of HR events were significantly lower ($p < 0.01$ and < 0.02, respectively).

Conclusions

The initial purpose of monitoring depth of anesthesia, i.e., the avoidance of undesired episodes of awareness [51], was greatly extended in the last decade.

Researchers reached the conclusion that there are different degrees of awareness, from recall to dreams and recollection under hypnosis. Soon afterwards, another goal was added, that of keeping the level of anesthesia as stable as possible in order to minimize undesired secondary effects of turbulent anesthesia, such as the sympathetic response to various stimuli during the anesthetic and surgical procedure [37, 47].

Clinicians are concerned about the fact that there is still no proved method or combination of methods for measuring depth of anesthesia. Neither indirect clinical signs of anesthesia nor the more modern tools of monitoring have solved the problem. No sign or parameter alone can give enough information to the anesthesiologist regarding the level of general anesthesia.

One evident obstacle is the way we bring about clinical anesthesia: "anesthesia is produced by a number of pharmacological effects that are not necessarily produced by all drugs or may be produced by different concentrations of the same drug" [48].

The use of EEG led to controversial results. Although a description of electrical activity during anesthesia emerged from early reports, it has been argued that in fact EEG does not seem to be directly related to the level of arousal, but rather to the "amount of work" done at a given moment by the cerebral cortex, and that synchronized EEG is found whenever this decreases below a certain threshold [53].

In the presence of so many unsolved problems, a single parameter such as SEF cannot offer all the data clinicians would expect. The combination between SEF and hemodynamic parameters can be viewed as a step forward.

Some unanswered questions still remain. One such question is the differentiation between the hypnotic and the analgetic effect of the various anesthetics used in clinical practice. In other words, when changing an anesthetic concentration or dosage, we are supposed to look for additional information before we decide whether more analgesia, more hypnosis, or both are required. A combination of data emerging from the concomittant use of processed EEG and evoked potentials (EP) could provide an answer to this problem. Meanwhile, the technical difficulties involved in using EP still prevent its routine use [45] in the operating rooms for this specific purpose.

A second interesting question concerns the proper level of sedation needed for supplementing a locoregional anesthetic procedure or for patients admitted to an intensive care unit [46]. Here, too, it seems that processed EEG correlated with hemodynamic behaviour could deliver a clinically acceptable method of monitoring.

These and other clinical questions which await a solution are possible new directions for future research.

References

1. Arden JR, Holley FO, Stanski AR (1986) Increased sensitivity to etomidate in the elderly: initial distribution versus altered brain response. Anesthesiology 65:19–27
2. Artusio JF Jr (1954) Di-ethyl ether analgesia: a detailed description of the first stage of ether anesthesia in man. J Pharmacol Exp Ther 111:343

3. Archibald JE, Drazkowski JE (1985) Clinical applications of compressed spectral analysis (CSA) in OR/ICU settings. Am J EEG 25:13–36
4. Bashein G, Nessly ML, Bledsoe SW et al. (1992) Electroencephalography during surgery with cardiopulmonary bypass and hypothermia. Anesthesiology 76:878–891
5. Bickford RG, Billinger TW, Fleming NI et al. (1972) The compressed spectral array (CSA) – a pictorial EEG. Proc San Diego Biomed Symp 11:365–370
6. Chang T, Dworsky WA, White PF (1988) Continuous electromyography for monitoring depth of anesthesia. Anest Analg 67:521–525
7. Chiappa KH, Burke SR, Young RR (1979) Results of EEG monitoring during 367 carotid endarterectomies. Stroke 10:381–388
8. Charlson ME, Mackenzie CR, Gold JP et al. (1990) Intraoperative blood pressure: what patterns identify patients at risk for postoperative complications. Ann Surg 212:567–580
9. Clark DL, Rosner BS (1973) Neurophysiologic effects of general anesthetics. I: the electroencephalogram and sensory evoked responses in man. Anesthesiology 38:564–582
10. Cullen DJ, Eger EI II, Stevens WC et al. (1972) Clinical signs of anesthesia. Anesthesiology 36:21–25
11. Donegan JH, Rampil IJ (1989) The electroencephalogram. In: Blitt CD (ed) Monitoring in anesthesia and critical care medicine, 2nd edn. Churchill Livingstone, New York, p 444
12. Evans JM, Fraser A, Wise CC (1983) Computer controlled anesthesia. In: Prakash O (ed) Computing in anesthesia and intensive care. Nijhoff, Boston, pp 279–291
13. Evans JM, Davies WL, Wise CC (1984) Lower oesophageal contractility: a new monitor of anesthesia. Lancet 1:1151–1154
14. Fleming RA, Smith NT (1979) Density modulation: a technique for display of three variable data in patient monitoring. Anesthesiology 50:543–546
15. Galla SJ, Rocco AG, Vandam LD (1958) Evaluation of traditional signs and stages of anesthesia: an electroencephalographic and clinical study. Anesthesiology 19:328–332
16. Gibbs GA, Gibbs El, Lennox WG (!937) Effect on the electroencephalogram of certain drugs which influence nervous activity. Arch Intern Med 60:154–166
17. Goddard GF (1982) A pilot study of changes in skin electrical conductance in patients undergoing general anesthesia and surgery. Anaesthesia 42:596–603
18. Goldman LJ, Goldman E (1989) Automated EEG analysis during recovery from isoflurane anesthesia in pediatrics. Anesth Analg 68:S105
19. Graff TD, Phillips OC (1959) Consciousness and pain during apparent surgical anesthesia. JAMA 170:2069–2071
20. Griffith H, Johnson GE (1942) The use of curare in general anesthesia. Anesthesiology 3:418–420
21. Guedel AE (1937) Inhalation anesthesia, a fundamental guide. Macmillan, New York
22. Gurman GM, Porath A, Fajer S, Pearlman A (1993) Correlation of EEG spectral edge frequency with hemodynamic stability during maintenance of general anesthesia. In: Sebel PS, Bonke B, Winograd E (eds) Memory and awareness in anesthesia. PRT Prentice Hall, NJ, USA, pp 265–274
23. Gurman GM, Schilly M, Porath A, Gdor S (1992) Spectral edge frequency during cesarian section and its influence on various intra- and postoperative parameters. Proceedings of the 10th World Congress of Anesthesiology, the Hague, no 283

24. Gurman GM, Glaser M, Korotkoruchko A et al. (1993) Continuous propofol infusion monitored by processed EEG as a supplement for epidural anesthesia. Acta Anesth Scand 37:228

25. Hameroff SR, Grantham CD (1989) Monitoring anesthesia depth. In: Blitt CD (ed) Monitoring in anesthesia and critical care medicine, 2nd edn. Churchill Livingstone, New York, pp 539–553

26. Hudson RJ, Stanski DR, Saidman LJ et al. (1983) A model for studying depth of anesthesia and acute tolerance to thiopental. Anesthesiology 59:301–308

27. Hugg CC (1990) Anesthesia for adult cardiac surgery. In: Miller RD (ed) Anesthesia, 3rd edn. Churchill Livingstone, New York, p 1646

28. Johnstone M (1974) Digital vasodilatation: a sign of anesthesia. Br J Anaesth 46:414–419

29. Levinson BW (1965) State of awareness during general anaesthesia. Br J Anaesth 37:544–546

30. Levy WJ, Shapiro HM, Maruchak G et al. (1980) Automatic EEG processing for intraoperative monitoring: a comparison of techniques. Anesthesiology 52:223–236

31. Long CW, Shah NK, Loughlin C et al. (1989) A comparison of EEG determinants of near-awakening from isoflurane and fentanyl anesthesia. Spectral edge, median power frequency and delta ratio. Anesth Analg 69:169–173

32. Meyer BC, Blacher RS (1961) A traumatic neurotic reaction induced by succinycholine chloride. NY State J Med 61:1255–1261

33. Mori K, Kawamata M, Miyajima S et al. (1972) The effects of several anesthetic agents on the neuronal reactive properties of talamic relay nuclei in the cat. Anesthesiology 36:550–557

34. Mori K, Winters WD (1975) Neural background of sleep and anesthesia. Int Anesthesiol Clin 13:67–108

35. Pearlman AL, Gurman GM (1985) Toward a unified monitoring system during anesthesia. Int J Clin Monit Comput 2:21–27

36. Rampil IJ, Sasse FJ, Smith NT et al. (1980) Spectral edge frequency – a new correlate of anesthetic depth. Anesthesiology 53:S12

37. Rampil IJ, Matteo RS (1987) Changes in EEG spectral edge frequency correlate with hemodynamic response to laryngoscopy and intubation. Anesthesiology 67:139–142

38. Reves JG, Smith LR (1990) From monitoring to predicting outcome. Ann Surg 212:559–560

39. Ross L, Matteo RS, Ornstein E et al. (1989) Dose-electroencephalographic (EEG) response to sufentanil in man. Anesthesiology 71:A263

40. Russell IF (1989) Conscious awareness during general anaesthesia: relevance of autonomic signs and isolated arm movements as guides to depth of anesthesia. In: Jones JC (ed) Clinical anesthesiology, Baillere Tindall, London, pp 511–532

41. Schaefer MK, Gurman GM, Moecke HP et al. (1993) Preliminary results of inhalatory and balanced anesthesia protocol on monitoring depth of anesthesia. In: Sebel PS, Bonke B, Winograd E (eds) Memory and awareness in anesthesia. PRT Prentice Hall, NJ, USA, pp 310–317

42. Schwilden H, Stoeckel H (1987) Quantitative EEG analysis during anesthesia with isoflurane in nitrous oxide at 1.3 and 1.5 MAC. Br J Anaesth 59:53–59

43. Schwilden H (1989) Use of median EEG frequency and pharmocokinetics in determining depth of anesthesia. In: Jones JG (ed) Depth of anesthesia. Clinical anesthesiology. Balliere Tindall, London, p 607

44. Schwilden H, Stoeckel H, Schuttler J (1989) Closed-loop feedback control of propofol anesthesia by quantitative EEG analysis in humans. Br J Anaesth 62:290–296
45. Sebel PS (1989) Somatosensory visual and motor evoked potentials in anaesthetized patients. In: Jones JG (ed) Depth of anesthesia. Clinical anesthesiology. Balliere Tindall, London, pp 587–602
46. Shearer ES, O'Sullivan EP, Hunter JM (1991) An assessment of Cerebro Trac 2500 for continuous monitoring of cerebral function in the intensive care unit. Anaesthesia 46:750–755
47. Sidi A, Halimi P, Cotev S (1990) Estimating anesthetic depth from computerized EEG during anesthetic induction and intubation in cardiac surgery patients. J Clin Anesth 2:101–105
48. Stanski DR (1990) Monitoring depth of anesthesia. In: Miller RD (ed) Anesthesia, 3rd edn. Lippincott, Philadelphia, pp 1001–1029
49. Stundt TM, Sharbrough FW, Piepgras DG et al. (1981) Correlation of cerebral blood flow and EEG changes during carotid endarterectomy. Mayo Clin Proc 56:533–543
50. Thornton C, Newton DEF (1989) The auditory evoked response: a measure of depth of anesthesia. In: Jones JC (ed) Depth of anesthesia. Clinical anesthesiology. Balliere Tindall, London, pp 559–585
51. Tunstall ME (1977) Detecting wakefulness during anaesthesia for caesarian section. Br Med J 1:1321
52. Volavka J, Matousek M, Feldstein S (1973) The reliability of electroencephalography assessment. Electroencephalogr Electromyogr 4:123–128
53. Webb AC (1983) Consciousness and the cerebral cortex. Br J Anaesth 55:209–211
54. Winterbottom EH (1950) Insufficient anaesthesia. Br J Anaesth 1:247–248
55. Withington PS, Morton J, Arnold R et al. (1986) Assessment of power spectral edge for monitoring depth of anesthesia using low rate methohexital infusion. Int J Clin Monit Comput 3:117–122
56. Woodbridge PD (1957) Changing concepts concerning depth of anesthesia. Anesthesiology 18:536–539

"Paradoxical Arousal" During Isoflurane/Nitrous Oxide Anesthesia: Quantitative Topographical EEG Analysis

P. Bischoff, E. Kochs, and J. Schulte am Esch

Electroencephalogram (EEG) recordings have been used for the evaluation of drug effects on brain electrical activity, and numerous studies suggest that EEG measures may be useful for the assessment of depth of anesthesia [16, 22]. It has been shown that increases in depth of anesthesia may be reflected by the appearance of EEG slow-wave activity, and decreases in fast-wave activity [20]. Arousal reactions during emergence from anesthesia have been found to be associated with EEG desynchronization with a shift to higher frequencies. Controversy exists on the EEG response indicating intraoperative arousal phenomena. In anesthetized patients a shift to EEG delta activity concurrent with cardiac and respiratory irregularities has been interpreted as an indicator of insufficient depth of anesthesia. These intraoperative electrophysiological arousal phenomena have been addressed as "reverse" or "paradoxical" arousal phenomena [1]. However, brain electrical activity is modulated not only by drug effects but also by changes in respiratory and hemodynamic parameters, body temperature, and exogenous stimuli. Previous studies have shown that EEG high voltage slow waves may also occur spontaneously or in response to auditory or painful stimulation [6, 25].

Only few data are available on alterations in brain electrical activity induced by the surgical procedure per se. The present study investigated the spatial distribution of quantitative EEG responses induced by noxious stimulation during steady-state anesthesia with 0.6% and 1.2% isofluranc in nitrous oxide. In addition, the effect of analgesic treatment on EEG responses to surgical manipulations was studied in additional patients.

Methods

Following institutional approval and written informed consent 48 patients [age 43 ± 10 years; American Society of Anesthiologists (ASA) score, I, II] without neurological disease scheduled for elective abdominal surgery were included in the study. After premedication with midazolam (7.5 mg orally) anesthesia was induced by etomidate (0.3 mg/kg), fentanyl (1.5–2.0 μg/kg), and vecuronium (0.1 mg/kg). Following tracheal intubation patients were mechanically ventilated [end-tidal carbon dioxide (P_{ETCO_2}) 35–38 mm Hg]. Anesthesia was maintained with 0.6% or 1.2% end-tidal isoflurane and 66% nitrous oxide in oxygen. The following parameters were recorded: mean arterial pressure (MAP; mmHg), heart rate (HR; beats/min), body temperature (°C; rectal), arterial oxygen saturation (SaO_2; %), end-tidal isoflurane (P_{ETISO}; percentage), and P_{ETCO_2} (mmHg). During surgery increases in MAP or HR of less than 40% from baseline were accepted; otherwise patients were treated with fentanyl and excluded from the study.

Original EEG tracings, recorded from 17 electrodes (reference Cz) placed according to the international 10–20 system (impedance > 20 MΩ; bandpass, 0.45–35 Hz) were displayed on a monitor, digitized (256/s), and stored on disk (CATEEM, Medisyst, Linden). Artifact control was by electro-oculogram and electrocardiogram. Following Fast Fourier Transformation (4-s period; bandpass 1.25–35.0 Hz) the topographical distribution of EEG power was calculated with respect to common average reference in selected frequency bands: delta (1.25–4.5 Hz), theta (4.7–6.8 Hz), alpha 1 (7.0–9.5 Hz), alpha 2 (9.7–12.5 Hz), beta 1 (12.7–18.5 Hz) and beta 2 (18.7–35.0 Hz). On a second monitor the topographical distribution of brain electrical activity was displayed as color maps (brain mapping).

Patients were randomly divided into four different treatment groups: In groups 1 ($n = 12$) and 2 ($n = 12$) anesthesia was maintained with 0.6% isoflurane in nitrous oxide in oxygen and in groups 3 ($n = 12$) and 4 ($n = 12$) with 1.2% isoflurane in nitrous oxide in oxygen. After establishment of steady-state anesthesia (P_{ETCO_2} 35–38 mmHg; P_{ETISO} 0.6% or 1.2%) all data were recorded over 20 min. Data collected during the first 6 min served as baseline values. In group 1 (P_{ETISO} 0.6%) and group 3 (P_{ETISO} 1.2%) no external stimulation was performed; these groups served as controls. In group 2 (P_{ETISO} 0.6%) and group 4 (P_{ETISO} 1.2%) surgical stimulation (skin incision with subsequent surgical procedures) was started after recording of baseline values.

Because it was unclear whether the EEG responses due to surgery may change with analgesic treatment, further patients with 0.6% isoflurane in 66% nitrous oxide in oxygen anesthesia and additional analgesic treatment (alfentanil infusion, epidural anesthesia) were studied according to a similar protocol. Provisional results (case reports) are presented. Alfentanil (bolus 100 μg kg^{-1}; infusion 1 μg kg^{-1} min^{-1}) in addition to 0.6% isoflurane in nitrous oxide in oxygen was used to achieve steady-state alfentanil plasma concentrations [9] before recording of baseline data. In further patients EEG recording was started after performance of lumbar epidural anesthesia (bupivacain 0.5%; 15 ml). Anesthesia was maintained with 0.6% isoflurane in nitrous oxide in oxygen.

In all patients EEG data from each electrode position were averaged over the following periods: 0–6 min (baseline), 1–2 min, 3–4 min, 5–6 min, 7–8 min, 9–10 min, 11–12 min, 13–14 min following the start of surgery. All data are given as median ($\mu V^2 \pm$ SD) or relative changes from baseline (% $\pm$ SD). Multivariate analysis of variance was performed to analyze changes over time in selected recording sites (F4, C4, T4, P4). The Wilcoxon test was used to test for significant differences between groups ($p < 0.05$, median $\pm$ SD).

Results

Demographical data were comparable in all groups (Table 1). $P_{ET}CO_2$, SaO_2, and body temperature did not change over time in any group.

Hemodynamics

During baseline (6 min) no differences in MAP or HR were noted among groups (groups 1–4) or in the patients given additional analgesic treatment. Following surgical stimulation (groups 2 and 4) MAP was increased by $30\% \pm 10\%$ and $28\% \pm 12\%$; versus baseline ($p < 0.05$), whereas HR did not change over time in either group. In patients treated with alfentanil or epidural anesthesia in addition to 0.6% isoflurane in 66% nitrous oxide in oxygen no changes in hemodynamics were seen over the whole observation period.

EEG

0.6% Isoflurane in 66% Nitrous Oxide in Oxygen. During steady-state anesthesia with 0.6% isoflurane and 66% nitrous oxide in oxygen (groups 1, 2) the EEG baseline pattern was dominated by alpha activity superimposed with low amplitude faster waves. After the start of surgery (group 2) the occurrence of slow-wave activity (delta-theta waves) was observed with a dominance at frontal leads (Fig. 1).

The topographical distribution (brain mapping) in selected frequency bands did not change during baseline recordings (6 min), and no differences between groups (group 1 and 2) were observed. Alpha 1/2 was dominant at frontal and central areas. The spatial distribution of delta and theta power was comparable,

Table 1. Patient characteristics

	Age (years)	Weight (kg)	Sex	ASA status
Group 1 ($n = 12$):				
0.6% isoflurane	39 ± 11	72 ± 17	5 M, 7 F	I, II
Group 2 ($n = 12$):				
0.6% isoflurane	45 ± 8	68 ± 15	6 M, 6 F	I, II
Group 3 ($n = 12$):				
0.2% isoflurane	41 ± 13	74 ± 12	6 M, 6 F	I, II
Group 4 ($n = 12$):				
1.2% isoflurane	44 ± 14	72 ± 13	5 M, 7 F	I, II

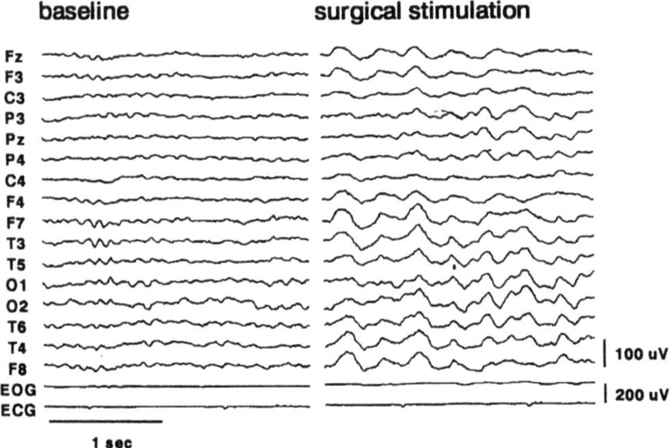

Fig. 1. Original tracings: 16-channel EEG, electro-oculogram, and electrocardiogram recordings (Cz reference) in one patient of group 2 before and after start of surgery. During baseline EEG patterns were dominated by alpha superimposed with fast-wave activity. The start of surgery produced abrupt EEG slowing, which was most pronounced at frontotemporal leads

with a relative maximum at frontal and occipital recording sites (Fig. 2). After the start of surgical stimulation (group 2) brain electrical activity was changed, and slow-wave activity became dominant. These increases in absolute and relative delta power were associated with decreases in fast-wave activity at identical cortical areas. EEG power was changed quantitatively, with a dominance at frontal leads (Fig. 2). At frontal areas (F4) delta activity was increased from 69.6 to 146.4 μV^2 during surgery. Fast-wave activity (alpha 2) was decreased from 25.0 ± 8.3 to $15.7 \pm 8.8 \mu V^2$. Relative decreases in alpha 1 and alpha 2 activities were $-37.2 \pm 14.2\%$ and $-50.9 \pm 33.2\%$, respectively (Fig. 3).

Differences in topographical distribution of EEG responses to surgery were observed at frontal areas (F4) consisting of changes in delta activity ($+111.8 \pm 34.8\%$) which were maximal 4–6 min after the start of surgery and more pronounced when compared to parietal areas (P4). This EEG response was associated with concurrent decreases in alpha 1 ($-37.2 + 14.2\%$) and alpha 2 activities ($-50.9 \pm 33.2\%$) with a maximum decrease 7–8 min after the start of surgery at identical cortical areas ($p < 0.05$; Fig. 4).

0.6% Versus 1.2% Isoflurane in 66% Nitrous Oxide in Oxygen. At baseline delta power was increased during 1.2% isoflurane (groups 3 and 4) when compared with 0.6% isoflurane (groups 1 and 2; Fig. 5). Maxima of slow waves were most prominent frontally in both groups (0.6% and 1.2% isoflurane) with a spread to central leads during high-dose isoflurane. Delta activity was increased by $+51\%$ (F4) during 1.2% isoflurane, and burst suppression periods

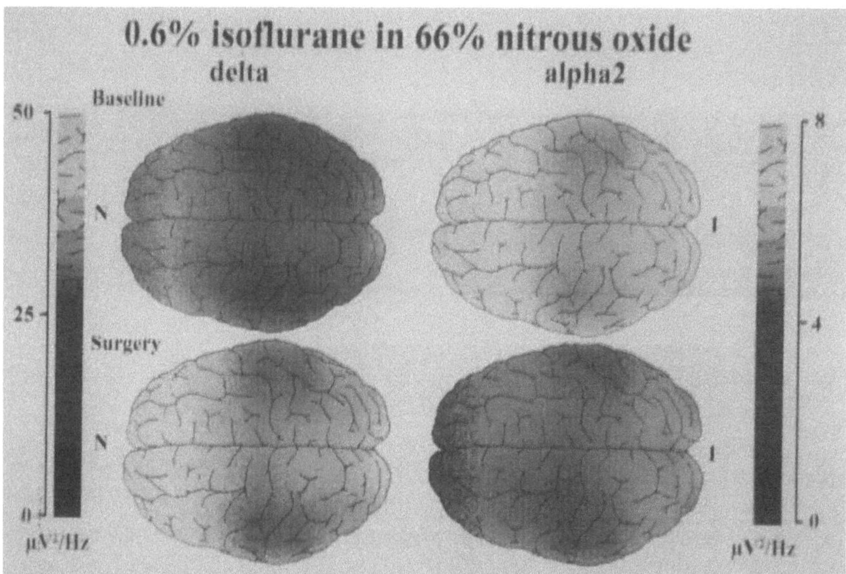

Fig. 2. Topographical distribution of absolute power densities in delta and alpha 2 were coded by colors (color scale): *black*, zero activity; *light blue*, maximal activity. EEG maps are given for baseline (6 min) and maximal changes (delta, 5–6 min; alpha 2, 7–8 min) following surgical stimulus. During baseline alpha 2 and delta were prominent frontally in a different degree. After the start of surgery changes in color indicate delta shift at identical cortical areas

(2–5 s) were noted in 8 of 11 patients. Surgical stimulation resulted in dose-dependent increases (0.6% versus 1.2% isoflurane) in median delta power at the frontal region F4 (group 2 $+111.8\% \pm 34.8\%$ versus group 4 $+40\% \pm 47.8\%$). The increase in delta power was also maximal at frontal leads. However, the delta shift with suppression in fast-wave activity was attenuated but not completely abolished during high-dose isoflurane (group 4; Fig. 5). In both isoflurane groups alpha 1/2 activities were decreased to a similar degree (groups 2 and 4).

0.6% Isoflurane in 66% Nitrous Oxide in Oxygen plus Analgesic Treatment. EEG responses to surgery as reflected by increases in delta and decreases in alpha activities (Fig. 6) were blocked in patients anesthetized with 0.6% isoflurane in 66% nitrous oxide in oxygen given additional analgesic treatment. During laparotomy all EEG frequency bands were unchanged over time when epidural anesthesia (0.5% bupivacain) was used in addition to 0.6% isoflurane in 66% nitrous oxide in oxygen (Fig. 7). Also, no delta shift was observed in patients treated with high doses of narcotics. When alfentanil (bolus 100 μg kg^{-1} continued by 1 μg kg^{-1} min^{-1}) was used to achieve steady-state plasma concentrations, EEG did not change in response to surgery (Fig. 7).

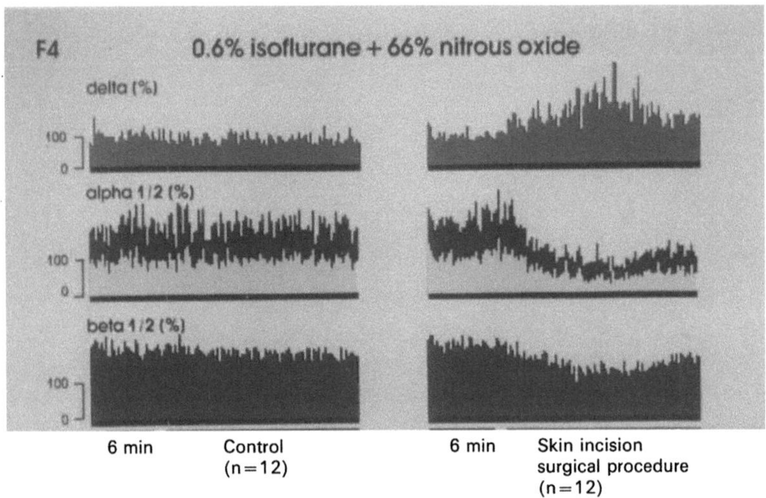

Fig. 3. Frontal lead F4. Time plot of relative changes in EEG power (%) from baseline: delta, alpha 1/2, and beta 1/2. *Left*, in group 1 (control; $n = 12$) all frequency bands were stabile over time; *right*, following surgical stimulation (group 2; $n = 12$) delta was enhanced. Increases of more than 100% occurred 5–6 min after the start of surgery and were associated with decreases in fast-wave activity (alpha 1/2, beta 1/2)

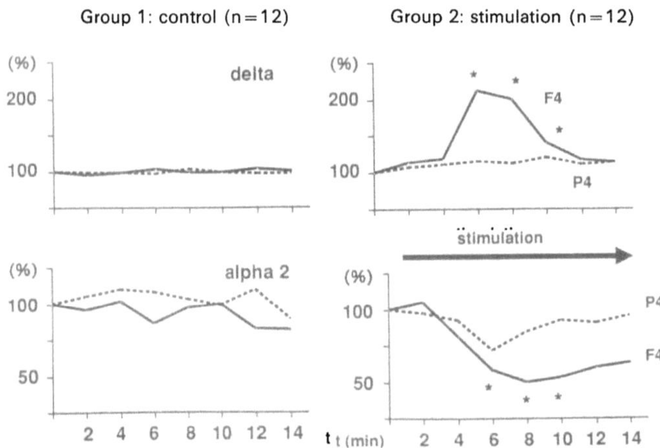

Fig. 4. *Left*, group 1 (control; $n = 12$) mean EEG data (%) of delta and alpha 2 did not change in F4 versus P4 over time; *right*, during surgical stimulation (group 2; $n = 12$) delta was enhanced maximally at F4($+ 111.8 \pm 34.8\%$) 5–6 min after the start of surgery, whereas parietal regions (P4) were only small affected. At the same time depression in alpha 2 activity was more pronounced at F4 when compared to P4 ($p < 0.05$), and maximum (F4, $- 50.9 \pm 33.2\%$) was reached 7–8 min after the start of surgical preparation. Relative changes (%) in F4 versus P4, $p < 0.05$

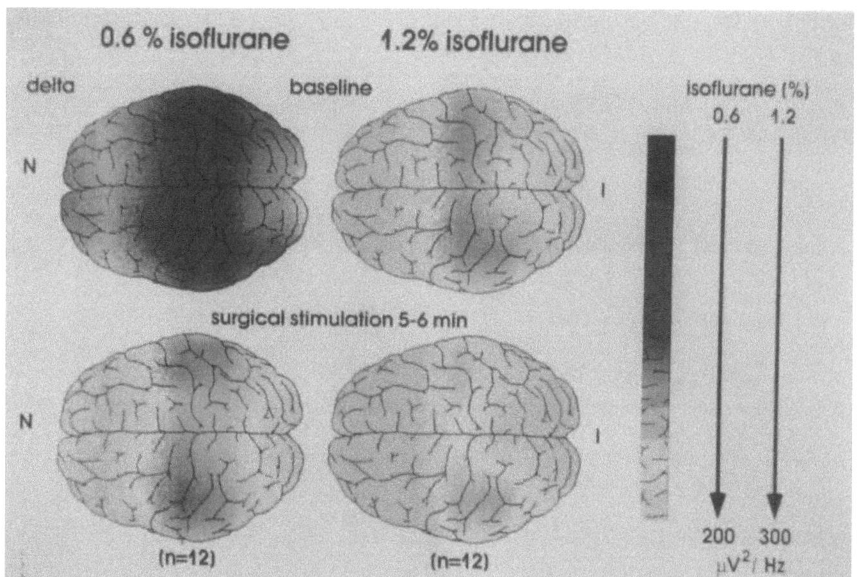

Fig. 5. Changes in topographical distribution of delta power (EEG maps) in patients treated with 0.6% ($n = 12$; *left*) and 1.2% isoflurane ($n = 12$; *right*); changes in color indicate that delta was enhanced using 1.2% isoflurane during baseline (*above*). Following surgical stimulation (*below*; maximal changes after the start of surgery) the delta shift was still noted in patients treated with high-dose (1.2%) isoflurane

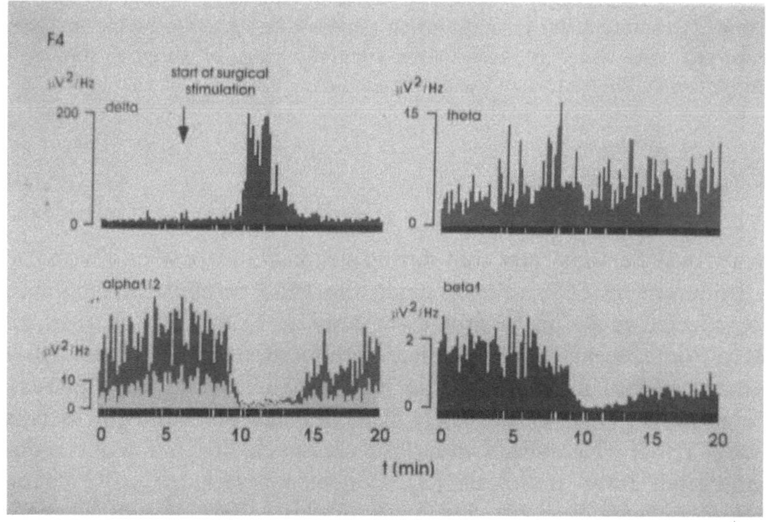

Fig. 6. Time plot of delta, theta, alpha 1/2, and beta 1/2 (F4 median, $\mu V^2/Hz$) of one patient anesthetized with 0.6% isoflurane in 66% nitrous oxide in oxygen (group 2; case report). Start of surgery (laparotomy) resulted in a large increase in delta activity associated with decreases in alpha 1/2 and beta 1/2 frequency band 5–6 min after the start of surgery. Theta band was affected only little

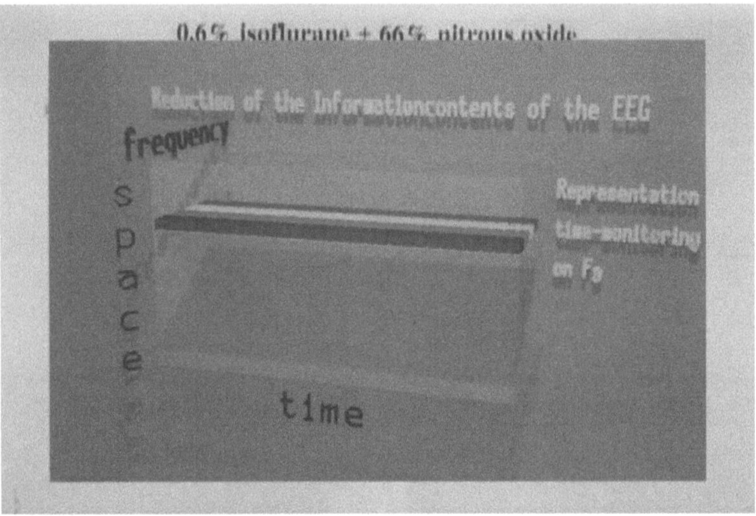

Fig. 7. Time plot of delta, theta, alpha 1/2, and beta 1/2 (F4 median, $\mu V^2/Hz$) recorded from two patients anesthetized with 0.6% isoflurane in nitrous oxide in oxygen and additional analgesic treatment (case reports). *Patient left*, EEG baseline recording was started after performance of epidural anesthesia (0.5% bupivacain 15 ml) and steady-state 0.6% isoflurane in nitrous oxide in oxygen anesthesia. *Patient right*, EEG recording was started during continuous alfentanial infusion (100 μg kg within 10 min and 1 μg kg^{-1} min^{-1}, continuously) in addition to steady-state 0.6% isoflurane in nitrous oxide in oxygen anesthesia. In both cases after the start of surgical stimulation (laparotomy) none of the frequency bands changed over time

Discussion and Conclusion

The present study demonstrates that during steady-state anesthesia with 0.6% or 1.2% isoflurane in 66% nitrous oxide the EEG response to abdominal surgery is reflected in a shift to slow-wave activity. These EEG changes were dominant at frontal areas. Similar findings with increases in EEG delta activity have been interpreted as characteristic EEG responses indicating increased depth of anesthesia [21, 22]. Conversely, EEG desynchronization and increases in fast-wave activity have been described as typical for imminent arousal during emergence from anesthesia [5]. Concurrent with the EEG changes seen here, the hemodynamic responses indicate intraoperative arousal phenomena induced by noxious stimulation. These results are in agreement with earlier findings [1] which reported so-called "reverse" arousal reactions during halothane anesthesia in man. The authors pointed out that either EEG desynchronization and increases in fast-wave activity or EEG synchronization with the appearence of high-voltage slow-wave activity was related to hemodynamic and respiratory irregularities or spontaneous movement in halothane anesthetized patients. However, in a later study no relationship between EEG changes induced by thiopental and clinically assessed depth of anesthesia were found [10]. From many studies performed so far it has been concluded that the attempt to define EEG parameters useful for the assessment of depth of anesthesia is confounded by the complexity of the EEG. In clinical practice, in addition to drug-induced effects, brain electrical activity is modulated by a variety of variables [17]. As a result, EEG descriptors which guarantee a precise definition of the level of anesthesia during surgery have never been agreed upon.

Our findings on the effects of isoflurane on brain electrical activity are in agreement with previous studies [4]. In patients without sensory stimulation delta activity was dominant frontally, spreading to central regions with increased isoflurane concentration. In accordance with previous studies in isoflurane anesthetized patients [7], slow-wave activity was prominent at frontal areas during baseline recordings in all groups. Following surgical stimulation frontal delta activity was enhanced and fast-wave activity (alpha 1/2) was decreased in both groups (0.6% and 1.2% isoflurane). The EEG responses to noxious stimulation were most prominent during lower levels of anesthesia (0.6% isoflurane). These findings are in agreement with previous studies using various anesthetic techniques. The intraoperative delta shift has been interpreted as electrophysiological reverse or paradoxical arousal reactions in the presence of clinical signs of insufficient anesthesia [1,3,23]. Most interestingly, the delta shift seen here was not abolished with 1.2% isoflurane in 66% nitrous oxide. Likewise, increases in MAP demonstrate that arousal phenomena were still present at this level of anesthesia.

It is unclear why these EEG phenomena have received scant attention in the literature so far. Only few studies report the effect of sensory stimulation on brain electrical activity during steady-state anesthesia [2]. The EEG changes seen here occurred predominantly at frontal areas. The frontal dominance of EEG changes induced by noxious stimulation may explain why similar EEG changes have not been found more often. Most studies using single- or two-channel recordings at parietal/temporal areas may not detect the significant EEG alterations in frontal delta power seen in the present study.

Concurrent with the EEG alterations seen here increases in MAP were observed. It has been shown previously that increases in cerebral blood flow may induce changes in brain electrical activity. However, the blood pressure response did not exceed the upper limit of cerebral autoregulation in all groups. In addition, autoregulation of cerebral blood flow is unaffected by isoflurane concentrations within the range used in the present study [11,14,24]. All other parameters measured (P_{ETCO_2}, SaO_2, body temperature, etc.) were stable over time. It can be concluded that the EEG slowing was not secondary to changes in blood pressure or brain perfusion, and that most likely neuronal arousal phenomena were involved. Since no general stress parameter is available so far, the EEG response to stress-induced alterations in humoral or neural homeostasis cannot be assessed by the present study. On the other hand, our preliminary investigations with adjuvant high doses of narcotics or epidural anesthesia indicate that EEG responses to nociceptive pain stimuli may be blocked by adequate analgesic treatment. This was supported by unchanged hemodynamics and lack of clinical signs of arousal for the whole observation period (baseline and surgical stimulation). These findings indicate that pain-related nociceptive transmission during general anesthesia with volatile anesthetics may be associated with a frontal EEG delta shift.

However, the underlying electrophysiological mechanisms of arousal are still unknown. We conclude that concurrent with clinical signs of insufficient depth of anesthesia two different types of intraoperative arousal reactions may be distinguished. Several studies suggest that arousal during emergence from anesthesia is indicated by EEG desynchronization and the appearance of fast-wave activity similar to findings in awake healthy subjects. These EEG changes my be related to clinical sings (respiratory and cardiac irregularities, movement) of a light level of anesthesia [18]. However, this arousal mechanism appears to be different from arousal phenomena following sensory stimulation in long-term sedated and head-trauma patients [8, 13, 19, 25]. It has been pointed out that in these patients the occurrence of slow-wave patterns following external stimulation may reflect an unconscious arousal reaction. Stimulus-induced delta rhythms in anesthetized patients are similar to delta rhythm during arousal seen during EEG sleep recordings [12]. These EEG patterns are thought to be generated in thalamic and subthalamic areas when the ascending reticular activation system is functionally blocked. The importance of the reticular activating system in the maintenance of the conscious state has been shown previously. Moruzzi and Magoun [15] reported in 1949 on arousal reactions originating from direct stimulation of the reticular formation of the brain stem. From these findings it may be concluded that the EEG response to surgery seen here is modulated by isoflurane in the brainstem reticular formation. There is ample evidence that the ascending reticular function is attenuated by anesthetic drugs. Thus EEG response to surgery is related to the actual level of anesthesia. Taken altogether, these findings may serve as an explanation of why different types of electrophysiological arousal (EEG desynchronization with shift to faster waves *or* synchronization with shift to slow waves) have been described and would partly explain the controversy about electrophysiological arousal reactions.

The Fast Fourier Transformation bandpass used in the present study may have influenced the results, It has been pointed out that during anesthesia most EEG activity may be found in the subdelta range (0.5–1.5 Hz) [21]. In the present study a Fast Fourier Transformation high-frequency bandpass of 1.25 Hz was used. Thus, it cannot be excluded that the increase in delta activity described here is actually a shift in the dominant frequency from the lower frequency range (e.g., 0.5–1.2 Hz), which could not be assessed by the setup used. However, the original EEG was recorded with a high-frequency bandpass of 0.45 Hz. Visual inspection of all EEG tracings before noxious stimulation revealed a dominant frequency in the theta-alpha bands. Following the start of the surgical procedure this EEG pattern was changed abruptly to slow-wave activity. Thus, it is unlikely that the high-frequency bandpass resulted in a false interpretation of the present data.

In conclusion, a general agreement that the appearance of slow-wave activity indicates increased depth of anesthesia cannot be accepted. The EEG changes with increases in slow-wave activity and decreases in fast-wave activity concurrent with changes in autonomic parameters indicate involvement of arousal mechanisms. Similar electrophysiological arousal reactions previously have been addressed as paradoxical or reverse arousal reaction. Our data indicate that topographical EEG monitoring is sensitive enough to detect nociceptive transmission induced by surgical manipulations. Using high-dose (1.2%) isoflurane for increased depth of anesthesia, the frontal delta shift was attenuated but not completely abolished. In contrast, using high-dose narcotics or epidural anesthesia in addition to 0.6% isoflurane in 66% nitrous oxide, no EEG responses were observed. Further studies are needed to quantify stimulus-induced EEG response with respect to changing levels of anesthesia and different anesthetic techniques.

References

1. Bimar J, Bellville JW (1977) Arousal reactions during anesthesia in man. Anesthesiology 47:449–454
2. Bischoff P, Kochs E, Droese D, Meyer-Moldenhauer WH, Schulte am Esch J (1993) Topographisch quantitative EEG-Analyse der paradoxen Arousalreaktion: EEG-Veränderungen bei urologischen Eingriffen unter Isofluran/N_2O Narkose. Anaesthesist 42:142–148
3. Brazier MA (1954) The action of anaesthetics on the nervous system with special reference to the brain stem reticular system. In: Delafresnaye JF (ed) Brain mechanisms and consciousness. Blackwell, Oxford, pp 163–199
4. Clark DL, Hosick EC, Adam N, Castro AD, Rosner BS, Neigh JL (1973) Neural effects of isoflurane (forane) in man. Anesthesiology 39:261–270
5. Drumond JC, Brann CA, Perkins DE, Wolfe DE (1991) A comparison of median frequency, spectral edge frequency band power ratio, total power, and dominance shift in the determination of depth of anesthesia. Acta Anaesthesiol Scand 35:693–699
6. Eger EI, Stevens WC, Cromwell TH (1971) The electroencephalogram in man anesthetized with forane. Anesthesiology 35:504–508
7. Engelhardt W, Carl G, Dierks T, Maurer K (1991) Electroencephalographic mapping during isoflurane anesthesia for treatment of mental depression. J Clin Moni 7:23–29
8. Evans BM (1976) Pattern of arousal in comatose patients. J Neurol Neurosurg Psychiatry 39:392–402
9. Geerts P, Noorduin H, Vanden Bussche G, Heykants J (1987) Practical aspects of alfentanil infusion. Eur J Anaesth [Suppl] 1:25–29

10. Hung OR, Varvel JR, Shafer SL, Stanski DR (1992) Thiopental pharmacodynamics. II. Quantitation of clinical and electroencephalgraphic depth of anesthesia. Anesthesiology 77:237–244

11. Kochs E, Hoffman WE, Werner C, Albrecht RF, Schulte am Esch J (1993) Cerebral blood flow velocity in relation to cerebral blood flow, cerebral metabolic rate for oxygen, and electroencephalogram analysis in dogs. Anesth Analg 76:1222–1226

12. Kubicki St, Haas J (1975) Elektro-klinische Korrelationen bei Komata unterschiedlicher Genese. Aktuel Neurol 2:103–112

13. Li CL, Jasper H, Henderson L (1952) The effect of arousal mechanisms on various forms of abnormality in the electroencephalogram. Electroenceph Clin Neurophysiol 4:512–526

14. Madsen JB, Cold GE, Hansen ES, Bardrum B (1987) The effect of isoflurane on cerebral blood flow and metabolism in humans during craniotomy for small supratentorial cerebral tumors. Anesthesiology 66:332–336

15. Moruzzi G, Magoun HW (1949) Brain stem reticular formation and activation of the EEG. Electroenceph Clin Neurophysiol 1:455–473

16. Pichlmayr, I, Lips U (1983) EEG Monitoring in anesthesiology and intensive care. Neuropsychobiology 10:239–248

17. Prys-Roberts C (1987) Anesthesia: a practical or impossible construct? (editorial). Br J Anaesth 59:1341–1345

18. Rampil IJ, Matteo RS (1987) Changes in EEG spectral edge frequency correlate with the hemodynamic response to laryngoscopy and intubation. Anesthesiology 67:139–142

19. Schwartz MS, Scott DF (1978) Pathological stimulus slow wave arousal responses in the EEG. Acta Neurol Scand 57:300–304

20. Schwilden H, Stoeckel H (1980) Untersuchungen über verschiedene EEG-Parameter als Indikatoren des Narkosezustands. Der Median als quantitatives Maß der Narkosetiefe. Anästh Intensivther Notfallmed 15:279–286

21. Schwilden H, Stoeckel H (1985) The derivation of EEG parameters for modelling and control of anaesthetic drug effect. In: Stoeckel H (ed) Quantitation, modelling and control in anaesthesia. Thieme, pp 160–168

22. Schwilden H, Schüttler J, Stoeckel H (1987) Closed-loop feedback control of methohexitone anesthesia by quantitative EEG analysis in humans. Anesthesiology 67:341–347

23. Shah NK, Long CW, Bedford RF (1988) Delta shift: and EEG sign of awakening during light isoflurane anesthesia. Anesth Analg 67:S206

24. Todd MM, Drummond JC (1984) A comparison of the cerebrovascular and metabolic effects of halothane and isoflurane in the cat. Anesthesiology 60:276–282

25. Zschocke S (1986) Diagnostic and prognostic value of EEG in brainstem lesions including spectral analysis. In: Kunze K, Zangemeister WH, Arlt A (eds) Clinical problems of brainstem disorders. Thieme, Stuttgart, pp 145–156

Central Nervous System Monitoring; Reduction of Information Content of Quantitative Electroencephalograms for Continuous On-Line Display During Anesthesia

W. Dimpfel and H.-C. Hofmann

Due to technical difficulties, monitoring of the main target organ influenced by the anesthetist – the central nervous system – has advanced rather slowly in comparison to, for example, monitoring of circulation parameters (for review see Pichlmayr et al. 1983 and Freye 1990). Recording microvolts in an environment where millionfold larger potential differences are induced by electromagnetic fields had to await a certain level of technical progress before monitoring of the brain's electrical activity became easy and reliable enough to be used in routine work. The separation of the amplifiers' power supply and the general power line by using rechargeable batteries followed by transmission of the signals via optical fibres (Hofmann et al. 1990) fulfilled one of the most important requirements for artefact-free recording. The second major technical breakthrough consisted in the feasibility of real-time, on-line recording and display of the frequency-analyzed signals from 17 electrode positions providing the base for true on-line monitoring with a high time resolution of 4 s (Dimpfel 1993). Finally, the problem of archiving the large mass of data obtained has only recently been solved by storage of computerized data on optical devices. The requirements for significant monitoring of electrical activity in the brain are as follows:

1. Artefact-free recording from 17 channels for topographical view
2. Artefact-free transmission of signals to computer by optical fibres
3. Continuous display of calculated information in real time
4. Storage of raw data on optical devices

The information content of the quantitative electroencephalogram (EEG) as the main representative of the brain's electrical activity can be described using three dimensions: space, time and frequency. In order to interpret the data in a meaningful way, data reduction is necessary. The question therefore arises which of the dimensions can be omitted or reduced in order to keep as much relevant information as possible.

We shall illustrate the consequences of data reduction of the quantitative EEG step by step, taking a recording obtained during bypass surgery by colleagues (courtesy of Dr. Gollnitz, Bad Oeynhausen, in 1992). For the sake of

 W. Dimpfel and H.-C. Hofmann

simplicity, we shall only concentrate on the temperature of the patient and try to compare the information provided by the EEG under four different temperature conditions. Part of the continuous electrographicial recording of surgery is depicted in Fig. 1 in combination with a documentation of the amount of data reduction adherent to a so-called time plot. The marked phases shown in Fig. 1 are used as examples throughout this chapter in order to document the changes in information content and its representation after computer-assisted coding.

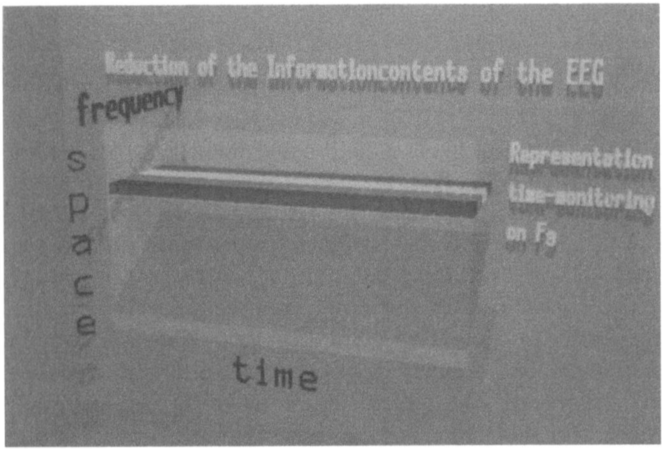

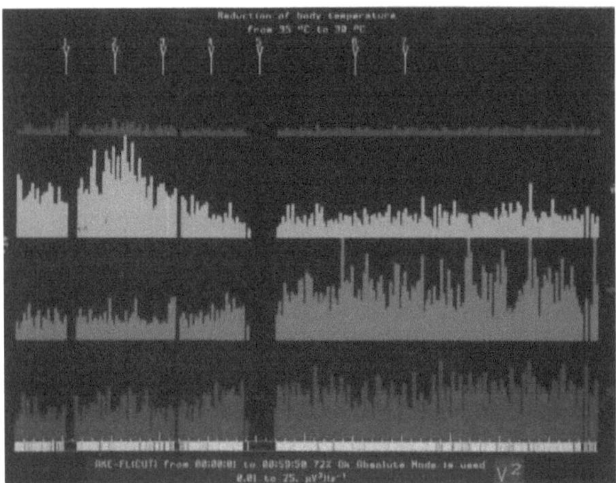

Fig. 1. Reduction of topographical information to one electrode position (F3), while preserving information about time and part of information about frequency (*top*). Example of on-line recording (*bottom*) of bypass surgery, documenting frequency changes during different phases of temperature changes (see *arrows*). *EEG*, electroencephalogram

The four phases of surgery under consideration depicted in Fig. 1 are as follows:

1. Normal body temperature (periods 1–2)
2. First part of transient fall of body temperature (periods 3–4)
3. Second part of transient fall of body temperature (periods 4–5)
4. Lowest phase of body temperature (periods 5–6)

All four episodes will be documented under four different conditions of data reduction. The first computer-assisted presentation mode saves the topographical and part of the time information, but averages the information on frequency content by incorporating 95% of the power (Fig. 2). The median of the frequency distribution as proposed by Schwilden and Stoeckel (1980) is based on the same type of data reduction (attenuating information on frequency content), but more attention to the fact that lower frequencies are more prone to changes during anesthesia. The computer-assisted representation in analogy to the previous 95% value is shown in Fig. 3.

In order to retain more information with respect to the frequency content of the signals, we propose a new computer-assisted representation mode consisting in the selection of three frequency bands from the computerized EEG and an additive colour mixture in analogy to the construction of a television picture using the three colors red, green and blue. The resulting colour represents the relationship and partial contribution of each of the frequency bands contained in the EEG. Using this mode, we are able to save the topographical and parts of the frequency information in an on-line display (Fig. 4).

There are circumstances in which one might be interested in only one particular frequency band. For example, if there is a strong correlation between one frequency band and a particular factor such as temperature or medication, one would like to trace this relationship in more detail. For such purposes, it is quite useful to document this feature by a representation mode called "glow mode" (Fig. 5) in analogy to the feature of glowing iron. In terms of physics, it represents the radiation of solid material which lightens from dark to red to white and blue depending on its temperature.

From the different presentations, it becomes clear that there is no ideal representation mode which is able to document all three features of EEG, namely, information on space, time and frequency, at the same time. It therefore depends very much on the question being asked by the user of this methodology and the on-line performance of the analysis system in use. If one wishes to preserve as much information of the signal as possible in an on-line approach during anesthesia, we propose that: (a) the analog signal is followed continuously in order to provide some kind of quality control and (b) an on-line trend monitor is used with information on the median of the topographical power density distribution for all six frequency bands and identification of the maximum and minimum power within each frequency band.

In this manner, the position with the highest and lowest local power density is shown, and by switching to the corresponding topographical chart one can

 W. Dimpfel and H.-C. Hofmann

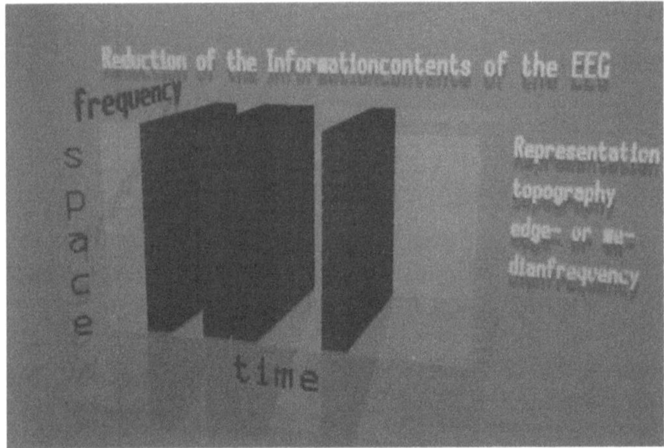

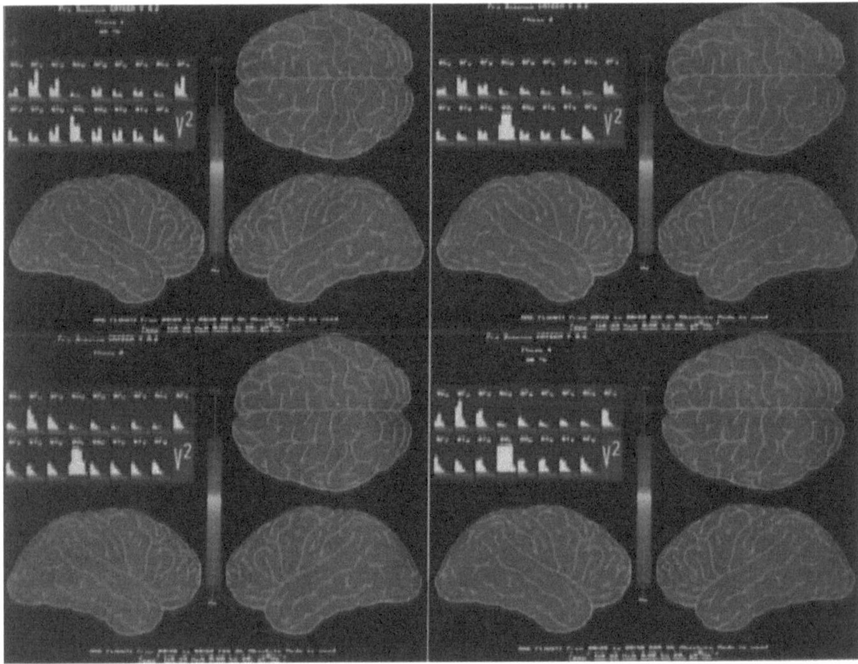

Fig. 2. Reduction of frequency content (*top*; note large, longitudinal *bricks* at four phases of surgery) by giving the 95% value of the frequency distribution (in Hz; Rampil et al. 1980). Computer-assisted colour coding (*bottom*) using the spectral colours from red to blue in analogy with the frequency content from low to high frequencies ("slowing" the electroencephalogram shifts the colour from blue to green or even yellow)

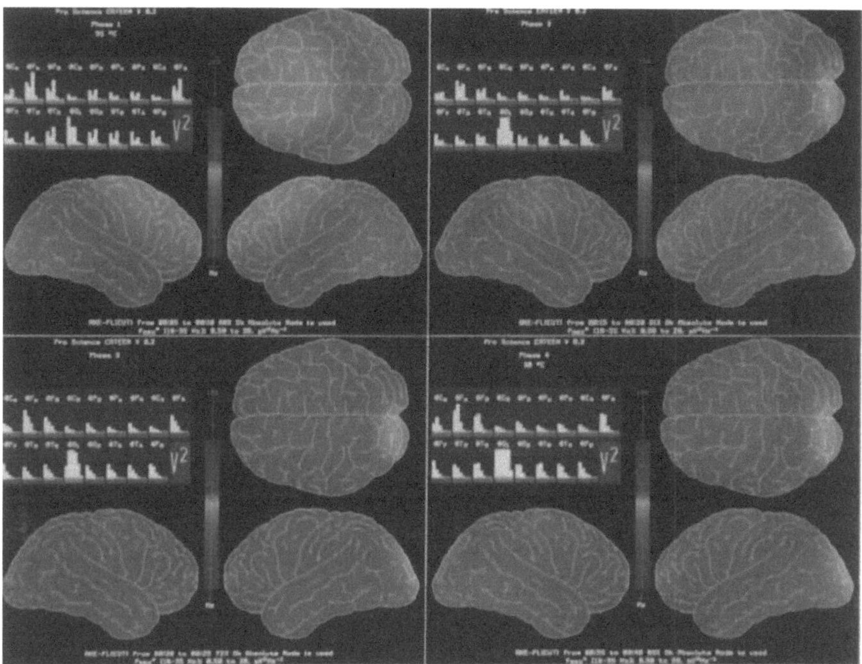

Fig. 3. Computer-assisted representation of the median of the frequency distribution for all four phases of cardiac surgery, reflecting the influence of temperature changes on topographical distribution

identify the locus of major change, for instance, under the conditions of a focal ischemia. This computer-assisted representation mode is depicted in Fig. 6 in order to clarify the combination of the information content of the signal according to the different demands.

In summary, neuromonitoring has become more attractive due to computer-based equipment now available. Battery-powered, high-impedance amplifiers (20 MΩ, no Faraday cage required) and fast laboratory computers based on motorola processors allow continuous on-line monitoring of the brain's electrical activity for the first time and provide the scientist with a tool to recognize drug effects, focal ischemia and effects of extracorporal circulation or temperature. Monitoring the patient's brain will at least give the anesthetist a better feeling of control.

                    W. Dimpfel and H.-C. Hofmann

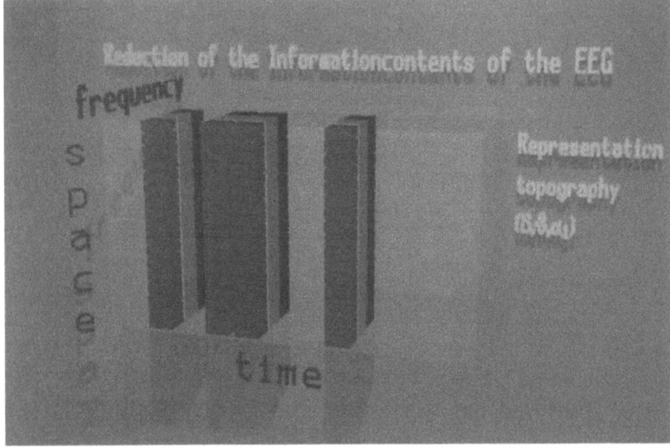

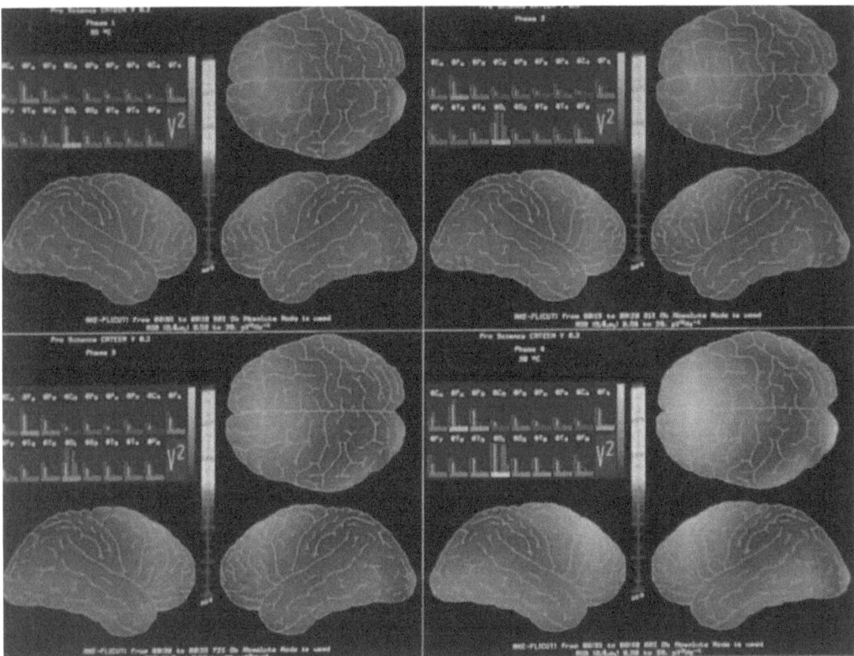

Fig. 4. The so-called Red–Green–Blue mode (top), which retains 50% of the information on frequency and 100% of the topographical information. A practical example (*bottom*) of cardiac surgery providing information on delta and theta frequencies. Shifts in colour represent shifts in contributions of slow frequencies among each other

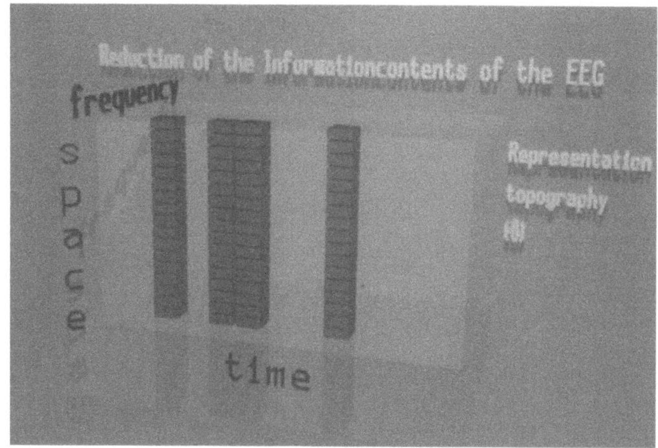

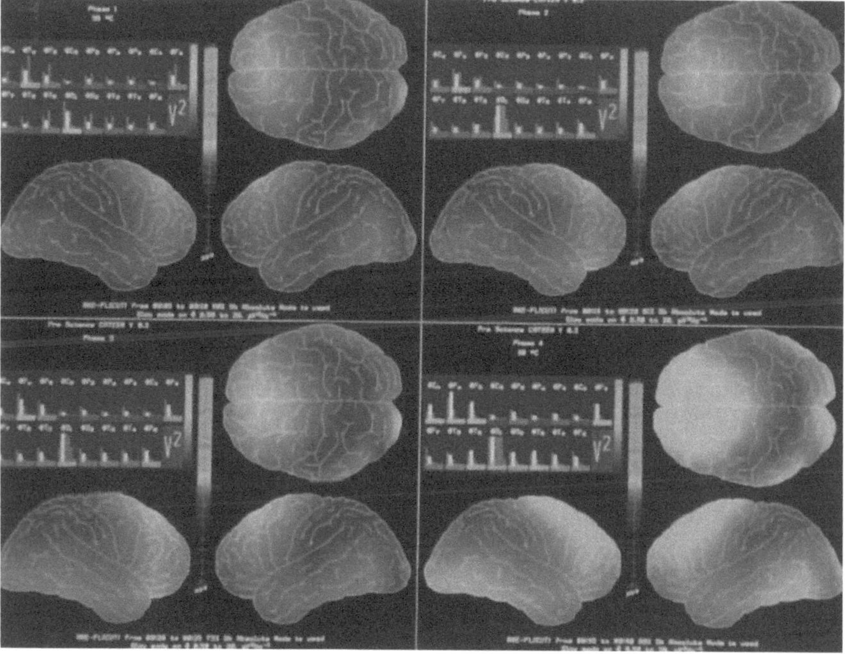

Fig. 5. "Glow mode" (*top*), showing the information content of one frequency band at one topographical position as one *brick* at four different time points. Reduction of information is quite considerable, but this is useful for particular purposes (see text). Computer-assisted representation (bottom) of the same situation concentrating on changes in theta frequencies. Note topographical distribution

W. Dimpfel and H.-C. Hofmann

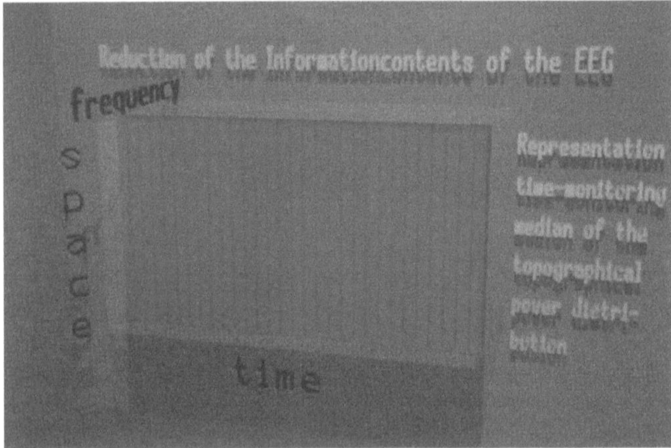

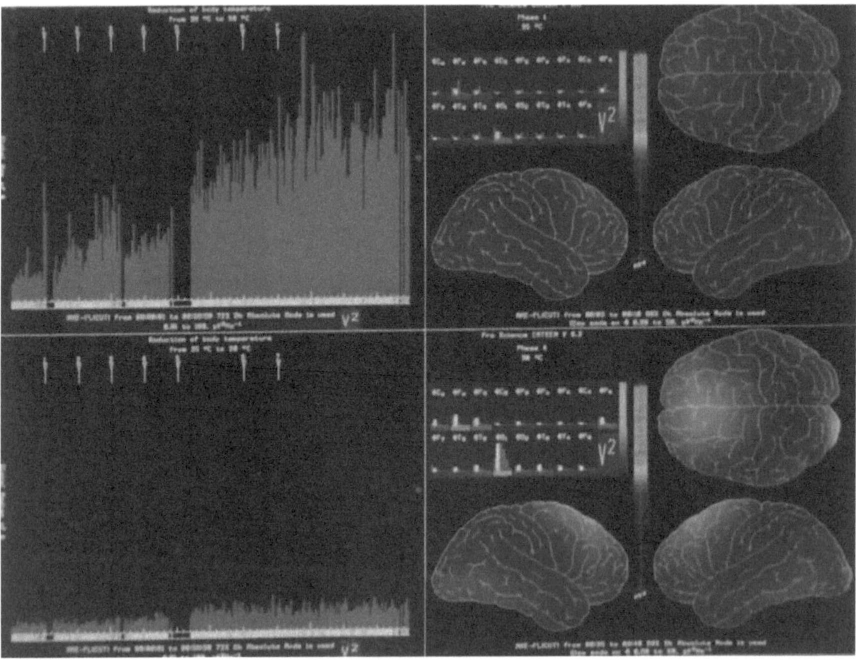

Fig. 6. Bar graph (*top*) representing power density distribution according to all positions of a 17-channel recording system (10/20 International System). Note projection of the topographical distribution of power on the *top y-axis*. Computer-assisted representation of trend monitor picture of maximum and average power density (*middle*) and corresponding topographical chart (*bottom*)

References

Dimpfel W (1993) The new route to brain mapping. Medical Technology International, pp 24–25

Freye E (1990) Cerebral monitoring in the operating room and the intensive care unit. Developments in critical care medicine and anesthesiology, vol 22. Kluwer Academic, Dordrecht

Hofmann H-C, Spüler M, Hellmann H, Dimpfel W (1990) Design and implementation of a real time system for computer aided topographical electroencephalometry. Paper on the international conference on medical information- international symposium on mathematical approaches to Brain Functioning and Diagnostics, 3–7 Sep, Prague

Pichmayr I, Lips U, Künkel H (1983) Das Elektroencephalogramm in der Anästhesie. Springer, Berlin Heidelberg New York

Rampil IJ, Sasse FI, Smith NT, Hoff GH, Flemming DC (1980) Spectral edge frequency: a new correlate of anesthetic depth. Anaesthesiology 53:S12

Schwilden H, Stoeckel H (1980) Untersuchungen über verschiedene EEG-Parameter als Indikatoren des Narkosezustandes. Anaesth Intensivther Notfallmed 15:279

Stoeckel H, Schwilden H, Lauwen PM, Schüttler J (1981) EEG indices for evaluation of depth of anesthesia. Br. J Anaesth 53:117

Part III
Monitoring of Stimulus Evoked Responses

Central Evoked Brain Potentials as Overall Control of Afferent Systems

B. Bromm

Introduction

The spontaneous electroencephalogram (EEG) is widely accepted to reflect states of consciousness; it is therefore increasingly used as a noninvasive tool to determine the depth of narcosis under general anesthesia. However, even the most sophisticated analytical methods are not able to differentiate cognitive states in spontaneous EEG. Especially in context with the issues covered in this book, no correlates have been found which are specific for pain; even under severest pain attacks at most some nonspecific dechronizations are visible, such as decreasing alpha and increasing beta activity, the well-known signs of states of mental stress, but nothing relevant for pain.

This chapter introduces another approach to access higher central nervous system (CNS) functioning by measuring electrical brain activity, namely, the evoked cerebral potentials as stimulus-induced changes of the spontaneous EEG. The appearance of an evoked cerebral potential provides evidence that the specific sensory system investigated does receive the applied stimulus and does mediate the activated peripheral impulse pattern towards the brain. We are all familiar with the usefulness of the visual (VEP), auditory (AEP), or somatosensory (SSEP) evoked potential in clinical routine, especially in the diagnosis of pathological alterations of the sensory channel controlled. This holds true for the investigation of the pain-processing nociceptive nervous system as well.

Figure 1 illustrates the principle. The CNS is more or less a black box activated by a certain stimulus. In pain research, the two most frequently used models are the intracutaneously administered electrical shock (Bromm and Meier 1984; see the following chapters by Kochs et al. and by Scharein) and the brief radiant heat pulse elicited by a CO_2 laser (for review see Bromm and Treede 1991). Both pain models predominantly activate the thinnest cutaneous nerve fibers belonging to the nociceptive system; these are A-delta and C-fibers. The elicited neuronal impulse pattern runs within dorsal and anterolateral tracts towards the thalamus and onto the cortex.

In the surface EEG, stimulus-locked changes occur, but these evoked potentials are small and masked by the spontaneous EEG. Therefore, averaging techniques are used to increase the signal-to-noise ratio. However, the spontaneous EEG is by no means "noise," as has already been discussed in the preceding

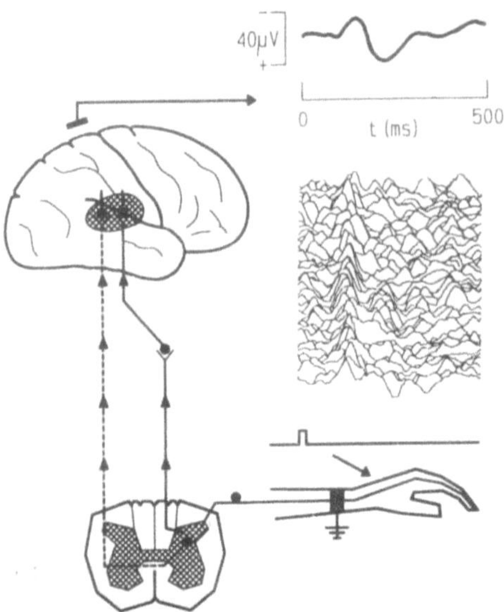

Fig. 1. The measurement of pain-related cerebral potentials. Pain-inducing stimuli, such as intracutaneous shocks or the laser heat pulses, activate specifically nociceptive afferents, which conduct information in anterolateral and dorsal tracts to the thalamus and from there to the cortex. In the surface electroencephalogram, stimulus-induced changes appear (*above*), which are visible after averaging over, e.g., 40 stimulus repetitions. The negativity (*upward deflection*) at 150 ms after stimulus onset and the positivity at 250 ms are late evoked potentials, which reflect the painfulness of the stimulus applied

chapters. In other words, the momentary state of spontaneous brain activity is not regarded in methods of signal averaging usually performed in clinical evoked potential applications. This is of special importance in the case of cognitive brain potentials evaluated in pain research which depend on the momentary arousal level of the brain (see below).

Analyses in the Time Domain

Evoked potentials measured over the scalp are usually divided into early and late components, defined by their polarity (P, positive; N, negative) and latency (in milliseconds) after stimulus onset. The early components, with peak latencies of, for example, less than 30 ms (depending on stimulus quality and site), are very small; thus, up to 1000 stimulus repetitions are needed to make them visible. Early brain potentials show an extremely small intraindividual latency jitter if stimulus conditions are well controlled. Also, the intraindividual variances in latencies are very small; the major source of variance from subject to subject is the neuronal distance to be conducted. Therefore, corrections for body size are necessary when general statements about latencies and site of generation of early potentials are to be made.

The earliest signals in the EEG are far-field potentials with a noncortical source of generation. Because of the short neuronal distance and high conduction velocity of the concerned nerves, earliest potentials can be observed in AEP: The first peak between 1 and 2 ms after stimulus onset is ascribed to peripheral nervous acusticus activity; the next four peaks reflect brain stem activity (nervous cochlearis, oliva, lemniscus lateralis, colliculus inferior), whereas the peaks between 8 and 10 ms are presumably already cortical, i.e., a near-field potentials (e.g. Picton 1988).

The near-field potentials are cortical and interpreted as summated electrical fields of postsynaptic potentials occurring synchronously with the earliest arrival of the nervous impulse pattern activated by the stimulus. Consequently, with scalp electrodes early potentials can best be recorded over the corresponding primary sensory cortex areas. The latencies of early potentials depend, of course, on stimulus quality. Earliest near-field potentials occur as early as 8 ms after stimulus onset in case of AEP (see above), whereas by median nerve stimuli SSEP do no show earliest cortical signals before 20 ms. In the case of selective C fiber activation, we would expect stimulus-induced cortical potentials not earlier than an entire second after stimulus onset (see below).

In contrast, the late components are unspecific and exhibit a considerable latency variance within and between subjects. They are thought to coincide with secondary mechanisms of processing of the received information, such as stimulus recognition, magnitude estimation, quality of sensation induced by the stimulus, e.g. its painfulness, or the cortical initiation of a movement in reaction to the event. The amplitudes of the late potentials are much larger; thus, about 40 stimulus repetitions may be sufficient for averaging. Typically, the late potentials consist of a negativity at about 140 ms and a subsequent positivity at about 240 ms (see, for example, Figs. 1, 2). Of course, the latencies again depend on stimulus quality and body site stimulated (see below).

The late potentials are preceded by so-called middle components, which are still maximal over the contralateral sensory cortex areas like the early ones, though they seem to reflect some secondary processing mechanisms as well. On the other end, the late components are followed by further cognitive potentials, such as the P300 or the contingent negative variations (CNV), which are both domains in psychological research. We will not discuss these very late components in this context (for more detail see Picton 1988).

In pain research, up to now only late potentials have been considered. Because of their unspecificity, they depend on the experimental surrounding, background noise, stress situations, on the subject's attention to the stimulus, distraction, stimulus expectancy, vigilance level, and many other sources of distortion (see Hillyard 1978; Desmedt 1979; for review, see Picton 1988). For these reasons, essential experimental conditions have to be fulfilled if late brain potentials are used to quantify pain. Most important is the necessity to randomize both stimulus intervals and intensities. With intensity randomization, switching unpredictably between nonpain and pain, the arousal state of the subject is kept high and constant, even in long experimental sessions (for details see Bromm 1985, 1989).

There are several reasons why early components have not been described in pain research: (a) because of the low signal to noise ratio, up to 1000 stimulus repetitions are necessary to extract early potentials from spontaneous EEG; this seems impossible in the case of pain-inducing stimuli; (b) because of the slow conduction velocity of nociceptive afferents, the earliest brain potentials in response to a noxious stimulus would appear in a latency range, in which secondary information processing of information to simultaneously coactivated fast-conducting A-beta fibers may occur; (c) with repeated stimuli we normally activate different fibers with slightly different conduction velocities; in the case of the slowly conducting nociceptive afferents, this means a latency jitter of single-trial brain responses, which are too large to apply averaging techniques (see below).

Figure 2 gives an example of the plasticity of late potentials in response to radiant heat stimuli elicited by the CO_2 laser. As already said, this stimulus activates both myelinated and unmyelinated nociceptors; the A-delta fibers exhibit mean conduction velocities in man of 18 m/s, and the C fibers of less than 1 m/s. As a consequence, one single laser stimulus applied to the back of the hand elicits a typical double pain sensation: first pain appears with a mean reaction time of 450 ms and is described as a sharp, stinging, well-localizable pinprick pain, induced by A-delta fiber activity. It is followed by a second, more diffuse burning component, with a mean reaction time of 1400 ms, which can be ascribed to C fiber conduction.

An unexperienced observer directs his attention only to the first painful sensation and thus misses the second. But with increasing experience, the volunteer is able to differentiate between both kinds of pain and to focus his attention more and more upon second pain. At this stage, he is able to create late

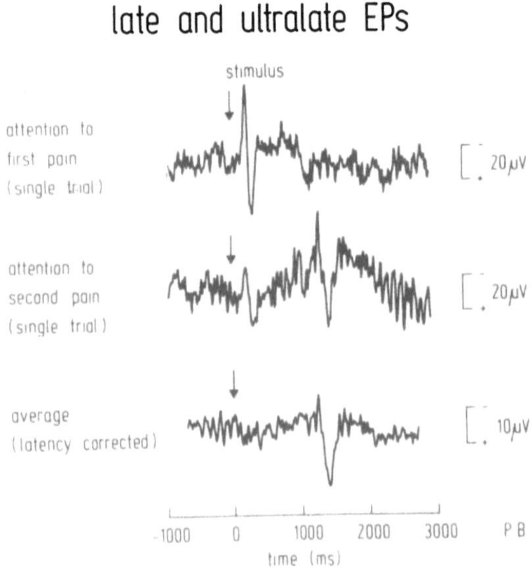

Fig. 2. First and second pain and their electrical brain correlates modulated by shift in attention. Late and ultralate cerebral potentials are given in response to CO_2 laser stimuli of 20 ms and 2.5 W/μm², applied to the back of the left foot. The stimulus activates both myelinated A-delta and unmyelinated C nociceptors and, accordingly, induces two pain sensations with a delay of about 1000 ms. If the subject focuses attention from first to second pain, the ultralate potential as correlate of C-fiber mediated second pain is increased (from Bromm 1994)

cerebral potentials in response to the C fiber input. The more the subject concentrates on the appearance of second pain, the more distinct these so-called ultralate cerebral potentials (for more details see Bromm 1994). The two pain sensations elicited by one stimulus depend on many factors and can emerge to a very different degree in patients, with a special loss in nerve conduction (for review see Bromm and Treede 1991).

Pain-related evoked potentials have increasingly been used for the quantitative evaluation of analgesic drugs, both nonsteroidal anti-inflammatory drugs (NSAID) and centrally acting nacroanalgesics (for review see Bromm et al. 1992). They are able to quantify analgesic potency and to differentiate between the degree of efficacy (see Chap. by Scharein). Late potentials in response to pain stimuli are furthermore used in those situations in which the patient is unable to report his pain, such as in coma or under general anesthesia (see Kochs et al. 1990). With well-controlled experimental conditions, the measurement of amplitude differences N150-P240 is in most cases sufficient for an estimation of pain, at least if pre–post comparisons in repeated measure designs are performed. For a more detailed analysis, multi variate statistical methods have been used, such as principal component analysis, by which pain-specific factors can be extracted (Bromm and Scharein 1982; see in Bromm 1989).

Analyses in the Frequency Domain

By definition, the evoked brain potential is the stimulus-induced change in the spontaneous electroencephalogram. For a detailed physiological analysis, we therefore have to include spontaneous EEG activity. One approach is the single trial analysis of both spontaneous and evoked EEG segments, pre-and post-stimulation. The activity of the spontaneous EEG is commonly described by spectral powers in discrete frequency bands. Since in evoked potential measurements analysis periods of usually only 500 ms are taken into account, parametric spectral estimates have to be adapted. Fourier transformation renders an insufficient frequency resolution for these brief EEG segments. The best success has been achieved by using parametric estimators, such as the autoregressive moving average filter or the maximum entropy method. These filters yield power density functions by modeling the data generating processes.

Figure 3 shows an essential observation concerning stimulus-induced EEG alterations. For simplification, in the upper line one single trial EEG of 500 ms prestimuli (spontaneous EEG) and 500 ms poststimuli (stimulus-induced change), in one subject, without medication, is given. Each of these single trial segments was subjected to maximum entropy spectral analysis; averages over blocks of 80 stimuli were then established for the 21 subjects participating in this study (second line). To the left, the power spectra of the spontaneous EEG prestimulation are shown with a clear α-peak and some power accumulated in the delta frequency range. To the right, we see the effects of the painful stimulus: there is an enormous increase of power in the delta band. This low-frequency

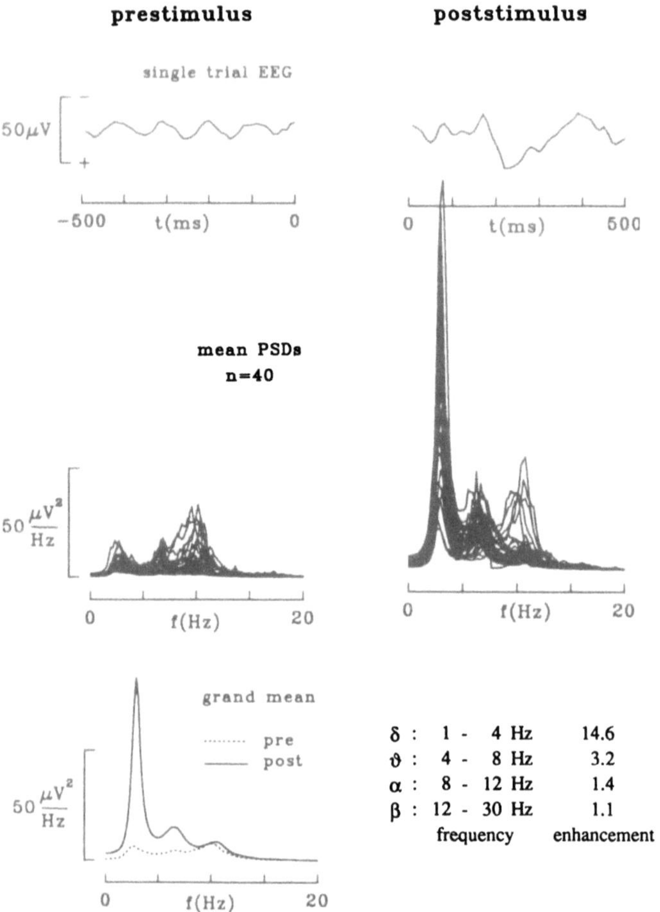

Fig. 3. The effect of the pain-inducing stimulus on electroencephalogram (EEG) activity. *Above*, one single trial peristimulus EEG segment of 500 ms before stimulus (spontaneous EEG) and 500 ms after stimulus (stimulus-induced change). *Second line*, mean power spectra density functions over 80 segments for the 21 subjects: *left*; sponataneous EEG, *right*, stimulus-induced EEG alterations; the stimulus increases low-frequency activity. Obviously, the same frequency bands, which were originally derived from spontaneous EEG, held true for the description of stimulus-evoked responses. *Lowest line*, grand mean averages over the 21 subjects (*left*) and frequency bands with stimulus-induced amplification factors (*right*)

activity might be deducable from the late potential waveforms. But precisely the same frequency bands, which were originally derived by Haus Berger from spontaneous EEG, obviously held true for the description of stimulus-evoked responses as well, with distinct borders between the frequency bands. We also know from single trial analysis of peristimulus EEG segments that the same delta generators responsible for immediate prestimulus activity are triggered by the stimulus. In other words, late potentials might be defined by the delta and theta generators of the spontaneous EEG immediately before stimulus onset.

To sum up, the stimulus enhanced activity of delta and theta frequencies, whereas alpha and beta activity was only slightly influenced by the event. This is once more demonstrated in the lower line of Fig. 3 by the grand mean power density functions pre- and poststimulation, averaged over all 21 subjects participating in the study. To the right, the mean factors are given by which the powers were enhanced by the pain-inducing stimulus.

These results are of special interest, since centrally acting drugs affect spontaneous and evoked EEG in different ways. An example with the strong narcoanalgesic pethidin (DolantinR) is shown in Fig. 4. The stimulus-induced increase in delta power was drastically attenuated by this analgesic with a time course which correlated well with the known pharmacokinetics of meperidine. The pain ratings of the stimuli were reduced in the same way under the analgesic. In fact, we know from many experiments that the delta power induced by pain-inducing stimuli is a good measure to quantify analgesic efficacy. On the other hand, we see an opposite effect in the spontaneous EEG: here, the delta power increased under meperidine. In other words, the same analgesic increases delta power in the prestimulus EEG and decreases it in the poststimulus EEG.

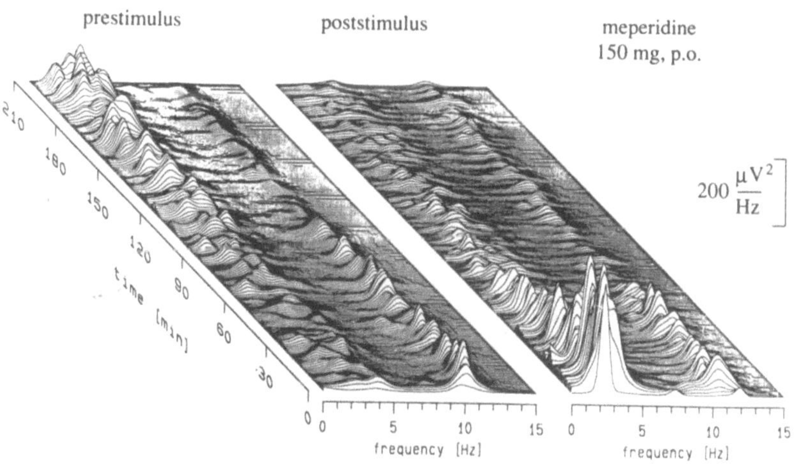

Fig. 4. Single trial power spectra before and after stimulation in one subject under the nacroanalgesic pethidin (150 mg, p.o.). The drug (given at time 0) exhibits different effects on spontaneous and evoked electroencephalograms (EEG). The stimulus-induced increase in delta power was drastically attenuated with a time course which correlates well with the known pharmacokinetics of the drug; analgesia developed with a similar time course. An opposite effect is seen in the spontaneous EEG: here the delta power increases slowly due to decrease in vigilance

From these and other experiments, we deduced that sedation is highly correlated with a decrease in alpha and an increase in delta activity of the spontaneous EEG, whereas analgesia correlates best with a decrease in stimulus-induced delta activity (for details and literature, see Chap. by Scharein). Obviously, spectra analysis of peristimulus EEG, both spontaneous and evoked, is a suitable tool for differentiation between drug-induced analgesia and drug-induced sedation.

Another approach of single trial analysis should briefly be mentioned in which single evoked responses have been estimated by means of parametric modeling and Kalman filtering procedures. The poststimulus EEG activity was separated into an estimated spontaneous activity predicted by the prestimulus segment and a stimulus-evoked change. In other words, the poststimulus EEG activity was assumed to be an additive superposition of an stimulus-evoked and spontaneous part. In this way, changes in evoked potentials under fast-acting drugs could be monitored with a high time resolution (von Spreckelsen and Bromm 1988).

Source Localization Procedures

Stimulus-evoked changes in the EEG indicate that somewhere in the brain the stimulus modifies activity. Many attempts have been made to localize the sites within the brain responsible for the evoked potential. One essential approach is the use of multielectrode recordings (up to 124; Gevins et al. 1990). Such procedures result in a three-dimensional topography of projections of brain sites activated by the stimulus.

Multilead data, however, are not easily interpreted. Normally, color maps are used, which visualize the spatial figures of the quantified recordings. An example is given in Fig. 4, in which potential maps of late and ultralate components in response to painful laser stimuli are given. Figure 4 shows a very similar potential distribution. Obviously A-delta and C fibers project into corresponding brain structures. If A-delta fibers are blocked, the C fibers activate these cortical generators, which causes a similar scalp potential projection, though more than 1000 ms later. In fact, convergence of myelinated and unmyelinated fiber input is known at all levels of the pain pathways, starting at the spinal cord dorsal horn cells; thus, it might be assumed that A-delta and C fiber input trigger a common generator.

Several efforts have been made to determine the sites of generators responsible for the evoked cerebral potential by combination with other imaging techniques; nuclear spin resonance spectrography seems promising as far as the quantification of local changes in phosphate and nitrogen metabolism concomitant with repeatedly applied pain stimuli and evoked potential measurements are concerned. First results correlating EEG and nuclear magnetic resonance (NMR) changes under cerebral ischemia in rats have been reported. Ingvar and

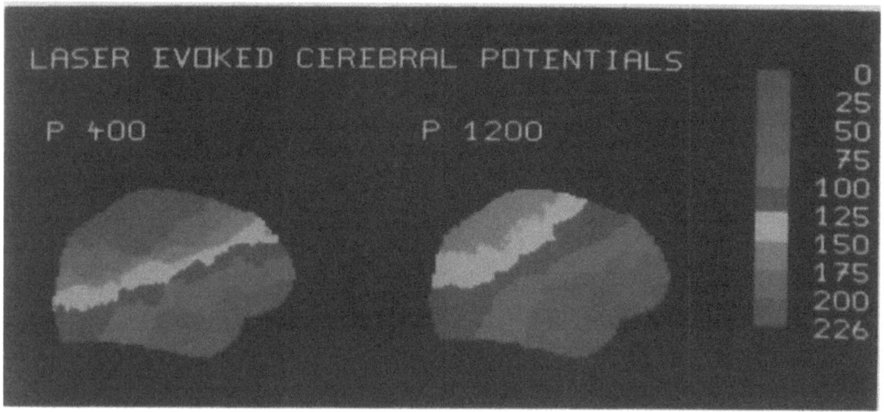

Fig. 5. Scalp topography of late (*P400*) and ultralate evoked potentials (*P1200*). The spatial distributions were interpolated between data from 12 leads, measured from standard 10–20 system positions over the right hemisphere versus linked earlobes. Frontal regions are to the right. The *gray scale* indicates amplitudes in $1/10$ μV for both maps. The *maps* indicate the distributions for the time points of maximal scalp positivity. Latency-corrected averages over 40 stimuli in one subject (from Treede and Bromm 1988)

his group investigated microcirculatory changes of the brain's blood flow and recorded the evoked potential at the same time (for review see Bromm 1985).

Positron emission tomography (PET) imaging of fluorodesoxyglucose uptake, in particular, has the ability to survey functional activity throughout the brain and thus to bring the disparate lines of neurochemical and behavioral approaches together. It has adequate resolution to view both individual gyri of the cortex and discrete portions of the basal ganglia and the limbic system. The metabolic information from the cortical surface may then be correlated with electrophysiological measurements providing collateral localization information, and the temporal information from the evoked potentials can enhance our ability for associate cognitive functions with metabolic increases. There are many reports about local cerebral metabolic changes and EEG variations in normal subjects and in patients suffering, for example, from schizophrenia (Guich et al. 1989).

This survey of stimulus-evoked brain signals is not complete without a look into the new technique of magnetoencephalography (MEG), the magnetic counterpart of brain electrical activity. Neuromagnetic recordings of tooth pulp evoked responses and somatosensory evoked responses have already yielded some interesting new data (Hari et al. 1983). The University Hospital in Eppendorf is one of the few places in Germany where SQUID (Supra Conducting Interference Device) technology is to be installed (Philips 38-channel DC SQUID system).

Preliminary results are shown in Fig. 6 (Bromm et al. 1992), in which we compared the late negativity and its magnetic counterpart for auditory (N100) and pain-related evoked components (N145). The MEG was measured with a

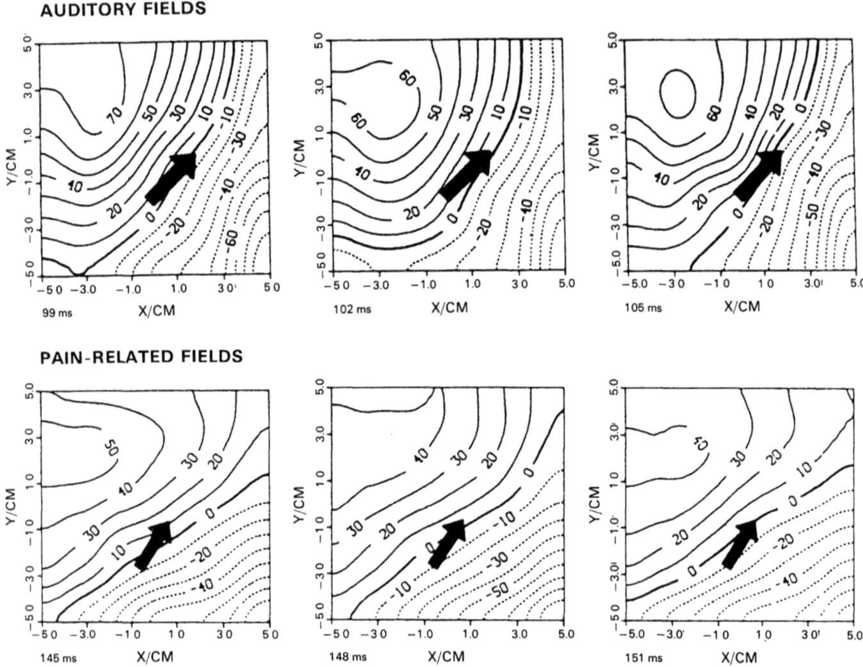

Fig. 6. The stability of single equivalent dipole reconstructions for late evoked components. The *upper row* illustrates auditory evoked fields around 100 ms following the stimulus, and the *lower row* pain-related somatosensory fields around 150 ms after intracutaneous stimuli. Magnetoencephalography with 19-channel Philips biomagnetometer, electroencephalography with Nicolett SM 2000; one subject. Same amplifier and filtering parameters were used: bandpass 0.5–250 Hz, sampling rate 1 kHz (from Bromm et al. 1992)

19-channel system at Philips Research Laboratories. From the raw data, isocontour maps for both auditory (above) and noxious stimuli (below) were computed on the basis of a 10 × 10 cm grid perpendicular to the axis of the SQUID system. The positioning of the cryostat for the two modalities was different, but the isocontour lines look quite similar. By using an inverse calculation procedure for a single equivalent current dipole and a homogeneous and isotropic sphere as reconstruction space, the generator for the N100 (AEP) was localized in the area of the right primary auditory cortex. The estimation of site, size, and direction of this current dipole turned out to be quite stable for at least 10 ms, here at 99 ms, 102, 105 ms. Then the amplitudes decreased and the localization became insufficient.

The lower row shows the pain-related somatosensory fields around 150 ms after painful stimulation. The generator for this component was localized in the contralateral secondary somatosensory cortex. The stability of the estimated dipole location around 150 ms was also good; the dipole is constant for some 10 ms at the same place and the same direction.

To sum up, the generators for late components of different stimulus modalities are located in different sites of the cortex. Whereas in evoked electrical potential analysis for all stimulus qualities the components appear similarly with a maximum over the vertex, the magnetic brain fields localize the generators in primary and secondary sensory cortex areas contralateral to the stimulated body site. However, other generators seem to appear in the ipsilateral hemisphere, though with a time delay of about 5 ms. Since late electrical potentials are usually recorded with a low sampling rate of only 100 Hz, generator fluctuations within 5 ms variations are smeared; thus, on average, vertex maxima are measured. In other words, MEG forces us to analyze more carefully the multilead EEG recordings measured so far. Nevertheless, with multichannel MEG, the localization of cortical generators of late evoked components has been enhanced and thus MEG adds to our understanding of the role of late components in sensory processing.

Acknowledgements. This work was supported by the Deutsche Forschungsgemeinschaft (DFG BR 310/19-3 and BR 310/20).

References

Bromm B (1989) Laboratory animal and human volunteer in the assessment of analgesic efficacy. Raven, New York (Advances in Pain Research and Therapy, vol 12)

Bromm B (1994) Consciousness, pain and the EEG. Raven, New York (Advances in Pain Research and Therapy, vol 18)

Bromm B, Meier W (1984) The intracutaneous stimulus: a new pain model for algesimetric studies. Methods Find Exp Clin Pharmacol 6(7):405–410

Bromm B, Scharein E (1982) Principal component analysis of pain related cerebral potentials to mechanical and electrical stimulation in man. Electroencephalogr Clin Neurophysiol 53:94–103

Bromm B, Treede RD (1991) Laser-evoked cerebral potentials in the diagnosis of cutaneous pain sensitivity. Rev Neurol (Paris) 147:625–688

Bromm B, Laudahn R, Tarkka I (1992) Late magnetic field components evoked by auditory and pain-inducing stimuli. In: Hoke N (ed) Biomagnetism Elsevier Science, Amsterdam

Desmedt JE (ed) (1979) Cognitive component in cerebral event-related potentials and selective attention. Karger, Basel (Progress in clinical neurophysiology, vol 6)

Gevins A, Brickett P, Costales B, Le J, Reutter B (1990) Beyond topographic mapping: towards functional-anatomical imaging with 124-channel EEGs and 3-D MRIs. Brain Topogr 3:53–64

Guich SM, Buchsbaum MS, Burgwald L,Wu J, Haier R, Asarnow R, Nuechterlein K, Potkin S (1989) Effect of attention on frontal distribution of delta activity and cerebral metabolic rate in schizophrenia. Schizophr Res 2:439–448

Hari R, Kaukoranta E, Reinikainen K, Huopaniemi T, Mauno J (1983) Neuromagnetic localization of cortical activity evoked by painful dental stimulation in man. Neurosci Lett 42:77–82

Hillyard SA (1978) Sensation, perception and attention: analysis using ERPs. In: Callaway E, Tueting P, Koslow SH (eds) Event-related brain potentials in man. Academic, London, pp 223–321

Kochs E, Treede RD , Schulte am Esch J, Bromm B (1990) Modulation of pain-related somatosensory evoked potentials by general anesthesia. Anesth Analg 71:225–230

Picton TW (ed) (1988) Human event-related potentials. Elsevier Science, Amsterdam (Handbook of electroencephalography and clinical neurophysiology, vol 3)

Von Spreckelsen M, Bromm B (1988) Estimation of single evoked potentials by means of parametric modelling and Kalman filtering. IEEE Trans Biomed Eng 35:691–700

Treede RD, Bromm B (1988) Reliability and validity of ultra-late cerebral potentials in response to C-fibre activation in situ. Proc Vth World Congress on Pain. Elsevier, Amsterdam, p 567

Indication for Evoked Potential Monitoring:
A Surgical View

J. Zentner and J. Schramm

Introduction

Evoked potentials (EP) have proven to be a useful diagnostic tool in clinical neurophysiology. They can be used to establish objective evidence of an abnormality when clinical signs and symptoms are equivocal or when the patient is not able to co-operate. They may also give evidence of "silent" or "subclinical" lesions, which are electrical abnormalities in a pathway when clinical function seems to be normal. Moreover, EP can help to define the anatomical level of a lesion. Although pathologic potentials do not reflect a certain disease, the relationship between clinical signs and changes in amplitudes, latencies, and wave shapes can give the clinician some hints to assess the general category of pathology, e.g., demyelinating versus compressive disorders. Finally, EP may be used on the intensive care unit for prognostic evaluation of comatose patients.

It has been found that EP are able to monitor changes in the nervous system both over long and short periods of time and even during acute clinical procedures including surgery. The last area of application – intraoperative monitoring – has been a neurosurgical goal for a long time. With the availability of modern neuroradiological imaging and microsurgical operation techniques, the spectrum of neurosurgical operations has enormously widened, and these procedures frequently involve functionally important nerval structures with a certain risk of neurological complications. This is also true for vascular and orthopedic surgery, where neurological complications are especially feared, since these patients usually present with an intact nervous system.

The concept of using EP for intraoperative monitoring was first formulated in 1972 [13]. Several international meetings, mainly addressing the use of somatosensory evoked potentials (SEP) for spinal cord monitoring, have followed [25, 49, 51, 66, 72]. In parallel, numerous reports have described results of intraoperative brain stem acoustic evoked potential (BAEP) monitoring for posterior fossa tumor and neurovascular decompression surgery. Visual evoked potentials (VEP) have been used for monitoring lesions along the optic pathways. Recently, the spectrum of intraoperative application has been widened with the introduction of motor evoked potentials (MEP) into the operating room [4, 45].

The main purpose of intraoperative monitoring is to help reduce neurological complications. According to this concept, changed potentials should allow identification of both local and systemic impairment. Unchanged potentials should provide reassurance to the surgeon that complications are unlikely to have occurred [22, 53, 57]. These paradigms of monitoring are based on several assumptions. First, EP alterations should occur before the lesion is irreversible. Second, an ideal monitoring method should have no false positive or false negative results. Third, the monitoring procedure itself should not harm the patient. Fourth, the warning or intervention criteria should be well defined [63].

The extensive body of literature on intraoperative EP monitoring shows promising results. However, while many questions have been answered, as many questions are arising during ongoing research in this field. It is the scope of this chapter to provide a brief overview on actual standards of intraoperative monitoring. We will try to give some statements about the possibilities and limitations of intraoperative monitoring from a surgical view with respect to the different modalities available.

Basic Remarks

Application of EP in the operating room confronts the monitoring team with several problems. First of all, we have to decide upon the adequate technique for a given lesion. This includes selection of the pathway to be monitored and of appropriate stimulation, recording sites, and parameters as well as data acquisition and processing. Another problem is the influence of anesthesia. Collaboration with the anesthesiologist is necessary both to decide upon an adequate anesthetic regime and to control physiologic and patient-related influences on EP such as body temperature, blood pressure, and blood gases. Having obtained baseline recordings after induction of anesthesia, we have to consider spontaneous fluctuation and variability of recordings and to decide upon acceptable limits for changes in potentials, i.e., to define "warning criteria" or "intervention criteria". Finally, the monitoring procedure itself must be safe and should not harm the patient.

Techniques

One principle limitation of intraoperative monitoring is that once we have decided upon a certain modality, we can only recognize impending complications affecting the respective pathway. Therefore, thorough knowledge of the topographical relationships of the lesion to be monitored is essential. For example, due to the anatomical relationships of acoustic pathways, BAEP monitoring cannot be useful in lesions localized below the IVth ventricle. Since the medial lemniscus crosses at the level of the obex, contralateral SEP have to be used for lesions above the obex, while lesions at the level of the foramen magnum can only be assessed by ipsilateral stimulation. The anatomical situation in aneurysm surgery is even more difficult. While in aneurysms of the carotid and middle cerebral artery, SEP elicited by contralateral median nerve stimulation are usually sufficient, both the leg and the arm area should be assessed in aneurysms of the anterior communicating artery, since the recurrent artery of Heubner supplies parts of the basal ganglia.

For details of stimulation and recording techniques and data acquisition, processing, and storing as well as special requirements for EP machinery in the operating room, we refer to the respective literature [25, 53, 66, 72]. While intraoperative recording of BAEP is standardized to a great extent, a variety of recording and stimulation techniques are available for SEP and MEP. SEP can be elicited by peripheral nerve, epidural, or intrathecal stimulation. Recording can be done from the scalp [44, 81] or the spinal cord [15, 30, 42]. For MEP, transcranial electrical or magnetoelectrical stimulation is available, and recordings are made epidurally, over nerve trunks, or from the muscles [6, 33, 43, 85]. The advantages of invasive techniques include a lower number of runs required per average, higher stimulus rates, shorter analysis times, and lower susceptibility of potentials to anesthetic drugs and patient-related influences. Although poor preoperative EP nearly always preclude useful monitoring, especially of the spinal cord, invasive techniques may improve intraoperative recordability of potentials in these cases [43, 63, 68, 85]. Possible complications, time-consuming pre- or intraoperative placement of electrodes, and susceptibility of electrodes to intraoperative dislocation are disadvantages of invasive techniques. With the use of different techniques, failure rates between 3% and 5% for SEP [7, 73] and between 5% and 25% for electrically evoked MEP [33, 38, 84] have been reported. It is impossible at present to judge whether invasive or noninvasive techniques are preferable. It is probably best to define individually the most suitable technique for a given patient depending on the location of the lesion, the quality of preoperative recordings, and the experience of the monitoring team.

Anesthesia

Signals obtained by modalities that evaluate centripetal or ascending pathways (SEP, VEP, BAEP) as well as signals obtained from the spinal cord or nerve trunks during motor tract stimulation represent neural activity, while MEP recorded from the extremity muscles reflect muscular activity. This is an important consideration, since there is a noticeable difference between anesthesia-related influences on neural activity and muscular activity [52, 56, 74, 77, 78, 80, 88].

Halogenated agents cause a dose-dependent reduction in amplitudes and an increase in latencies of cortical neurogenic potentials, but much less so for brain stem potentials. However, monitoring is still compatible with concentrations of up to 1.0 MAC of halogenated agents. Although the effect of nitrous oxide is similar, it can be used during EP monitoring in concentrations of up to 70 vol%. Drugs commonly used not showing significant influence on neural activity include fentanyl, midazolam, dehydrobenzperidol, and thiopental. Both balanced anesthesia based on moderate concentrations of halogenated agents as well as neuroleptic anesthesia based on nitrous oxide are compatible with intraoperative recording of neural activity. Total intravenous anesthesia based on propofol has recently been found to provide useful SEP with only a low number (50–150) of averages necessary [76]. Koht [34] gives an excellent overview of the influence of anesthesia on neural activity.

The situation concerning the recording of myogenic MEP is quite different. Using inhalational anesthetics, we found motor responses to be abolished at concentrations beyond 0.5 MAC both in humans and in animal experiments with rabbits [87]. Despite noticeable suppression, myogenic MEP have been described as sufficient for regular monitoring procedures during anesthesia with nitrous oxide [16, 62] or propofol [28, 29, 32]. In our experience, continuous infusion of intravenous narcotics such as fentanyl and midazolam and allowing the patient to breathe an oxygen – air mixture is the type of anesthesia most suitable for intraoperative recording of muscular activity [86]. The main disadvantage of this sophisticated anesthetic regime, however, is that in these patients ventilation usually has to be controlled for 1-2 h after the end of the operation. Besides modification of anesthesia, intraoperative recordability of myogenic MEP may possibly be improved by use of facilitation [14, 77, 78] and likely by repetitive stimulation [74]. Both techniques promise to make MEP monitoring feasible during total intravenous anesthesia or balanced anesthesia.

Both in the recording of neural and muscular activity, bolus administration of anesthetic drugs should be avoided. It is important that the anesthesiologist controls patient-related influences on EP such as changes in blood pressure, blood gases, body temperature, and muscular relaxation, since dramatic potential changes due to these factors have been reported.

Warning or Intervention Criteria

Due to the well-known variability of intraoperative recordings, characteristic changes which are thought to be clinically significant ("warning criteria" or "intervention criteria") have mostly been determined retrospectively and arbitrarily [64]. Depending on the patients' clinical condition and the recording technique used, reduction in amplitudes between 20% [81] and 60% [31] and increase in latencies between 4% [61] and 10% [59] were found to be associated with additional postoperative neurological deficits in spinal surgery. The smaller the degree of change used as a warning criterion, the greater the chance that cases without neurological sequelae will be included and vice versa. So far, the only reliable warning criterion seems to be a nontechnical loss of a previously good SEP lasting at least 15 min or a potential loss that is not reversed until the end of the surgery [65].

Although generally accepted warning criteria are still lacking, several principles are clear. Before warning the surgeon, the typical sources for technical, anesthesiological, and patient-related influences have to be eliminated. Moreover, changes must be recorded over a period of time sufficiently long to ensure that the change is different from the many spontaneous fluctuations, especially in spinal cord monitoring. It seems to be justified to use different interventional criteria depending on the preoperative neurological condition and the recording site, since in spinal patients, for example, changed potentials in neurologically intact cases are more significant than in cases with impaired function. Invasive recording sites provide much more stable potentials; thus, changes are more significant than in noninvasive recording sites [65].

For correlating postoperative neurological findings with intraoperative recordings, terms such as "correct detection" and "false positive" and "false negative" results are commonly used. False negative results with only minor or transient neurological deficits have been reported with a frequency of up to 3.5% in spinal SEP monitoring [8] and up to 6% in AEP monitoring [82]. Major false negative cases that are of great consequence seem to be rare. False positive recordings are observed in up to 25% of cases for spinal MEP and AEP monitoring [82,85].

Safety

During a period of nearly 2 decades, no undesirable side effects have been reported with the intraoperative use of SEP and BAEP. Transcranial motor tract stimulation, however, needs separate consideration, due to the relatively high charges applied. Agnew and McCreery [1] found in animal experiments that the likelihood of tissue damage at the site of stimulation was related to the charge density per phase and the total charge delivered. It was concluded that with the present state of knowledge, charge density per phase should not exceed $40\ \mu C/cm^2$ at scalp stimulating electrodes and $10\ \mu C/cm^2$ in neural tissue. Barker et al. [3] calculated that the magnetic stimulator induces a brain charge density of about $0.5\ \mu C/cm^2$ per phase. Estimates of charge density delivered by the electrical stimulator are hampered by the uncertain degree of shunting of current through low resistance scalp pathways. However, charge density at the brain can be assumed to be below $10\ \mu C/cm^2$ per phase. The results of animal experiments imply that if stimulus frequency is kept low, there is no danger of kindling [21]. The clinical magnetic stimulator failed to induce ventricular fibrilation in a single experiment with an anesthetized dog in which the coil was placed directly over the precordium [3]. However, the possibility that the stimulator could cause permanent damage to the functioning of a cardiac pacemaker by inducing currents in the implanted circuitry cannot be discounted.

Altogether, both electrical and magnetoelectrical stimulation seem to be safe when single shocks are used. Cumulative experience now involves many hundreds of patients. To date, no serious side effects have been reported. However, further work is necessary to provide safety criteria for the use of repetitive stimulation. Although no valid data exist, we would advise caution with motor tract stimulation in patients with a history of epilepsy, previous surgical procedures, or with cardiac pacemakers.

Somatosensory Evoked Potential Monitoring

SEP have now reached the stage of widespread application in neurosurgery [15, 69], orthopedic surgery [30, 42], vascular surgery [44], and interventional neuroradiology [23]. Most experience with the use of SEP exists in assessment of spinal cord function. Several reports have described the results of intraoperative SEP monitoring for posterior fossa tumor surgery [65], carotid endarterectomy [48], and aneurysm surgery [67]. We will briefly discuss the current status of SEP monitoring in spinal and aneurysm surgery.

Spinal Cord Monitoring

Neurosurgical domains for the use of SEP as a monitor of spinal cord function include intramedullary tumorous or vascular lesions and metastases. Clinical studies reported up to now have been very heterogenous regarding techniques, patient selection, and results. In consequence, the data reported are fluctuating [50, 68]. The main problem is that in these patients, baseline recordings are frequently poor due to preoperative neurological deficits. Therefore, monitoring is often unsuccessful in those problem patients where intraoperative electrophysiological data are more desirable. This is a major limitation of spinal cord monitoring in neurosurgical patients.

The situation in other domains such as scoliosis surgery is quite different, since only a small proportion of these patients have preexisting neurological deficits. In these cases, normal potentials are usually available and valid recordings can be obtained. Impending neurological complications can be expected to be well recognized by significant changes in SEP. This is also true for spinal cord monitoring in vascular surgery and interventional neuroradiology. Since we still know too little about the variable degree of tolerance to vessel occlusion, SEP monitoring seems to be particularly useful for detecting spinal cord impairment during these procedures. During intraoperative embolization, SEP have been described to disappear rapidly when embolization material or even contrast dye enters the anterior spinal artery. After aortic cross-clamping in humans, SEP usually take between 5 and 15 min to disappear. McWilliam et al. [44] mentioned that in six of 13 patients, SEP did not change for up to 23 min of clamping. They observed that recovery of SEP without neurological sequelae was possible after responses had been absent for 28 min, and they discuss another case with 58 min of absent SEP, followed by postoperative anterior spinal artery syndrome.

In summary, several aspects have to be considered regarding the value of spinal cord monitoring with SEP. The technical reliability both for spinal and cortical recordings is satisfying, the requirements for the conduction of anesthesia are well known, and monitoring has become possible with hardly any anesthesiologic problems. The empirical criteria for warning the surgeon or changing the procedure have often been used successfully to reverse potential deteriorations with no or little neurological deficits. Minor deficits are, however, occasionally seen without significant potential changes. Severe neurological deficits with unchanged potentials are not impossible, but remain rare events. However, the clinical relevance of changed potentials needs further clarification in order to reduce the high number of false positive recordings. In our opinion, SEP monitoring seems to be particularly useful in patients with a pre-operative intact spinal cord who run a definite risk of deterioration, while its value in many neurosurgical patients is limited due to poor baseline recordings, as frequently observed in patients with severe or noticeable neurological impairment.

Somatosensory Evoked Potential Monitoring in Aneurysm Surgery

The surgical therapy for cerebral aneurysms carries a risk which stems from manipulation, accidental or intentional vessel occlusion, or bleeding due to premature rupture. These events usually cause reduction of cerebral blood flow. It has been well established both in humans and in animal experiments that there is a close relationship between reduced cerebral blood flow and SEP changes [12, 47]. This relationship justifies the use of SEP as a monitor of cerebral blood flow during surgical treatment of aneurysms.

We observed significant intraoperative SEP changes in 32 of 282 (11.3%) surgically treated aneurysms. These changes were mainly related to accidental or intentional vessel occlusion. Response to these changes included reapplication of aneurysm clips, repositioning of retractors, or removal of temporary clips in 23 cases (8.1%). In six of 18 patients (33.3%), in whom vessel occlusion coincided with SEP changes, we encountered an additional postoperative neurological deficit, while the outcome was uneventful in 12 patients (66.7%). On the other hand, 26 of 28 patients (92.9%) without significant SEP changes during vessel occlusion had an uneventful outcome. However, in two cases (7.1%) neurological status had deteriorated postoperatively [67].

Little et al. [40] and Friedman et al. [19] have pointed out that in aneurysms of the basilar artery neurological deficits may occur after temporary clipping despite normal SEP. The reason is that in these cases the vascular territory in question often does not include sensory pathways. Therefore, if a temporary vessel clip is needed in surgery for aneurysms located in the posterior circulation, the surgeon should not feel reassured if SEP persist during clipping of the vessel. However, the surgeon should be alarmed if SEP disappear.

Altogether, provided that the vascular territory in question includes the sensory pathways, SEP seem to be a reliable indicator of ischemia. In our experience, SEP monitoring during aneurysm surgery has proven to be helpful in many cases. It has proven to be particularly useful and has influenced the course of surgery in more complicated cases such as multilobed aneurysms, giant aneurysms, trapping procedures, and procedures requiring long-term temporary or even permanent vessel occlusion.

Brain Stem Auditory Evoked Potential Monitoring

The main area for intraoperative application of BAEP is in the assessment of acoustic nerve function during surgical treatment of cerebellopontine angle lesions including acoustic neurinomas in hearing patients [18, 41, 46, 57, 58, 65, 70, 82]. With the availability of gadolinium-enhanced magnetic resonance imaging, a non-negligible proportion of patients now present with a useful degree of hearing at time of diagnosis, and this number will become higher in the future. Consequently, surgeons are more frequently confronted with the problem of hearing preservation. BAEP are also used during neurovascular decompression procedures [18, 82] as well as in posterior fossa tumor [65] and vascular [41] surgery. Another approach for intraoperative assessment of the acoustic nerve is the use of compound nerve action potentials (CNAP), which are recorded from the VIIIth nerve [46].

Acoustic Nerve Monitoring with Brain Stem Auditory Evoked Potentials

Obviously, only patients with preserved waves I or V are suitable candidates for intraoperative monitoring, since hearing preservation cannot be achieved in patients who had no BAEP preoperatively. However, the significance that has to be attributed in particular to waves I and V and their intraoperative change in amplitudes, latencies, or both still remains unclear.

From our findings in 103 patients with posterior fossa lesions in whom preoperative hearing was preserved, we conclude that both waves I and V seem to be relatively reliable for prediction of postoperative hearing. Loss of wave I coincided in 15 of 19 cases (79%) and loss of wave V in 16 of 20 cases (80%) with postoperative deafness. Hearing loss was also observed in four of 70 cases (6%) despite preserved wave I and in two of 75 cases (2.7%) with preserved wave V. Thus, evaluating wave I, we encountered 21% false positive and 6% false negative cases versus 20% false positives and 3% false negatives with wave

V [82]. Therefore, the predictive value of preserved waves I and V is not an absolute one, but they strongly suggest preserved postoperative hearing. The dilemma remains that once wave I or V are lost during surgery, there is no certainty as to the postoperative preservation of hearing. Wave V changes were reversible or irreversible with nearly the same frequency, while wave I changes were mostly irreversible. In wave I, amplitude changes alone were more frequent than in wave V, where isolated latency changes were more often observed. However, no absolute guidelines exist about whether the surgeon should be more worried by an amplitude or by a latency change [82].

The fact that a large proportion of patients with acoustic neurinomas have either no or poor preoperative BAEP, which precludes ipsilateral monitoring, is an obvious limitation of the technique. In our experience, the use of contralateral monitoring should not be advocated in these cases. Contralateral monitoring may only be useful in particularly dangerous lesions such as angiomas or tumors involving the brain stem.

In summary, although no absolute criteria for prediction of postoperative hearing exist, BAEP have proven to be useful as a monitor of acoustic nerve function in small cerebellopontine angle tumors with preserved hearing. BAEP are particularly useful during microvascular decompression procedures for trigeminal neuralgia or hemifacial spasm, since in these cases postoperative hearing loss was found in 2.8%–8% [46]. Contralateral monitoring seems only to be useful in lesions affecting the brain stem and is now definitely considered unnecessary for small or middle-sized acoustic neurinomas without a useful degree of preoperative hearing. In our opinion, BAEP monitoring seems to be a typical field where electrophysiology can help the surgeon to learn much about the pathophysiology of perioperative cranial nerves lesions and the brain stem. The information derived from monitoring will concentrate the surgeon's thoughts on a more functional and not purely morphological aspect of tumor surgery [82].

Acoustic Nerve Monitoring with Compound Nerve Action Potential

Two factors limit the value of BAEP monitoring even with preserved responses. First, a high number of averages (1500–5000) is necessary to obtain distinct potentials, which takes a relatively long time (2–5 min). This means a considerable delay during critical periods of surgery before a serious injury to the hearing system is revealed. Second, BAEP monitoring includes many false positive (about 20%) and some false negative (about 5%) results. In order to improve both availability of potentials and reliability of monitoring, Moller [46] introduced recording of the CNAP, a near-field potential that can be obtained directly from the VIIIth nerve.

Our experience with the simultaneous use of CNAP and BAEP in 24 patients with acoustic neurinomas shows that CNAP can be recorded 10–15 times faster than BAEP, since only 10–500 averages are necessary. Moreover, in several cases of reversible impairment of the VIIIth nerve, CNAP were obtained after loss of BAEP. We have not encountered any false negative or false positive results with CNAP so far.

In summary, the essential advantage of CNAP is that this technique allows nearly real-time monitoring of the VIIIth nerve function. This is desirable, since deterioration often occurs suddenly and a delay in alarming the surgeon reduces the chance of altering the operative strategy adequately before further irreversible damage to the hearing systems occurs. No special technique is necessary for starting CNAP recording except for a tiny wick electrode [46, 75]. Although our limited number of cases up to now does not allow any definite conclusions, CNAP seem to be an accurate monitor of VIIIth nerve function. However, the major limitation for CNAP is that monitoring can only be started after exposure of the VIIIth nerve. Therefore, CNAP recording is restricted to small or intrameatal acoustic neurinomas, to microvascular decompression procedures, and all tumors that displace the acoustic nerve posteriorly. Further experience is necessary to decide whether simultaneous recording of CNAP and BAEP might increase the accuracy of the monitoring [75].

Motor Evoked Potential Monitoring

The need for motor tract monitoring is obvious, since the sensivity of SEP in detecting lesions along the descending pathways has been found to be limited. Although acute spinal cord impairment usually affects both motor and sensory pathways, a few isolated cases have been documented in the literature in which motor impairment occurred when SEP remained stable [20, 36, 37, 60, 90]. This is not surprising, since motor and sensory pathways travel along separate tracts, each with its own vascular supply. Therefore, it is possible to injure one while leaving the other intact. In supratentorial lesions, isolated motor deficits that may be missed by SEP can be expected to occur more often than in brain stem and spinal cord lesions. Animal studies of spinal cord trauma and ischemia confirm a close relationship between changes in motor potentials and the neurological condition [17, 54, 71, 89]. These results encourage the use of MEP for intraoperative assessment of the descending pathways.

Our experience with MEP monitoring in 122 patients during neurosurgical operations on the posterior fossa and spinal cord using different recording sites along the spinal cord and the extremity muscles after transcranial electrical stimulation suggests that MEP may be a reliable tool for assessment of motor pathway function. If at the end of an operation we had amplitudes with 50% of baseline values obtained after induction of anesthesia, a good correlation between this degree of MEP change and missing motor deficit was found. On the other hand, intraoperative loss of potentials as observed in five patients coincided in every case with severe postoperative deficits. We encountered false negative recordings in one patient. Permanent reduction in amplitudes did not necessarily indicate poor outcome, since we found false positive results in 11.1%–22% of recording situations [84, 85].

Some reports describe the use of MEP for intraoperative monitoring during orthopedic, vascular, and neurosurgical operations [6, 16, 33, 37, 38, 80]. Several technical variations exist: stimulation can be performed transcranially or the motor cortex may be stimulated directly. Single or repeated stimuli may be used, and responses can be recorded from the muscles of the extremities, peripheral nerves, or the epidural space along the spinal cord and cauda equina. Electrical stimulation is usually preferred, especially when spinal evoked responses (D waves) are recorded. Magnetic stimulation has been tried intraoperatively by several authors [5, 33, 79]. However, the overall experience with MEP monitoring is limited and only rough statements on its value are possible at this time. First, both spinal cord, peripheral nerve, and muscular responses seem to be sensitive for detection of impending neurological complications. There is no evidence that serious deficits were missed with any technique. However, minor and transient motor deterioration may coincide in rare cases with unchanged potentials mainly obtained from epidural recording sites. Second, changes and even loss of potentials have found to be reversible in principle, and this is usually followed by an uneventful outcome. Third, the nontechnical intraoperative loss of MEP lasting until the end of surgery undoubtedly seems to indicate serious complications.

Although existing reports suggest that MEP may be a promising tool for intraoperative assessment of the descending pathways, MEP monitoring is still at an experimental stage. The current situation is characterized by many technical problems in obtaining intraoperative signals, due to the influence of anesthesia. Repetitive stimulation [74] seems to be a useful approach in order to improve intraoperative recordability. Only when more experience has been gained in managing these problems may MEP monitoring become a feasible method for routine intraoperative use. Other open questions concern evaluation of results and the definition of warning criteria. Further clinical studies are necessary to show the range of false positive and false negative results. Moreover, the sensitivity of different stimulation and recording techniques with respect to different localizations of lesions has to be defined. It still remains unclear whether MEP are as sensitive for supratentorial lesions as they seem to be for spinal cord lesions.

As mentioned previously, it is possible in theory to injure motor pathways while leaving the sensory ones intact and vice versa. The question arises as to whether the combination of two monitoring techniques (SEP and MEP) would contribute to a better prediction of the postoperative outcome. Experimental studies indicate that MEP and SEP indeed reflect the functional status of the respective pathways [39, 55, 89]. The few clinical studies that are available with combined SEP–MEP monitoring suggest that both modalities may provide supplementary information [24, 43]. However, further studies are required to justify this optimism.

Visual Evoked Potential Monitoring

Neurosurgical procedures in perisellar tumors carry a high risk of postoperative visual deterioration, which has been found to be around 5% for pituitary adenomas and somewhat higher for meningiomas [26]. Some reports suggest that VEP monitoring might be useful to avoid additional damage to the visual system [35, 83].

In our series of 35 patients with tumors along the visual pathways, we found an intraoperative VEP loss in 25 cases. Of 20 patients who were evaluated postoperatively, visual function improved in 12 cases, while it was unchanged in seven patients. Thus, 19 of 20 patients (95%) showed false positive recordings. Two of three patients who deteriorated postoperatively had a homonymous hemianopsia which did not coincide with intraoperative potential changes. Another patient had a decrease in visual acuity from 0.4 to 0.3, which coincided with an intraoperative potential loss. Although this might be called a correct detection, a decrease in visual acuity in such a low range usually would not be considered to be significant [9]. Similar results have been reported by others [2, 57].

Further studies showed a significant influence of anesthetics, in particular of nitrous oxide and inhalational agents [11] on VEP. Surgical maneuvers far away from visual pathways such as trephination and dura opening even on the contralateral side were also followed by essential changes or loss of VEP [10].

Altogether, our results show a high variability of VEP due to the combined effects of anesthesia, surgical manipulation, and compression of the visual pathways. As long as the high proportion of potential loss and the large variability of VEP cannot be reduced, flash VEP seem not to be helpful in management of lesions close to the visual pathways [10]. Flash VEP have also been found to be variable in the awake patient, both among large numbers of subjects and within the same subject on multiple trials. It is well known in clinical neurophysiology that pattern-shift VEP are a powerful diagnostic tool [27]. However, currents methods of presenting pattern-shift stimuli require visual fixation. As this is not possible during operation, we currently have no effective approach to provide stable VEP in the operating room.

General Conclusions and Outlook

The idea of using EP as a monitor for intraoperative assessment of the integrity of various pathways during operations affecting the nervous system is a stimulating one. This is especially true since apart from the wake-up test, no other efficient way exists for intraoperative evaluation of the actual neurological condition. The concept of intraoperative EP monitoring was started in 1972 with the use of SEP for assessment of spinal cord function. Based on experimental data, different modalities have been introduced into the operating room, the most recent being MEP. Although considerable progress has since been made, there are still as many questions that remain unsolved as there are answers that have been found. Therefore, it is impossible at this time to make general statements on the value of intraoperative monitoring. Differentiating statements are required with respect to modalities, type of surgery, and patient selection.

The use of BAEP in patients with small acoustic neurinomas with preserved hearing and during neurovascular decompression procedures has been well established. It has been shown that critical situations were recognized and managed using BAEP and–more recently–CNAP. On the other hand, BAEP cannot be recommended for patients without a useful degree of hearing, neither with ipsilateral nor with contralateral stimulation. Exceptionally, BAEP may be useful in more complicated posterior fossa tumors for assessment of brain stem function.

In our experience, SEP have proven to be useful as a monitor of cerebral blood flow during aneurysm surgery. We found that SEP are definitely useful in more complicated cases. This is especially true, since baseline recordings are usually normal in these cases and since changes in SEP in these patients are rapid as a rule, if they occur at all. The situation in neurosurgical patients treated for spinal cord lesions is quite different. SEP monitoring is often unsuccessful due to poor baseline recordings in problem patients in whom electrophysiological data would be desirable. However, spinal cord monitoring with SEP seems to be useful in domains such as scoliosis surgery, vascular surgery, and interventional neuroradiology, since these patients usually have an intact spinal cord.

To date, motor tract monitoring is at an experimental stage. There are still too many intraoperative recording difficulties – mainly related to the influence of anesthesia – which considerably limit routine use of this technique. Further experience is necessary to improve intraoperative recordability of potentials, and both facilitation and repetitive stimulation seem to be promising approaches. Only when technical problems have been solved can other open questions such as the significance of changed and unchanged potentials be addressed adequately.

The intraoperative use of flash VEP has been disappointing. The high variability of flash VEP due to anesthesia and surgical manipulation far away from the visual system precludes useful intraoperative monitoring. Pattern-shift VEP could be expected to provide more stable potentials. However, to date an appropriate intraoperative stimulation technique is not yet available.

While many questions concerning intraoperative monitoring have been answered, as many remain unsolved. Further work is necessary to define generally accepted techniques for monitoring a given lesion. Spontaneous variability of EP as well as the influence of anesthesia and of patient-related factors should be clarified more precisely in order to define more reliably the significance of changed and unchanged potentials. Finally, safety problems in the context of motor tract monitoring have to be considered. We are still far away from the stage where EP monitoring is necessary in every operation affecting the nervous system. However, EP monitoring is definitely useful in selected cases. Further clinical studies are required to define more precisely the patient group that can be expected to most probably benefit from intraoperative monitoring.

References

1. Agnew WF, McCreery DB (1987) Considerations for safety in the use of extra-cranial stimulation for motor evoked potentials. Neurosurgery 20:143–147
2. Allen A, Starr A, Nudleman K (1981) Assessment of sensory function in the operating room utilizing cerebral evoked potentials: a study of fifty-six surgically anesthetized patients. Clin Neurosurg 28:457–482
3. Barker AT, Freeston IL, Jalinous R, Jarratt JA (1988) Magnetic and electrical stimulation of the brain: safety aspects. In: Rossini PM, Marsden CD (eds) Non-invasive stimulation of brain and spinal cord. Liss, New York, pp 131–144
4. Barker AT, Jalinous R, Freeston IL (1985) Non-invasive magnetic stimulation of the human motor cortex. Lancet 1:1106–1107
5. Beradelli A, Inghilleri M, Cruccu G, Manfredi M (1991) Corticospinal potentials after electrical and magnetic stimulation in man. In: Levy WJ et al. (eds) Magnetic motor stimulation. Basic principles and clinical experience (EEG Suppl 43). Elsevier, Amsterdam, pp 147–154
6. Boyd SG, Rothwell JC, Cowan JMA, Webb PJ, Morley T (1986) A method of monitoring functions in corticospinal pathways during scoliosis surgery with a note on motor conduction velocities. J Neurol Neurosurg Psychiatry 49:251–257
7. Breitner S, Matzen KA (1985) Scalp recorded somatosensory evoked potentials during spinal surgery. In: Schramm J, Jones SJ (eds) Spinal cord monitoring. Springer, Berlin Heidelberg New York, pp 173–178
8. Brown RH, Nash CL (1984) Implementation and evaluation of intraoperative somatosensory cortical potential – procedures and pitfalls. In: Homma S, Tamaki T (eds) Fundamental and clinical application of spinal cord monitoring. Saikon, Tokyo, pp 373–384
9. Cedzich C, Schramm J, Fahlbusch R (1987) Are flash-evoked visual potentials useful for intraoperative monitoring of visual pathway function? Neurosurgery 21:709–715
10. Cedzich C, Schramm J, Mengedoht CF, Fahlbusch R (1988) Factors that limit the use of flash visual evoked potentials for surgical monitoring. Electroencephalogr Clin Neurophysiol 71:142–145
11. Costa e Silva J, Wang AD, Symon L (1985) The application of flash visual evoked potentials during operations on the anterior visual pathways. Neurol Res 7:11–16
12. Coyer PE, Simeone FA, Michele J (1981) Latency of the cortical component of the somatosensory evoked potential in relation to cerebral blood flow measured in the white matter of the cat during focal ischemia Neurosurgery 4:497–502
13. Croft TJ, Brodkey JS, Nulsen FE (1972) Reversible spinal cord trauma: a model for electrical monitoring of spinal cord function. J Neurosurg 36:402–406

14. Date M, Schmid U, Hess CW, Schmid J (1991) Influence of peripheral nerve stimulation of the responses in small hand muscles to transcranial magnetic cortex stimulation. In: Levy WJ, Cracco AT, Barker AT, Rothwell J (eds) Magnetic motor stimulation: basic principles and clinical experience. Elsevier, Amsterdam, pp 212–223
15. Dinner DS, Lueders H, Lesser RP, Morris HH, Barne HG, Klem G (1986) Intraoperative spinal somatosensory evoked potential monitoring. J Neurosurg 65:807–414
16. Edmonds HL, Paloheimo MPJ, Backman MH, Johnson JR, Holt RT, Shields CB (1989) Transcranial magnetic motor evoked potentials (tcMEP) for functional monitoring of motor pathways during scoliosis surgery. Spine 14:683–686
17. Fehlings MG, Tator CH, Linden RD, Piper IR (1987) Motor evoked potentials recorded from normal and spinal cord-injured rats. Neurosurgery 20:125–130
18. Friedman WA, Kaplan BJ, Gravenstein D, Rhoton AL (1985) Intraoperative brain-stem auditory evoked potentials during posterior fossa microvascular decompression. J Neurosurg 62:552–557
19. Friedman WA, Kaplan BL, Day AL, Sypert GW, Curran MT (1981) Evoked potential monitoring during aneurysm operation: observations after fifty cases. Neurosurgery 20:678–687
20. Ginsburg HH, Shetter AG, Raudzens PA (1985) Postoperative paraplegia with preserved intraoperative somatosensory evoked potentials. J Neurosurg 63:296–300
21. Goddard GV, McIntyre DC, Leech CK (1969) A permanent change in brain function resulting from daily electrical stimulation. Exp Neurol 25:295–330
22. Grundy BL (1983) Electrophysiologic monitoring: EEG and evoked potentials. In: Newfield P, Cottrell J (eds) Manual of Neuroanesthesia. Little Brown, Boston, pp 28–59
23. Hacke W, Hündgen R, Zeumer H, Ferbert A, Buchner H (1985) Überwachung der therapeutischen neurologischen Untersuchungs- und Therapieverfahren mittels evozierten Potentialen. Z EEG EMG 16:93–100
24. Hicks RG, Burke DJ, Stephen JPH (1991) Monitoring spinal cord function during scoliosis surgery with Cotrel–Dubousset instrumentation. Med J Aust 154:82–86
25. Homma S, Tamaki T (eds) (1984) Fundamental and clinical application of spinal cord monitoring. Saikon, Tokyo
26. Horwitz NH, Rizzoli HV (1982) Postoperative complications of intracranial neurological surgery. Williams and Wilkins, Baltimore, pp 55–100, 132–141
27. Hughes JR, Fino J, Sagnon L (1984) A comparison of flash and pattern evoked potentials in patients with demyelinating disease and in normal controls. In: Nodar RH, Colin B (eds) Evoked potentials II. Butterworth, Boston, pp 302–309
28. Jellinek D, Jewkes D, Symon L (1991) Noninvasive intraoperative monitoring of motor evoked potentials under propofol anesthesia: effects of spinal surgery on the amplitude and latency of motor evoked potentials. Neurosurgery 29:551–557
29. Jellinek D, Platt M, Jewkes D, Symon L (1991) Effects of nitrous oxide on motor evoked potentials reduced from sceletal muscle in patients under total anesthesia with intravenously administered propofol. Neurosurgery 29:558–562
30. Jones SJ, Carter L, Edgar MA, Morley T, Ransford AO, Webb PJ (1985) Experience of epidural spinal cord monitoring in 410 cases. In: Schramm J, Jones SJ (eds) Spinal cord monitoring. Springer, Berlin Heidelberg New York, pp 215–220

31. Jones SJ, Howard L, Shakwat F (1988) Criteria for detection and pathological significance of response decrement during spinal cord monitoring. In: Ducker TL, Brown RH (eds) Neurophysiology and standards in spinal cord monitoring. Springer, Berlin Heidelberg New York, pp 201–206
32. Keller BP, Haghighi SS, Oro JJ, Eggers GWN (1992) The effects of propofol anesthesia on transcortical electric evoked potentials in the rat. Neurosurgery 30:557–560
33. Kitagawa H, Itoh T, Takano H, Takekuwa K, Yamamoto N, Yamada H (1989) Motor evoked potential monitoring during upper cranial spine surgery. Spine 14:1078–1083
34. Koht A (1988) Anesthesia and evoked potentials: an overview. Int J Clin Monit Comput 5:167–173
35. Koshino K, Kuroda R, Mogani H, Takimoto H (1978) Flashing diode evoked responses for detecting optic nerve function during surgery. Med J Osaka Univ 29:39–47
36. Lesser RP, Raudzens PA, Lueders H, Nuwer MR, Goldie WB, Morris HH, Dinner DS, Klem G, Hahn JF, Shetter AG, Ginsburg HH, Gurd AR (1986) Postoperative neurological deficits may occur despite unchanged intraoperative somatosensory evoked potentials. Ann Neurol 19:22–25
37. Levy W (1983) Spinal cord potentials from the motor tracts. J Neurosurg 58:38–44
38. Levy WJ (1987) Clinical experience with motor and cerebellar evoked potential monitoring. Neurosurgery 20:169–182
39. Levy WJ, McCaffrey M, York D (1986) Motor evoked potential in cats with acute spinal cord injury. Neurosurgery 19:9–19
40. Little JR, Lesser RP, Lueders H (1987) Electrophysiological monitoring during basilar aneurysm operation. Neurosurgery 20:421–427
41. Little JR, Lesser RP, Lueders H, Farlan AJ (1983) Brain stem auditory evoked potentials in posterior circulation surgery. Neurosurgery 12:496–502
42. Maccabee PJ, Levine DB, Pinkhasov EI, Cracco RQ, Tsairis P (1983) Evoked potentials recorded from scalp and spinous processes during spinal column surgery. Electroencephalogr Clin Neurophysiol 56:569–582
43. Matsuda H, Shimazu A (1989) Intraoperative spinal cord monitoring using electric responses to stimulation of caudal spinal cord or motor cortex. In: Desmedt JE (ed) Neuromonitoring in surgery. Elsevier, Amsterdam, pp 175–190
44. McWilliam RC, Conner AN, Pollock JCS (1985) Cortical somatosensory evoked potentials during surgery of scoliosis and coarctation of the aorta. In: Schramm J, Jones SJ (eds) Spinal cord monitoring. Springer, Berlin Heidelberg New York, pp 167–172
45. Merton PA, Morton HB (1980) Stimulation of the cerebral cortex in the intact human subject. Nature 285:227
46. Moller AT, Jannetta PJ (1991) Compound action potentials recorded intracranially from the auditory nerve in man. Exp Neurol 74:862–874
47. Momma F, Wang AD, Symon L (1987) Effects of temporary arterial occlusions on somatosensory evoked response in anerysm surgery. Surg Neurol 27:343–352
48. Narayan PV, Gilmour MP, Lloyd AJ, Dahn MS, King SD (1985) An assessment of the variability of early scalp-components of the somatosensory evoked response in uncomplicated, unshunted carotid endarterectomy. Clin Electroenc 3:157–160
49. Nash CL, Brodkey JS (eds) (1977) Proceedings: clinical application of spinal cord monitoring for operative treatment of spinal diseases. Case Western Reserve University, Cleveland

50. Nash CL, Brown RH (1989) Current concepts review. Spinal cord monitoring. J Bone Joint surg 71:627–630
51. Nash CL, Brown RH (eds) (1979) Proceedings: spinal cord monitoring workshop. Case Western Reserve University, Cleveland
52. Nogueira MC, Brunko E, Vandesteen A, De Rood M, Zegers de Beyl D (1989) Differential effects of isoflurane on SEP recorded over parietal and frontal scalp. Neurology 39:1210–1215
53. Nuwer MR (1986) Evoked potential monitoring in the operating room. Raven, New York, pp 1–4
54. Owen JH, Jenny AB, Naito M, Weber K, Bridwell KH, McGhee R (1989) Effects of spinal cord lesioning on somatosensory and neurogenic-motor evoked potentials. Spine 914:673–682
55. Owen JH, Naito M, Bridwell KH, Oakley DM (1990) Relationship between duration of spinal cord ischemia and postoperative neurological deficits in animals. Spine 15:846–851
56. Pelosi L, Caruso G, Balbi P (1988) Characteristics of spinal potentials to transcranial motor cortex stimulation: intraoperative recording. In: Rossini PM, Marsden CD (eds) Non-invasive stimulation of brain and spinal cord. Liss, New York, pp 297–304
57. Raudzens PA (1982) Intraoperative monitoring of evoked potentials. Ann NY Acad Sci 388:308–326
58. Raudzens PA, Shetter AG (1982) Intraoperative monitoring of brain-stem auditory evoked potentials. J Neurosurg 57:341–348
59. Romstöck J, Watanabe E, Schramm J (1988) Variability of epidural SEP from below and above spinal cord lesions – the effect of the lesion on spinal SEP. In: Ducker TL, Brown RH (eds) Neurophysiology and standards of spinal cord monitoring. Springer, Berlin Heidelberg New York, pp 261–267
60. Rosenberg JNC (1991) Somatosensory and magnetic evoked potentials in a postoperative paraparetic patient: case report. Arch Phys Med Rehabil 72:154–156
61. Salzman SK, Dobney KW, Mendez AA, Beauchamp JT, Daley JC, Freemen GM (1988) The somatosensory evoked potential predicts neurologic deficits and serotonergic pathochemistry after spinal distraction injury in experimental scoliosis. J Neurotrauma 5:173–186
62. Schmid UB, Boll J, Liechti S, Schmid J, Hess CW (1992) Influence of some anesthetic agents on muscle responses to transcranial magnetic cortex stimulation: a pilot study in humans. Neurosurgery 30:85–92
63. Schramm J (1985) Spinal cord monitoring: current status and new developments. CNS trauma 2:207–225
64. Schramm J (1988) Acute evoked potential changes in operative treatment: a summary. In: Ducker TL, Brown HR (eds) Neurophysiology and standards of spinal cord monitoring. Springer, Berlin Heidelberg New York, pp 268–273
65. Schramm J (1989) Intraoperative monitoring with evoked potentials in cerebral vascular surgery and posterior fossa surgery. In: Desmeth JE (ed) Neuromonitoring in surgery. Elsevier Science, Amsterdam, pp 243–262
66. Schramm J, Jones SJ (eds) (1985) Spinal cord monitoring. Springer, Berlin Heidelberg New York
67. Schramm J, Koht A, Schmid G, Pechstein U, Taniguchi M, Fahlbusch R (1990) Surgical and electrophysical observations during clipping of 134 aneurysms with evoked potential monitoring. Neurosurgery 26:61–70

68. Schramm J, Kurthen M (1992) Recent developments in neurosurgical spinal cord monitoring. Paraplegia 30:609–616
69. Schramm J, Romstöck J, Watanabe E (1986) Intraoperatives Rückenmarkmonitoring: eigene Ergebnisse and Bestandsaufnahme. Z Orthop 124:671–682
70. Schramm J, Watanabe E, Strauss C, Fahlbusch R (1989) Neurophysiologic monitoring in posterior fossa surgery. I. Technical principles, applicability and limitations. Acta Neurochir (Wien) 98:9–18
71. Shiau JS, Zappulla RA, Nieves J (1992) The effect of graded spinal cord injury on the extrapyramidal and pyramidal motor evoked potentials of the rat. Neurosurgery 30:76–84
72. Shimoji K, Tamaki T (eds) (1991) Spinal cord monitoring and electrodiagnosis. Springer, Berlin Heidelberg New York
73. Tamaki T, Takano H, Takakuwa K, Tsuji H, Nakayawa T, Imai I, Inooe S (1985) Assessment of the use of spinal cord evoked potentials in prognosis estimation of injured spinal cord. In: Schramm J, Jones SJ (eds) Spinal cord monitoring. Springer, Berlin Heidelberg New York, pp 221–226
74. Taniguchi M, Cedzich C, Schramm J (1993) Modification of cortical stimulation for motor evoked potentials under general anesthesia: technical description. Neurosurgery 32:219–226
75. Taniguchi M, Müller K, Rödel R, Schramm J (1992) Intraoperative monitoring of hearing function during acoustic neurinoma surgery using BAEP and CNAP from the eighth nerve. In: Tos M, Thomsen J (eds) Proceedings of the first international conference on acoustic neurinoma. Kugler, Amsterdam, pp 549–556
76. Taniguchi M, Nadstawek J, Pechstein U, Schramm J (1992) Total intravenous anesthesia for improvement of intraoperative monitoring of somatosensory evoked potentials during aneurysm surgery. Neurosurgery 31:891–897
77. Taniguchi M, Schramm J (1991) Motor evoked potentials facilitated by an additional peripheral nerve stimulation. In: Levy WJ et al. (eds) Magnetic motor stimulation: basic principles and clinical experience (EEG Suppl 43). Elsevier, Amsterdam, pp 202–211
78. Taniguchi M, Schramm J, Cedzich C (1991) Recording of myogenic motor evoked potentials (mMEP) under general anesthesia. In: Schramm J, Moller AR (eds) Intraoperative electrophysiological monitoring. Springer, Berlin Heidelberg New York, pp 72–87
79. Thompson PD, Day BL, Crockard HA, Caldler I, Murray NMF, Rothwell JC (1991) Intraoperative recording of motor tract potentials at the cervico-medullary junction following scalp electrical and magnetic stimulation of the motor cortex. J Neurol Neurosurg Psychiatry 54:618–623
80. Tsubokawa T, Yamamoto T, Hirayama T, Maejima S, Katayama Y (1986) Clinical application of corticospinal evoked potentials as a monitor of pyramidal function. Nikon Univ J Med 28:27–37
81. Veilleux M, Daube JR, Cucchiara RF (1987) Monitoring of cortical evoked potentials during surgical procedures on the cervical spine. Mayo Clin Proc 62:256–264
82. Watanabe E, Schramm J, Strauss C, Fahlbusch R (1989) Neurophysiologic monitoring in posterior fossa surgery. II. BAEP-waves I and V and preservation of hearing. Acta Neurochir (Wien) 98:118–128
83. Wilson WB, Kirsch WM, Neville H, Stears J, Feinsod M, Lehrmann RAW (1976) Monitoring of visual function during parasellar surgery. Surg Neurol 5:323–329

84. Zentner J (1989) Noninvasive motor evoked potential monitoring during neurosurgical operations on the spinal cord. Neurosurgery 24:709–712
85. Zentner J (1991) Motor evoked potential monitoring during neurosurgical operations on the spinal cord. Neurosurg Rev 14:29–36
86. Zentner J (1991) Motor evoked potential monitoring in operations on the brain stem and posterior fossa. In: Schramm J, Moller AR (eds) Intraoperative electrophysiological monitoring. Springer, Berlin Heidelberg New York, pp 95–105
87. Zentner J, Albrecht T, Heuser D (1992) Influence of halothane, enflurane, and isoflurane on motor evoked potentials. Neurosurgery 31:298–305
88. Zentner J, Ebner A (1989) Nitrous oxide suppresses the electromyographic response evoked by electrical stimulation of the motor cortex. Neurosurgery 24:60–62
89. Zileli M, Taniguchi M, Cedzich C, Schramm J (1989) Vestibulospinal evoked potential versus motor evoked potential monitoring in experimental spinal cord injuries in cats. Acta Neurochir (Wien) 101:141–148
90. Zornow MH, Drummond JC (1989) Intraoperative somatosensory evoked responses recorded during onset of the anterior spinal artery syndrome. J Clin Monit 5:243–346

Anesthesia and Somatosensory Evoked Responses

E. Kochs and P. Bischoff

Introduction

Intraoperative monitoring of somatosensory evoked potentials (SEP) is used increasingly to improve monitoring of neural tracts at risk during scoliosis surgery, surgical procedures involving the thoracic aorta, or cerebral perfusion such as carotid endarterectomy and intracranial aneurysm surgery. Changes in SEP latency and waveform may indicate impaired transmission in the pathway monitored. In addition to SEP changes due to surgical trauma and cerebral ischemia, SEPs are also modulated by a variety of other factors including drug-induced changes of the evoked electroencephalogram (EEG). The effects of anesthetics on spinal, subcortical, and cortical SEP have been extensively studied over the past 20 years. Anesthetic and sedative agents are known to exert their effects primarily on the association areas in the cortex and second- or third-order neurons involved in sensory signal processing and evaluation of information transmitted by the somatosensory system. Because drug-induced changes in SEP are predictable and appear to be non-agent-specific, the use of SEP monitoring for assessment of depth of anesthesia has been proposed.

SEPs are changes in brain electrical activity induced by electrical, mechanical, thermal, or tactile stimulation of large mixed peripheral nerves and represent the activities of a rather large number of subcortical and cortical neural generators. For perioperative monitoring the most commonly stimulated nerves are the median and ulnar at the wrist, the peroneal nerve at the leg, and the tibial nerve at the ankle. The voltages generated are of much lower amplitude than the spontaneous EEG. The amplitude of the EEG may reach $100\,\mu V$ whereas scalp-recorded SEPs range in amplitude from about $0.5\,\mu V$ to 10–$20\,\mu V$ (in rare instances). Electrical signals from a variety of different sources in the human body, such as heart, skeletal muscle, and different parts of the central nervous system (CNS), may also interfere with EEG and SEP recordings. SEPs recorded over the spinal cord are generated in the dorsal horn and dorsal columns of the spinal cord. The early cortical SEPs reflect generators activated via the dorsal column (lemniscal pathway) [19].

Soon after the introduction of SEP monitoring techniques in clinical medicine it became clear that the electrical activity of the brain can be used as an indicator of critical cerebral perfusion. A threshold relationship between the depression of SEP and cerebral blood flow has been established. SEPs become depressed at a cerebral blood flow of 15–18 ml 100^{-1} g min^{-1} [5, 35]. Recovery of evoked potentials occurs on reperfusion if residual blood flow can be maintained above levels to disturb ionic homeostasis (< 10 ml 100^{-1} g min^{-1}), and the ischemic period is too short to impair cellular integrity [40, 45, 91]. These findings serve as a rationale for using SEP as an indicator of imminent or ongoing intraoperative ischemia. As a consequence SEPs have been used primarily to monitor patients whose major afferent nerve tracts are at risk during surgical procedures. However, evoked responses are not only subject to critical brain perfusion but are also modulated by a variety of other factors such as blood pressure, body temperature, arterial blood gases, hematocrit, age, gender, neurological disease, and anesthetics. Dose-dependent increases in latency and reductions in amplitude of cortical SEPs have been demonstrated with both inhalational and intravenous anesthetics [33, 97]. Late cortical SEP have been found to correlate with the painfulness of noxious stimuli [7–11]. From this it was concluded that evoked responses hold promise as measures of depth of anesthesia.

Methodological Considerations

Stimulation

Although SEP monitoring is used increasingly for intraoperative monitoring, no definite criteria regarding choice of stimulus intensity have been agreed upon. The electrical stimulus delivered via stainless-steel needle or skin-surface electrodes is usually a monophasic square wave pulse of 100 μs–1 ms in duration. Stimulator output must be both resistively and capacitatively isolated from ground. To produce a synchronized volley accurate placement of the cathode over the nerve is extremely important for recruiting most fibers in the nerve. Care must be taken to assure a relatively constant electrode impedance to minimize fluctuations in stimulus intensity. Stimulators maintaining the output current constant may help to avoid gross stimulus artifacts and changes in the evoked responses. The interstimulus interval must be chosen long enough to allow recovery of the SEP components. For the assessement of cortical components (< 50 ms) the stimulation rate should not exceed 5/s. To minimize short-term habituation and cumulative interactions between SEP subcomponents and successive data irregular stimulation patterns should be used [17]. This is of particular importance for the assessment of long-latency cortical components (> 100 ms). SEP may also be recorded using mechanical or tactile stimulation techniques. SEP following electrical and mechanical stimulation have been found to be rather similar [63]. For perioperative monitoring electrical stimulation is usually preferred to the more natural mechanical stimulation.

However, by mixed nerve stimulation a heterogeneous population of afferent fibers from skin, joints, deep tissue, and even efferent motor neurons are activated [21]. Mechanical stimulation more selectively stimulates skin and joint afferents avoiding interference from muscle innervation.

Most studies have been performed using stimulus intensities of either motor threshold or just above motor threshold. In awake volunteers motor threshold stimulation gives submaximal responses [55]. In anesthetized patients SEP amplitude reaches a plateau when stimulus intensity is gradually increased to about 20 mA [67, 68]. In the case of stimulation of peripheral somatosensory nerves SEPs are generated which modulate the ongoing background EEG activity. Methods to enhance the signal-to-noise ratio (SEP versus EEG) are summation and time-locked averaging. The assumption is that identical SEP responses occur at fixed time intervals after stimulus delivery. Thus, the SEP is regarded as the summation of an exactly defined stimulus-related signal and an unrelated background activity. Using the trigger pulse for the averaging process, only the stimulus-locked response "survives" when more and more randomly distributed poststimulus EEG intervals are summed and averaged (Fig. 1).

SEPs can be recorded along the afferent pathways stimulated (Fig. 2). SEP recorded from the scalp comprise early components generated in peripheral nerves and in the brainstem, as well as short-, middle-, and long-latency

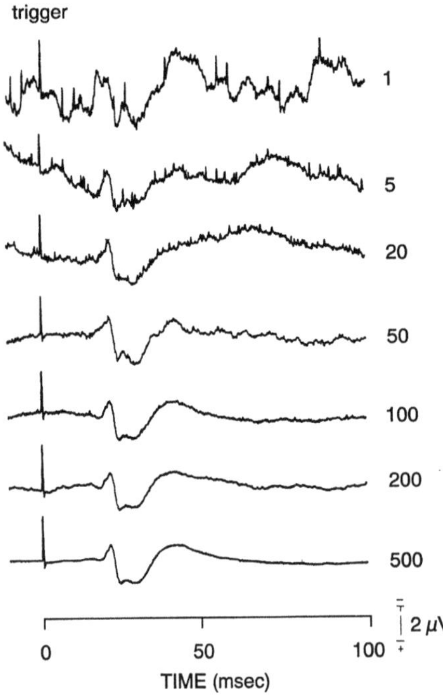

Fig. 1. Signal-to-noise enhancement by signal averaging of median nerve somatosensory evoked responses. The traces present the average of an increasing number of trials (1–500) with the same amplification

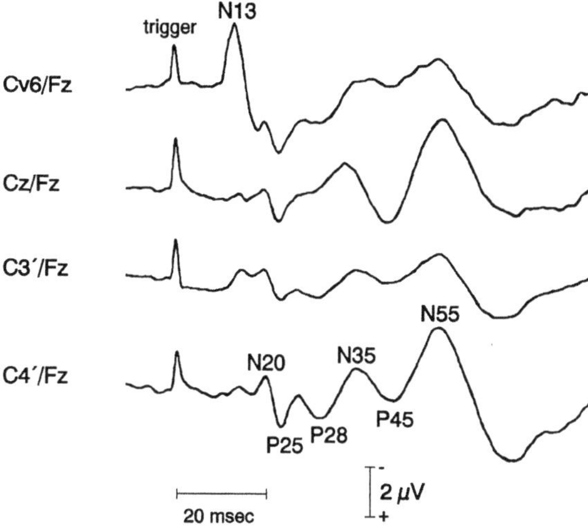

Fig. 2. SEP recordings following median nerve stimulation at the vertebral column at the level of the 6th vertebra (*Cv*6), at the vertex (*Cz*), the ipsilateral (*C3'*) and contralateral somatosensory projection area (*C4'*) with frontal references (*Fz*)

components with origins in subcortical and cortical areas. In anesthesiological practice usually brainstem (latency 13–15 ms) and early cortical components (latency 18–50 ms) following median or posterior tibial nerve stimulation are evaluated.

Bandpass filtering

Critical for waveform analysis is the bandpass used for SEP recordings. Restrictive filtering may introduce unwanted distortions in the SEP waveform [18]. Slow SEP components are affected by high-pass filtering and high-frequency transients by low-pass filtering. For diagnostic procedures the high-pass cutoff is frequently set to 1 Hz (time constant 0.16 s). In the unshielded environment of an operating room with a variety of electrical noise this setting may result in distorted waveforms due to baseline shifts and low-frequency artifacts. For intraoperative monitoring filtering at 10 Hz or higher may be helpful to minimize these artifacts. However, using these filter settings the SEP profile is modified and slow SEP components such as the subcortical N13 or cortical components with latencies longer than 25 ms may be modulated [18]. As a result, comparisons with normative data from SEP components recorded with other filter setting are made more difficult. On the other hand, intraoperative SEP recording makes use of intraindividual comparisons. Thus, changes in SEP waveform due to an ischemic event or due to changes in depth of anesthesia may still be assessed in the individual subject. Low-pass filtering may also affect SEP components. Because early SEP components include high-frequency transients, the use of standard EEG amplifiers with a high-pass cutoff of 50 Hz or lower is precluded. The low-pass filter should be set at 1 kHz or higher. Smoothing by low-pass filtering may be applied to the original waveform, but this affects data information. The minimum sampling rate must be more than twice the highest frequency present in the sampled data (Nyquist theorem) to avoid spurious low-frequency components that are not present in the original signals. For the assessment of early SEP components the bin width should be set a 250 μs or lower.

Nomenclature

As recommended by an international committee components of evoked responses are identified by the polarity (P = positive; N = negative) and the peak latency of the individual SEP component following the trigger pulse [22]. The label of the same SEP component may be different for stimulation at different levels along the sensory afferent pathway. Latencies may also be different when peripheral conduction times (depending on arm length or limb temperature) are different. SEP waveforms vary among subjects but are rather stable in one subject. Without general agreement poststimulus (median nerve) SEPs may be arbitrarily subdivided into short-latency (less than 25–30 ms), intermediate-latency (30–100 ms), and long-latency (> 100 ms) components. Short-latency SEPs are generated in the peripheral nerves, spinal cord, subcortical and primary cortical structures. Subcortical-generated potentials are of lower amplitude than the spontaneous EEG. Intermediate-latency SEP components are generated in the cerebral cortex. They are subject to the effects of anesthetic agents and changes in physiological variables such as arterial oxygen concentration, blood flow, and body temperature. Long-latency SEP are thought to be generated in the association areas of the cerebral cortex and may reflect information processing. They are very sensitive to the effects of anesthetics but are also subject to changes in vigilance, attention, and emotional state.

Nearfield, Farfield Recordings

The human head represents a volume conductor which allows recordings of evoked responses at almost all surface areas. The terms nearfield and farfield recordings do not necessarily imply different origins of the generated signals. The original reason for introducing the concept of nearfield versus farfield components was to distinguish between scalp-recorded fields resulting from generators in the cortex and sources in subcortical areas. Depending on the recording site the signals generated must travel varying distances. If the recording electrode is close (2–3 cm) to the neural generator of the individual SEP component a "nearfield" potential can be recorded. The waveform is changed if the electrode is moved away only short distances. Using scalp electrodes, near- and farfield potentials can be recorded. According to Eccles [25] cortical potentials of brain electrical activity are related to postsynaptic potentials of the apical dendrites of pyramidal neurons and can thus generate volume-conducted potentials which are recordable distant to the origin. Farfield potentials from several neural generators at different distances may be recorded at the same electrode position via volume conduction and are less affected by electrode displacement. However, the electrical field strength has an inverse relationship varying with the distance of the generator to the recording electrode. Thus, in general fearfield potentials are of small amplitude ($< 1\ \mu$V).

Reference Electrode

Recording electrode montages should be designated in conjunction with established evidence about the underlying brain generators. In clinical SEP monitoring a frontal scalp reference is most often used because it appears to be less noisy than non-cephalic references, although some SEP components may not be distinguishable with this montage. Noncephalic reference was a breakthrough in identifying early subcortical SEP components. Noncephalic reference recording is best used if the true SEP waveform is of concern because the neural generators with origins in the spinal cord and the brain are located at greater distances. Using a frontal scalp reference, the widespread farfield components with latencies less than 18–20 ms usually cancel out because the neural generators have approximately the same distance to the reference and the "active" scalp electrode. To obtain subcortical components in addition to scalp recordings an electrode must be placed at the level of the C2–C7 vertebrae in the neck.

Temperature

In anesthetized patients limb temperature can be below 28°C. For 1°C the nerve conduction velocity decreases by about 2.5 m/s, affecting latencies of SEP components. Long-term intraoperative SEP recordings may thus be compromised by lack of control of limb temperature. The effects of hypothermia on SEP latencies have been well documented. Hypothermia appears to decrease conduction velocity and to delay synaptic transmission. A linear relationship between latency and tympanic temperature has been calculated for the temperature range of 25°–35°C [75]. The central conduction time (CCT; difference in latencies; N20–N13) has been shown to vary as a logarithmic function [53] or to increase exponentially with decreasing temperature [38]. The spinal conduction time also increases exponentially but less steeply than the CCT [38].

Miscellaneous

Studies in animals and humans have shown that during isoflurane anesthesia acute hypocapnia with arterial carbon dioxide tensions in a range of 20–35 mmHg exerts small or no changes in subcortical and cortical SEP components [31, 79]. Similar findings with no significant changes in posterior tibial nerve SEP have been reported for variations in arterial carbon dioxide tensions in the range of 20–50 mmHg during alfentanil/nitrous oxide anesthesia [44]. From this it can be concluded that in the clinical setting monitoring sensitivity should not be compromised during hyper- or hypoventilation. Hypotension induced by hemorrhagic shock causes graded increases in SEP latencies and decreases in amplitude [34]. Cortical SEPs are depressed at a mean arterial blood pressure below 40 mmHg. Spinal SEPs show more resistance and disappear at lower levels of hypotension. Sequential recovery of SEPs upon restoration of blood flow and mean arterial blood pressure is dependent on the length of hypotension. When 15 min elapsed between loss of responses and reinfusion of blood, cortical SEPs do not resume within 1 h after infusion. No restoration of SEP is noted when more than 30 min has elapsed between loss of SEPs and blood reperfusion.

SEP Components for Intraoperative Monitoring

For intraoperative SEP recording following upper limb stimulation, usually the negative deflection with a latency of approximately 20 ms (N20) is evaluated. This represents the first major early thalamocortical SEP component, which is best recorded at the postcentral cortex contralateral to the stimulation site. Using non-cephalic reference, a widespread bilateral component N18 may be distinguished from the later N20 component. The N18 component appears to be generated below the thalamus whereas the N20 seems to be generated in the thalamus or by thalamic-cortical radiation [14]. The N20 is followed by a contralateral positivity with varying latency (P23–P27). Using neck recordings with noncephalic reference a spinal nearfield component N13 may be detected. This SEP component is distinct from the P15 component, which is recorded with frontal reference. The neck N13 potential is generated below the foramen magnum. This most probably reflects initial intraspinal postsynaptic activity generated in neurons of the dorsal horn. This component has no farfield representation at the scalp [20]. The P14 SEP farfield potential is generated above the foramen magnum between the lower medulla and the thalamus probably by the afferent volley in the medial lemniscus. The difference between the major cervical SEP N13 and the first negative thalamocortical peak N20 recorded over the scalp represents the CCT. This reflects the time interval for the evoked response to travel through the intracranial portion of the somatosensory pathway. Depending on the filter settings and electrode locations the CCT of healthy subjects is 5.8 ± 0.5 ms. The CCT has been shown to correlate with critical levels of brain perfusion [37, 47, 92]. However, the CCT is also subject to changes in body temperature [59] and drug effects.

Lower limb SEPs may be assessed during surgical procedures which may compromise spinal cord function. Following posterior tibial nerve stimulation at the ankle a large negative potential is recorded over the upper lumbar region (N20). This component is probably generated at the entry of the spinal roots into the spinal cord. Over the entire spinal column negative potentials representing the traveling waves with increases in latencies more rostrally may be recorded. At the level of the cervical vertebral column a P27 component appears which represents the arrival of the stimulus at the cervicomedullary junction. With noncephalic reference recording (at the shoulder) the scalp SEP of posterior tibial nerve stimulation shows a lemniscal P30 farfield followed by a contralateral frontal negativity N37 and a large midline parietal positivity P40. Using frontal reference montage, the parietal SEP is shifted downward [20, 57]. Similar SEPs with shorter latencies are elicited by peroneal nerve stimulation at the knee [68].

Effects of Anesthetics

Hypnotics

In contrast to spinal or subcortical SEP components, later SEPs (latencies > 20 ms) with origins in thalamocortical or cortico-cortical projection systems are subject to the effects of hypnotics. Only minimal changes in the amplitude of the early cortical response but significant increases in latencies are seen with 4 mg/kg thiopentone [89]. Latencies of later SEP components are even more prolonged consistent with an effect of barbiturates on synaptic transmission [23]. High-dose barbiturates result in dose-dependent reductions in amplitudes and increases in latencies of the primary cortical response (N20/P25). Later SEP potentials may be completely abolished [1]. Most interestingly, it has been shown that median nerve SEPs can be recorded in the presence of a thiopental-induced isoelectric EEG [23]. The CCT is significantly increased at 6 mg/kg intravenous thiopental [74]. In contrast to the depressing effects of thiopental on median nerve SEPs, administration of etomidate results in a tremendous increase in the early cortical SEP amplitude (Fig. 3) [48, 74, 62]. Latencies including CCT are also prolonged. The increase in amplitude indicates suppression of inhibitory neural mechanisms in thalamocortical pathways [48]. A later study has questioned the enhancement of the very first cortical component N20/P25 and demonstrated instead an increase of the P25/N30 component [24]. It has been suggested that the amplitude enhancing effect of etomidate may be beneficial for intraoperative monitoring in patients with small SEP amplitudes [52].

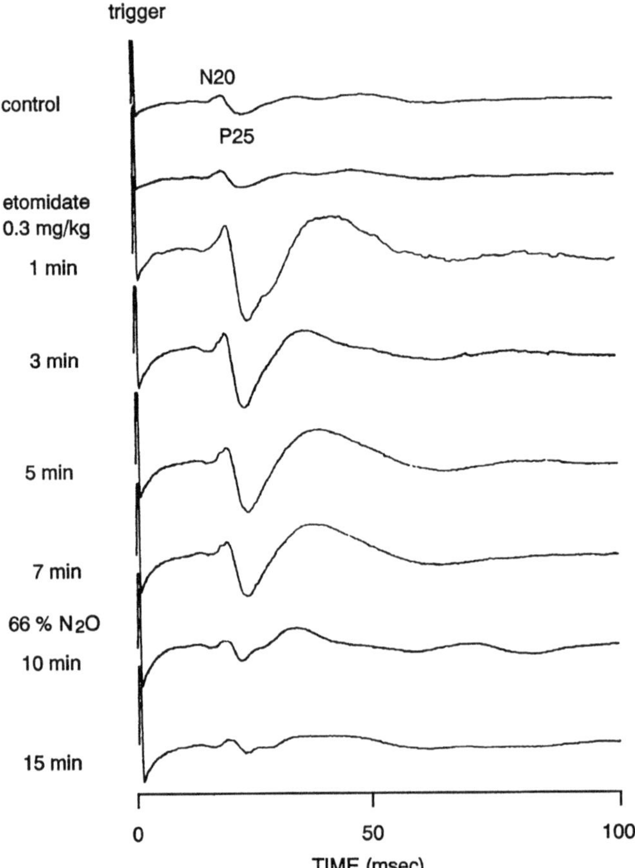

Fig. 3. Effect of etomidate on somatosensory evoked responses following median nerve stimulation. The amplitude of the SEP component N20/P25 was enhanced following induction of anesthesia with etomidate (0.3 mg/kg). Administration of 66% nitrous oxide in oxygen resulted in SEP depression. Recording: contralateral somatosensory projection area (*C4'*) versus frontal reference; stimulation rate: 3 Hz; bandpass: 10–2000 Hz

Propofol given for induction (2–2.5 mg/kg) and for maintenance of anesthesia (9 mg/kg) does not suppress subcortical and early cortical SEP [50, 78]. Moderate increases in CCT and decreases in amplitude of later components are comparable to the effects of thiopental [23]. However, during anesthesia with propofol (bolus, 2 mg/kg; infusion, 6 mg/h for the first hour and 3 mg/kg subsequently) in combination with 50% nitrous oxide early cortical SEP responses have been shown to be suppressed [60]. These findings have been confirmed in later studies which show that posterior tibial nerve SEPs are almost twice as large with propofol infusion (10 mg/kg for 10 min, $8 \, \mathrm{mg\,kg^{-1}\,h^{-1}}$ for the next 10 min and $6 \, \mathrm{mg\,kg^{-1}\,h^{-1}}$ thereafter) in combination with alfentanil (100 μg/kg bolus followed by an 2 μg kg^{-1} min^{-1} infusion) when compared to 66% nitrous oxide [42].

Ketamine (2 mg^{-1}kg^{-1} intravenous bolus followed by 30 μg kg^{-1}h^{-1}) enhances the early cortical SEP component [81]. The effect of ketamine on SEP amplitudes may be dose dependent because in a previous study no changes in early cortical SEPs following low-dose ketamine (0.5 mg/kg intravenously) were found [51] (Fig. 4). In the same study the middle-latency component N35 was suppressed during the period when subjects were unconscious. Because ketamine is a racematic mixture of the S-(+)- and the R-(−)-isomers it is not clear whether the effects of ketamine are mediated by the synergistic action of the two isomers or to a predominant effect of one of the isomers. Administration of S-(+)-ketamine results in dose-dependent increases in latency and decreases in amplitude of the N50 component of median nerve SEP in dogs [26]. The highly specific μ-receptor antagonist cyprodime is able to partially restore this component. It has been concluded that in part S-(+)-ketamine induces an opioid receptor mediated blockade of impulses in the sensory nervous pathways. Because restoration of SEP following cyprodime is not complete, additional

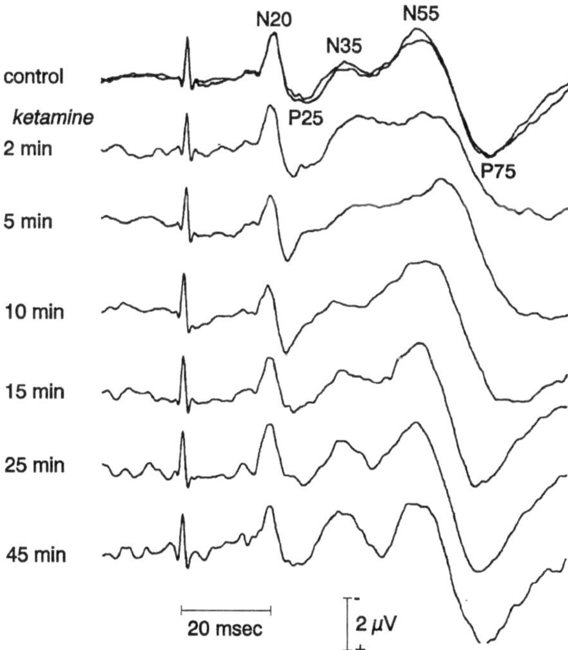

Fig. 4. Effect of low-dose ketamine (0.5 mg/kg) on median nerve SEP. The primary cortical component N20/P25 is not changed whereas the component N35 is depressed during the period when the subject was unconscious. Recording: contralateral somato-sensory projection area (*C4'*) versus frontal reference; stimulation rate: 3 Hz; bandpass: 10–2000 Hz

interaction of S-(+)-ketamine with other binding sites (i.e., NMDA-receptors) was postulated.

Diazepam has been shown to decrease amplitudes of cortical SEPs in a dose-dependent manner, with late cortical SEP components completely abolished [32]. These findings were not supported in a later study showing that 20 mg diazepam has no effect on median nerve SEP [56]. Midazolam (bolus 0.2 mg/kg followed by an infusion at 5 mg/h) has been reported to reduce the amplitudes of cortical peaks by approximately 60% [90]. It has been concluded that with unchanged systemic variables major latency changes following administration of midazolam may be indicative of other deleterious factors. However, these findings are at variance with previous results showing that midazolam (bolus 0.3 mg/kg followed by an infusion 0.2 $mg\,kg^{-1}\,h^{-1}$) may also increase SEP latencies [52]. Because benzodiazepines depress SEP amplitudes by approximately 50%–60%, these drugs may not be the best choice for SEP monitoring in the case of initially small SEP amplitudes. However, in contrast to ketamine, no additive effect of midazolam and nitrous oxide on SEP amplitudes has been reported.

In summary, using intravenous hypnotics the early cortical SEP component (N20/P25) may be used for intraoperative monitoring. Polysynaptic SEP components with latencies of 25 ms or more are depressed by all intravenous hypnotics. In contrast, subcortical SEP components are not significantly affected.

Narcotic Analgesics

Morphine, Fentanyl. The administration of morphine and fentanyl results in changes in cortical SEPs elicited by posterior tibial nerve stimulation [70]. Latencies of N1 (47.25 ± 3.31 ms), P2 (56.51 ± 3.35 ms), and N2 (65.78 ± 3.47 ms) are increased to a similar degree when morphine (250 μg/kg) or fentantyl (2.5 μg/kg) are given in combination with 60% nitrous oxide in oxygen after an induction dose of thiopental (3 mg/kg). Likewise, continuous infusion of both narcotics (fentanyl, 1.5–2.5 μg$\,kg^{-1}\,h^{-1}$; morphine, 150–250 μg$\,kg^{-1}\,h^{-1}$) produce similar increases in N1, P1, and N2 latencies. However, SEP pharmacodynamics may depend on the anesthetic technique used. Bolus injections may result in a greater increase in N1 latencies compared to an infusion. The N1/P2 amplitude is depressed with both drugs whereas the component P2/N2 is affected differently. Morphine has been found to depress, while fentanyl produces variable changes in amplitudes. It is concluded that increases in latencies and depression of amplitudes seen with opioids reflect inhibition of velocity and amplitude transmission of neural action potentials. In contrast, when given intrathecally, morphine (15 μg/kg) does not affect SEP following posterior tibial nerve stimulation [82]. These data indicate that opioid-activated spinal pathways do not interfere with transmission of afferent impulses resulting from stimulation of peripheral somatic nerves. The N20 and P24 components following median nerve stimulation are depressed by bolus doses of 25 μg/kg fentanyl [62]. Similar findings have been reported with bolus injections (200 μg) [56] and continuous high-dose fentanyl infusion (bolus 53.2 μg/kg following by a continuous infusion 10–20 μg kg^{-1} h^{-1}) [80]. During hypothermic cardiopulmonary bypass administration of 75 μg/kg fentanyl does not affect SEP [38].

Sufentanil. According to a rapid onset of action sufentanil (5 μg/kg) decreases cortical SEPs with minimal changes in latency within 1 min after intravenous administration [46].

Alfentanil. Cumulative doses of alfentanil (3–120 μg/kg) have been shown to decrease the amplitude (N100) and to increase latency (N140) of late cortical SEP following upper limb stimulation in dogs [27]. From the differential effects of alfentanil on cortical SEP (N100, decrease in amplitude; N140, increase in latency) an interaction of alfentanil with different opioid receptor subtypes (μ-receptor, decrease in amplitude; κ-receptor, increase in latency) which can be assessed by electrophysiological methods has been postulated [27]. In intensive-care patients long-term sedation (3–14 days) with alfentanil (0.6–2 mg/h) in combination with midazolam (1.5–5 mg/h) was found to decrease the late cortical SEP N100 concurrent with decreases in EEG beta-activity [30]. A close correlation between alfentanil dose and depression of amplitude was found ($r = 0.94$). The SEP was restored within 24–48 h after termination of sedation. It is concluded that changes in the N100 amplitude reflect blockade of nociceptive transmission. Propofol-alfentanil anesthesia (TIVA) has been found to be superior to enflurane or isoflurane in 66% nitrous oxide anesthesia for median and posterior tibial nerve SEP monitoring during surgical procedures on the spine [43] and intracranial aneurysma surgery [94]. In the TIVA group the amplitudes of cortical SEPs (N20/P25; P40/N50) were significantly higher than in the group with inhalational anesthesia. The N30 component was more frequently observed with TIVA. It is concluded that this component is a more sensitive indicator of cortical hypoperfusion than the N20/P25 complex. High-dose alfentanil given for cardiac surgery does not increase latencies of early cortical posterior tibial nerve SEPs [42]. Amplitudes are decreased to 60% of control.

In summary, the changes on upper or lower extremity SEPs induced by narcotic analgesics are smaller compared to the effects of intravenous or inhalational anesthetics. An opioid-based anesthetic technique allows adequate SEP signal acquisition in most instances.

Inhalational Agents

Nitrous Oxide. Increasing concentrations of up to 50% nitrous oxide cause graded reductions in amplitudes without changes in latencies of the median nerve evoked early cortical responses [83]. These findings are consistent with earlier findings during Harrington rod insertion and posterior fossa surgery which show reduction in SEP amplitude during administration of nitrous oxide [41, 61]. In latter studies it has been demonstrated that the SEP component N20/P25 is reduced by approximately 50% without changes in latencies during 66% nitrous oxide in oxygen anesthesia [43, 73]. These results are similar to other findings showing that the addition of 50% nitrous oxide to fentanyl (10–20 μg/kg followed by 20–50 μg/kg as needed) produces a consistent 50% decrease of the early cortical SEP [88]. Changes in latency are variable. Earlier studies have shown that 50% nitrous oxide depresses late cortical responses (100–250 ms) elicited by painful electrical stimulation of the pulp by approximately 50% [4]. Interestingly, naloxone (0.4 mg) has been found to be efficacious in restoring nitrous oxide depressed late cortical SEP amplitude (N100) and the negative peak latency at 175 ms [10]. The authors concluded that these findings are consistent with the hypothesis that the negative peak at 175 ms reflects primarily the analgesic effects of nitrous oxide, and that the mechanism of nitrous oxide analgesia involves the action of endorphins at selected sites along somatosensory pathways. The depressing effect of nitrous oxide on SEP amplitude is more pronounced with reduced stimulus intensity and extreme hyperventilation [93]. The decrease in SEP amplitude seen with nitrous oxide may result in deterioration of the signal-to-noise ratio especially in patients with initial small SEP components. Taken together these findings demonstrate that nitrous oxide exerts a general cortical depressant effect [10].

Volatile Anesthetics

Subcortical SEP Components. Volatile anesthetics (halothane, enflurane, isoflurane) have been shown to depress cortical SEP and, quantitatively differently from the effects of nitrous oxide, to increase latencies in a dose-dependent manner. There is controversy on the effects of volatile anesthetics on spinal and subcortical SEP. Halothane does not change spinal SEP in sheep [3]. Quite differently from the effects of enflurane and isoflurane, halothane does not interfere with spinal synapses but with synaptic transmission rostral to the medial lemniscus [99]. However, at concentrations of 2% halothane or greater a significant attenuation occurs [3]. These findings are at variance with previous studies in cats showing no change in spinal SEPs up to 4% halothane [66]. In humans the prolongation of CCT during halothane anesthesia is explained by depression of synaptic transmission between spinal cord and cortex [73]. At the spinal level enflurane induced slowing through interaction with dorsal horn cells giving rise to a prolongation of the N11–N13 interval [99]. Similarly to the effects of isoflurane, the cuneate synapse appears not to be affected by enflurane. Isoflurane and enflurane have been demonstrated to increase the N13 peak latency and to decrease the interval between the N14 peak and the P14 peak of median nerve SEPs [99]. The authors concluded that these findings implicate interference with synaptic transmission at the dorsal horn cells and cannot be explained by slowing of axonal conduction in the spinal cord. In humans anesthetized with enflurane or isoflurane in 66% nitrous oxide no significant changes in latencies of subcortical SEPs have been found [77, 96]. However, using higher concentrations of isoflurane (> 1.0%), the spinal N13 and the subcortical P14 have been shown to be prolonged in latency and decreased in amplitude [99, 100].

Cortical SEP components

Halothane. Reproducible cortical SEPs can be recorded during 0.5% halothane in 66% nitrous oxide [73]. At 1.0 MAC halothane in 60% nitrous oxide SEP components following posterior tibial nerve stimulation may be so small that neither amplitudes nor latencies can be measured [70]. In comparison to enflurane and isoflurane, halothane administration (1 MAC) results in the greatest depression of the cortical component P35/N45. Qualitative similar results showing dose related SEP depression have been obtained for median nerve evoked responses [71]. A quantitative difference is that in comparison to halothane and isoflurane, enflurane produces the greatest, and halothane the least, depressing effect on early cortical SEP. Similar results showing dose related SEP depression have been reported for intraoperative posterior tibial nerve SEP recording during spinal fusion surgery [76]. Using 0.25%–2.0% halothane, reproducible posterior tibial nerve SEPs were obtained throughout the surgical procedure in 91% of the patients ($n = 116$) studied. Small but significant decreases in the N25 and P30 amplitudes and significant increases in the latency of the P53 peak were found. The authors concluded that the use of halothane does not interfere with intraoperative SEP recordings. Late farfield components and nearfield cortical potentials are substantially altered by increments in halothane doses [36]. It has been concluded that early farfield SEP recorded from vertex to neck, together with lumbar spinal cord potentials may be the preferred monitoring technique for halothane anesthesia [36]. The amplitudes of subcortical farfield potentials measured from the scalp to noncephalic reference do not decrease as much of the nearfield cortical potential with increasing halothane concentrations [85].

Enflurane. Increasing concentrations of enflurane also depress cortical SEPs and increase latencies [62, 70, 71, 85]. In humans 0.5 MAC enflurane has been reported to abolish later cortical SEP whereas peripheral, spinal, subcortical, and early cortical generated SEP demonstrate minimal amplitude depression and increased latencies [16]. In monkeys 0.25–1 MAC enflurane causes marked amplitude depression and small increases in cortical SEP latencies [90]. In contrast to human studies, no changes in latencies of subcortical SEPs were found. Quite differently to the effects of other anesthetics [76], later cortical components were reproducible with 1 MAC enflurane. Enflurane seems to exert a biphasic effect on cortical SEP with depression of amplitude at lower and enhancement at higher concentrations. Increased SEP amplitudes have been reported for high-dose enflurane [15, 28]. These findings have been explained by enflurane-induced synchronization of cortical neurons and excitation [54]. The increase in SEP amplitudes can be reversed by administration of 66% nitrous oxide [96]. Similar to the EEG spike activity seen with enflurane, it has been argued that the enhancement of cortical SEPs may reflect "epileptic" activity induced by enflurane. However, this characterization has been questioned because no increase in cerebral demand for oxygen has been noted [95].

Isoflurane. Isoflurane results in more pronounced SEP depression and increases in latencies compared to equipotent doses of halothane [95]. In contrast to 2% halothane, administration of equipotent concentrations of isoflurane in combination with 66% nitrous oxide may result in complete loss of cortical SEPs. In contrast, reproducible median nerve SEPs have been demonstrated with an anesthetic technique using isoflurane in concentrations up to 1%–1.5% [77]. However, no nitrous oxide was administered in this study. Isoflurane has dissociated effects on short-latency cortical SEPs over the parietal as opposed to the central and precentral cortex. Multichannel SEP recordings have demonstrated isoflurane-induced increases in the precentral SEP P22 and depression of the postcentral peak N20 [100]. The focal amplitude increase of P22 has been attributed to facilitation, i.e., inhibition of inhibitory synapses at the level of thalamocortical somatosensory projection to the precentral cortex. CCT was increased significantly whereas the spinal conduction was not delayed by increasing levels of isoflurane [39, 65, 99].

Sevoflurane. Recent data suggest that sevoflurane given at increasing concentrations (0.5, 1.0, 1.5 MAC) has similar effects on median nerve SEPs as isoflurane [64].

Anesthetics are known to attenuate early cortical somatosensory evoked responses. The shorter latency components appear to be more stable during anesthesia than the later components, demonstrating more inter- and intraindividual variability. The high variability of late cortical components may preclude their use for intraoperative monitoring [33]. Recent studies are at variance with suggestions that advocate total avoidance of volatile anesthetics for SEP monitoring [33]. However, opioid-based anesthetic techniques may provide better SEP monitoring conditions.

Depth of Anesthesia

With the exception of etomidate anesthetic-induced effects on SEP latencies and amplitudes are not agent specific. From this it may be concluded that under certain conditions (i.e., no changes in systemic variables such as blood pressure, blood gases, hematocrit, temperature, or cerebral blood flow) changes in SEPs may be a reflection of depth of anesthesia rather than the specific anesthetic used. The requirements of such an indicator of depth of anesthesia have been defined previously [84]. The most important criteria are independence of anesthetic technique, graded, easily quantifiable responses to changes in depth of anesthesia, and changes to surgical stimulation when depth of anesthesia is inadequate. EEG measures have been found to be more or less agent specific, with large interindividual variability. These are therefore of limited applicability for the measurement of depth of anesthesia. In contrast, cortical components of visual, auditory, and somatosensory evoked potentials demonstrate graded non-agent-specific alterations with changes in depth of anesthesia. However, the lack of consistency in the effects of the various techniques on the cortical SEP components may be a drawback [97] although, with the exception of etomidate, increasing concentrations of anesthetics all produce reductions in SEP amplitude and increases in latency. Accordingly, it has been argued that SEP hold promise as an indicator of depth of anesthesia [84]. These theoretical considerations have been proved to be useful during a "balanced" anesthetic technique using 66% nitrous oxide supplemented by halothane or fentanyl [86]. Induction of anesthesia resulted in 62% decrease of median nerve SEP N20/P25 and increase in latency from 19.2 ± 1.3 to 20.0 ± 1.5 ms. On arousal the SEP returned toward normal. In a different study, tracheal intubation was accompanied by a decrease in latency and an increase in amplitude [86]. Intraoperative arousal produced by surgical stimulation was accompanied by a small decrease in latency and an increase in amplitude. However, these changes did not correlate with hemodynamic changes during surgery. This may be explained in part by the time periods (80–100 s) needed to average subsequent SEPs whereas heart rate and blood pressure were evaluated at different time points. Using propofol ($100\ \mu g\,kg^{-1}\,min^{-1}$) and 66% nitrous oxide in oxygen noxious stimulation has been shown to increase the amplitude and decrease latency of the SEP N100 [29]. This was explained by increases in the amount of afferent nerve potentials from the median nerve induced by surgical stimulation (traction of the mesentery). The authors concluded that generalized activation of the central nervous system results in SEP changes which offset the effects of anesthesia/analgesia. After terminating drug administration restoration of SEPs coincided with the patients becoming oriented in time and space. However, restoration of SEP after surgery may not always be observed because noxious stimulation may induce long-lasting changes in sensory threshold and amplitudes of early cortical SEPs [58]. The decrease in SEP amplitudes seen after surgery suggests modified transmission of stimuli, which has also been demonstrated during ischemic pain in volunteers [13].

It may be that SEPs reflect the analgesic rather than the hypnotic action of anesthesia [97]. This would be consistent with the findings that nitrous oxide depresses SEP more than volatile anesthetics when given at equipotent doses, and that in comparison to narcotic analgesics hypnotics such as etomidate and propofol without analgesic potency fail to depress the SEP response. There is increasing evidence that the amplitudes of late cortical event-related SEPs during painful laboratory stimulation relate to the individual pain relief [6–12] after administration of narcotic analgesics [6–8, 11]. However, electrical median or posterior tibial nerve stimulation employed for intraoperative SEP monitoring not only stimulates Aδ- and C-fibers but also thick myelinated nerves not involved in nociceptive transmission. Therefore, specific pain models developed for pain research in humans have been used for the assessment of analgesic treatment. These models assess late SEP components elicited by intracutaneous stimulation using weak electrical currents or thermal laser-induced activation of intracutaneous nociceptors. Amplitude changes in pain-related SEPs have been validated as indicators of analgesia in awake human subjects [6,7,9,11]. It has been shown that late SEPs due to noxious stimulation can be recorded during halothane anesthesia (Fig. 5) [51]. These SEP are sensitive to analgesic treatment. In contrast, EEG and late auditory evoked potentials were unspecific in response to painful stimulation. Laser-induced pain has been

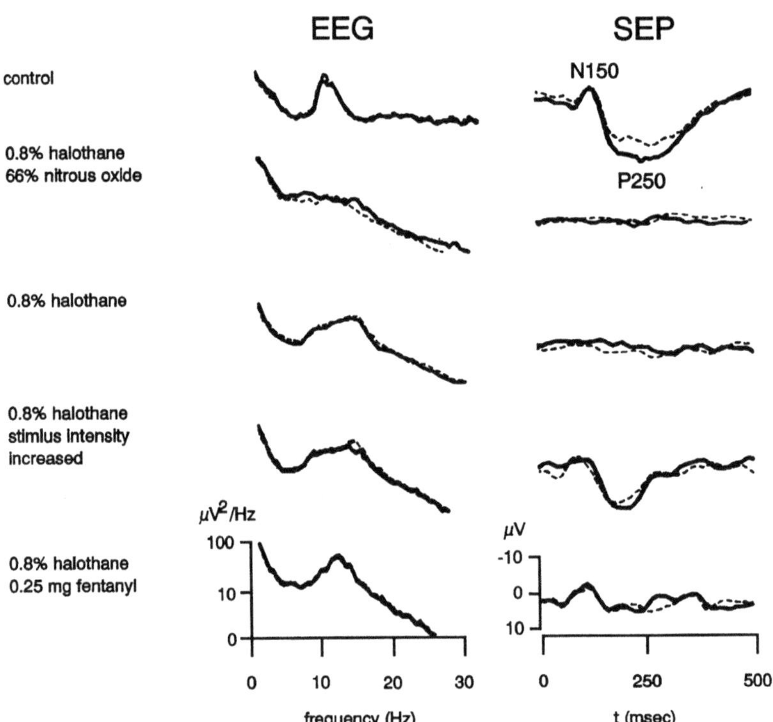

Fig. 5. Effect of 0.8% halothane with and without administration of 66% nitrous oxide on prestimulus EEG and SEP following intracutaneous noxious stimulation. *Control,* after premedication with 5 mg midazolam before induction of anesthesia. EEG was dominated by alpha-activity. SEP consisted of a biphasic deflection with a vertex negativity at 150 ms and a vertex positivity at 250 ms. *Second row,* 0.8% halothane in 66% nitrous oxide in oxygen. EEG was shifted to slow-wave activity, and SEPs were completely suppressed. *Third row,* termination of nitrous oxide administration did not result in restoration of SEP; EEG was shifted toward faster wave activity. *Fourth row,* increased stimulus intensity at the same anesthetic depth restored SEP but did not change EEG. *Last low,* SEP were attenuated after administration of 0.25 µg fentanyl

shown to be reduced by 30 μg/kg alfentanil [2]. In this study the SEP amplitude (N300), which reflects activity mediated by the thin myelinated Aδ-fibers, correlated with the intensity of the pain perceived. Naloxone (0.4 mg) antagonized the analgesic effect of alfentanil on both pain threshold and SEP.

In conclusion, specific pain models hold promise for the assessment of analgesic treatment. However, the techniques developed so far make use of late cortical SEP with latencies of 100 ms and more. These components are very vulnerable to the effects of anesthetics. More studies must be performed to asses the usefulness of this monitoring technique for the assessment nociceptive transmission in anesthetized patients.

Summary

Middle and long-latency SEP components correspond to the functional integrity of cortical projection areas and are modulated by the ascending reticular activating system. SEPs with an origin in thalamocortical or cortico-cortical projection systems may change with variations in vigilance, attention, and physiological variables such as temperature, blood pressure, and hematocrit. In addition, late SEPs are affected in amplitude and/or latencies by virtually all anesthetics in a dose-dependent manner. In contrast, similar to brainstem auditory evoked responses spinal and subcortical SEPs are very resistant to the effects of anesthetics.

SEP monitoring has been used increasingly because of (a) noninvasive measurement techniques from surface scalp electrodes, (b) reproducible and relatively stable waveforms, (c) spinal, subcortical and cortical SEP having a predictable dose-dependent relationship to the effects of anesthetics, and (d) the ability to control most of the physiological variables with an effect on SEP such as temperature, age, gender, stimulus pattern, and intensity. In addition, these effects can be assessed by using the contralateral sensory pathway in each individual as control [68]. SEP monitoring may provide information on the functional integrity of specific neuronal tracts at risk during surgery or in patients with central nervous system trauma. However, there is still no consensus on the indications of such monitoring in many types of surgical procedures. Recently the use of intraoperative SEP monitoring has expanded to the assessment of depth of anesthesia. Nonspecific dose-dependent drug effects have been shown for most anesthetics studied so far. This would fulfill one criterion for an indicator of depth of anesthesia. However, contrary to middle-latency auditory evoked potentials, which have been demonstrated to indicate intraoperative arousal or awareness [97] only few studies have shown that SEP monitoring provides information during inadequate depth of anesthesia. Studies in healthy volunteers and patients indicate that SEP monitoring may be useful for the assessment of subjective pain experience and nociceptive blockade. Further comparative studies must determine whether SEP or auditory evoked potential monitoring is more adequate for the assessment of drug effects and the patient's response to noxious stimulation.

Most studies report the average effects of anesthetics on SEP on populations of patients. One point not rigorously studied so far is the possibility of a differential sensitivity of individuals to anesthetics. It has been pointed out that anesthetics given in identical doses may result in interindividually differing SEP changes [76]. In addition, the interactions between anesthetics employed intraoperatively and the compound electrophysiological effects are poorly understood.

Appendix

Anesthetic	Stimulation	SEP component	Effect	Reference
Thiopental (4 mg/kg)	Median nerve	Early cortical response N20/P25	Minimal changes	[1, 89]
Thiopental (1–15 mg/kg)				[74]
Thiopental (5 mg/kg; $2\,mg\,kg^{-1}h^{-1}$)				[52]
Thiopental (1.25 mg $kg^{-1}\,min^{-1}$)			Dose-related depression of cortical responses	[23]
Etomidate (0.3 mg/kg)	Median nerve	Early cortical response N20/P25	Amplitude increased Latencies prolonged	[48, 62, 74]
Etomidate (0.3 mg/kg; $2\,mg\,kg^{-1}\,h^{-1}$)	Median nerve	Early cortical responses	Amplitude increased Latencies prolonged	[52]
Etomidate (0.3 mg/kg)	Posterior tibial nerve	Cortical P25/N30	Amplitude increased latencies polonged	[24]
Propofol (2–2.5 mg/kg)	Median nerve	Early cortical response N20	Latencies increased Minimal changes in amplitude	[47]
Propofol(2.5 mg/kg; $9\,mg\,kg^{-1}\,h^{-1}$)	Median nerve	Early cortical responses N20	Increased in latency Minimal changes in amplitude	[78]
Propofol ($6\,mg\,kg^{-1}\,h^{-1}$)/50% N_2O	Posterior tibial nerve	P40	Amplitude decreased Latency prolonged	[60]
Propofol ($6\,mg\,kg^{-1}\,h^{-1}$)/ alfentanil ($2\,\mu g\,kg^{-1}\,min^{-1}$ versus 66% N_2O/alfentanil	Posterior tibial nerve	Early cortical response P42/N51	Marked depression by nitrous oxide	[43]
Propofol (60 mg; 15 mg/min; 6 mg/min) + alfentanil (1 mg; 0.3 mg/min; 0.1 mg/min)	Median nerve Posterior tibial nerve	Early cortical responses N20/P25, P40/N50	Amplitude: $\pm$ 25% of control	[94]

Appendix (Continued)

Anesthetic	Stimulation	SEP component	Effect	Reference
Propofol ($6\ \text{mg kg}^{-1}\ \text{h}^{-1}$)/50% N_2O	Median nerve	Late cortical response N100	Amplitudes decreased amplitudes increased with noxious stimulation	[27]
Ketamine (0.5 mg/kg)	Median nerve	Early cortical response N20/P25	No changes	[50]
Ketamine (0.5 mg/kg)	Median nerve	Middle latenct N35	Suppression	[50]
Ketamine (2 mg/kg; $30\ \mu\text{g kg}^{-1}\ \text{h}^{-1}$)	Median nerve	Early cortical response N20/P25	Amplitude increased No changes in latency	[81]
S-(+)-ketamine (2–20 mg/kg)	Median nerve	Late cortical response N50	Amplitude decreased Latencies increased	[26]
Benzodiazepines Diazepam (20 mg)	Median nerve	cortical response N20/P25	Minimal changes	[56]
Midazolam (0.2 mg/kg; 5mg/h)	Median nerve	Cortical response N20/P25	Amplitude decreased	[90]
Midazolam (0.3 mg/kg; $0.2\ \text{mg kg}^{-1}\ \text{h}^{-1}$)	Median nerve	Cortical response N20/P25	Latencies increased	[52]
Opioids Morphine (250 μg/kg)	Posterior tibial nerve	Cortical responses N1, P2	Latencies increased Amplitudes decreased	[70]
Fentanyl (2.5 μg/kg)	Posterior tibial nerve	Cortical responses N1, P2	Latencies increased Amplitudes decreased	[70]
Fentanyl (63.6 $\pm$ 10.1 μg/kg)	Median nerve	Cortical	Latencies increased Changes in amplitude variable	[80]
Fentanyl (75 μg/kg)	Median nerve	Subcortical, cortical	No change	[38]
Fentanyl (2.5 μg/kg) + 0.8% halothane	Finger (pain stimuli)	Late cortical responses N150/P250	Amplitudes decreased	[51]

Alfentanil (100 μg/kg; $2\,\mu\text{g}\,\text{kg}^{-1}\,\text{min}^{-1}$) + propfol (10 mg/kg; $8\,\text{mg}\,\text{kg}^{-1}\,\text{h}^{-1}$; $6\,\text{mg}\,\text{kg}^{-1}\,\text{h}^{-1}$)	Posterior tibial nerve	Early cortical response	Early cortical response P1N1 twice as high when compared to alfentanil/66% N_2O	[42]
Alfentanil	Median nerve	Late cortical response N100	Amplitude decreased Latencies increased	[27]
Alfentanil (30 μg/kg)	Hand (C7 dermatome), laser stimulus	Late cortical response N300	Amplitude decreased	[2]
Alfentanil (0.6–2.0 mg/h) + midazolam (1.5–5.0 mg/h)	Median nerve	Late cortical evoked response N100	Amplitudes decreased	[30]
Sufentanil (5 μg/kg)	Median nerve	Early cortical response N20/P25	Amplitude decreased Latencies increased	[46]
Inhalational anesthetics 10–50%N_2O	Median nerve	Early cortical response N20/P35	Dose-dependent decrease in amplitude	[83]
50%N_2O	Median nerve Posterior tibial nerve	Early cortical responses	amplitudes decreased	[61]
66% N_2O	Median nerve	Early cortical response N20/P25	Amplitude decreased Latency increased	[73]
50%N_2O;fentanyl (20–70) μg/kg	Median nerve Posterior tibial nerve	Early cortical response N20/P25; N45	Amplitude decreased	[88]
66%N_2O + 0.8% halo-thane 33% N_2O	Finger (pain stimuli)	Late cortical responses N150, P250	Amplitudes abolished	[51]
	Tooth pulp	Late cortical response P100, N175, P265	Latency (N175) increased Amplitudes decreased	[4, 10]
Halothane (1%–3%)	Sciatic nerve	Spinal responses	Minimal changes	[3]
Halothane (1%–4%)	Posterior tibial nerve	Spinal responses	No changes	[66]
Halothane (0.5–2.0%)	Median nerve	Early cortical response N20/P25	Amplitude decreased Latencies increased	[73]

Appendix (Continued)

Anesthetic	Stimulation	SEP component	Effect	Reference
Halothane (0.5–1 MAC) + 60% N_2O	Posterior tibial nerve	Early cortical response P20/N45	Amplitude decreased Latencies increased	[70]
Halothane (0%–2.2%)	Median nerve	Early cortical responses N20/P27 P22/N30	Amplitudes decreased, Latencies increased Amplitudes unchanged	[99]
Halothane (0.5%–2.0%)	Median nerve	Farfield N13 Early cortical response N18 N20	No change Amplitude decreased Amplitude decreased Latency increased	[85]
Halothane (0.25%–2.0%) + 66% N_2O	Posterior tibial nerve	Early cortical response P30/N40	Latencies increased Small decrease in amplitude	[76]
Enflurane (0.25%–1.0%)	Median nerve Posterior tibial nerve	Early cortical responses	Amplituded decreased Amplituded unchanged	[61]
Enflurane (0.25–1.0 MAC)	Median nerve Posterior tibial nerve	Early cortical response	Amplitude decreased Minimal changes in latency	[91]
Enflurane (0.5%–1.5%)	Median nerve	Early- and middle-latency cortical responses	Dose-dependent increase in latencies Amplitude N55 decreased at 1.5%	[99]
Enflurane (0.5%–2.0%)	Median nerve	Farfield N13 Early cortical response N18 N20	No change Amplitude decreased Amplitude decreased Latency increased	[85]
Enflurane (1.1%–4.3%)	Median nerve	Early cortical response N20/P25	Dose-dependent increase in latency Amplitudes increased with high-dose enflurane	[96]

Enflurane (0%–2.2%)	Median nerve	Early cortical responses N20/P27 P22/N30	Amplitude unchanged, latencies increased Amplitude increased	[99]
Enflurane (1%–2%) + 50% N$_2$O	Median nerve	Late cortical response N100	Amplitude increased at 2% vs. 1% enflurane Latency increased	[28]
Enflurane (0.5–1 MAC) + 60% N$_2$O	Posterior tibial nerve	Early cortical response P25/N45	Amplitude decreased Latencies increased	[70]
Isoflurane (0.25%–1.0%)	Median nerve Posterior tibial nerve	early cortical responses	Amplituded decreased, Latency increased Amplitudes unchanged	[67]
Isoflurane (0.5%–1.5%)	Median nerve	Early cortical response	Amplitude decreased Latency increased	[77]
Isoflurane (0.25%–2.0%)	Median nerve	Subcortical responses Cortical responses	No change Dose-dependent increases in CCT N20 decreased	[38]
Isoflurane (0.8%–3.3%)	Median nerve	Early cortical response N20/P25	Amplitude decreased Latency increased	[95]
Isoflurane (0%–1%)	Median nerve	Early cortical responses N20/P27 P22/N30	Amplitude increased Unchanged < 2%	[65, 99]
Isoflurane (1–2 vol%) + 50% N$_2$O	Median nerve	Late cortical response N100	Amplitude decreased Latency increased	[28]
Isoflurane (0.5–1 MAC) + 60% N$_2$O	Posterior tibial nerve	Early cortical response P25/N45 Precentral evoked response P22	Amplitude decreased Latency increased Amplitude increased	[70] [99, 101]
Sevoflurane (0.5–1.5 MAC)		Early cortical evoked Response	Dose-dependent decrease in amplitude Dose-dependent increase in latency	[64]

References

1. Abrahamian HA, Allison T, Goff WR, Rosner BS (1963) Effects of thiopental on human cerebral evoked responses. Anesthesiology 24:650–659
2. Arendt-Nielsen L, Oberg B, Bjerring P (1990) Analgesic efficacy of i. m. alfentanil. Br J Anaesth 65:164–168
3. Baines DB, Whittle IR, Chaseling RW, Overton JH, Johnson IH (1985) Effect of halothane of spinal somatosensory evoked potentials in sheep. Br J Anaesth 57:896–899
4. Bendetti C, Chapman CR, Colpitts TH, Chen AC (1982) Effect of nitrous oxide concentration on event-related potentials during painful tooth stimulation. Anesthesiology 56:360–364
5. Branston NM, Symon L, Crockard HA (1976) Recovery of the cortical evoked response following temporary middle cerebral artery occlusion in baboons: relation to local blood flow and PO_2. Stroke 7:151–157
6. Bromm B (1985) Evoked cerebral potential and pain. In: Fields HL, Dubner R, Cervero F (eds) Advances in pain research and therapy, vol. 9. Raven, New York, pp 305–329
7. Bromm B, Scharein E (1982) Principal component analysis of pain related cerebral potentials to mechanical and electrical stimulation in man. Electroencephal Clin Neurophysiol 53:305–329
8. Buchsbaum MS, Davis GC, Naber D, Pickar D (1983) Pain enhanced naloxone-induced hyperalgesia in humans as assessed by somatosensory-evoked potentials. Psychopharmacology 79:99–103
9. Carmon A, Dotan Y, Sarne Y (1978) Correlation of subjective pain experience with cerebral evoked responses to noxious thermal stimulations. Exp Brain Res 33:445–453
10. Chapman CR, Benedetti C (1979) Nitrous oxide effects on cerebral evoked potentials to pain: partial reversal with a narcotic analgesic. Anesthesiology 51:135–138
11. Chapman CR, Chen ACN, Colpitts YM, Martin RW (1981) Sensory decision theory describes evoked potentials in pain discrimination. Psychophysiol 18:114–120
12. Chapman CR, Colpitts YM, Benedetti C, Butler S (1982) Event-related potential correlates of analgesia; comparison of fentanyl, acupuncture and nitrous oxide. Pain 14:327–337
13. Chen ACN, Treed RD, Bromm B (1986) Modulation of pain evoked cerebral potentials by concurrent subacute pain. In: Bromm B (ed) Pain measurement in man. Neurophysiological correlates of pain. Elsevier, Amsterdam, pp 301–310
14. Chiappa KH, Choi SK, Young RR (1980) Short-latency somatosensory evoked potentials following median nerve simulation in patients with neurological lesions. In: Desmedt JE (ed) Clinicial uses of cerebral, brainstem and spinal somatosensory evoked potentials, vol 7. Prog Clin Neurophysiol, Karger, Basel, pp 264–281
15. Clark DL, Rosner BS (1973) Neurophysiologic effects of general anesthetics. I. The electroencephalogram and sensory evoked responses in man. Anesthesiology 38:564–582
16. De Cosmo G, DiLazzaro V, Testuccia D, Lo Monaco M, Primeri P, Bosco M, Villani A (1990) Effects of enflurane on human median nerve somatosensory evoked potentials. Eur J Anaesth 7:177–183

17. Desmedt JE (1977) Some observations on the methodology of cerebral evoked potentials in man. In: Desmedt JE (ed) Attention, voluntary contraction and event-related cerebral potentials, vol. 1. Progress in clinical neurophysiology, Karger, Basel, pp 12–19

18. Desmedt JE (1988) Somatosensory evoked potentials. In: Piction TW (ed) Human event-related potentials. Handbook of electroencephalography and clinical neurophysiology, vol 3. Elsevier, Amsterdam pp 245–360

19. Desmedt JE (1989) Somatosensory evoked potentials in neuromonitoring. In: JE Desmendt (ed) Neuromonitoring in surgery. Elsevier, pp 1–20

20. Desmedt JE, Cheron G (1980) Central somatosensory conduction in man: neural generators and interpeak latencies of the far-field components recorded from neck and right or left scalp and earlobes. Electroenceph Clin Neurophysiol 50:382–403

21. Desmedt JE, Brunko E, Debecker J (1976) Maturation of the somatosensory evoked potentials in normal infants and children, with particular reference to the early N1 component. Electroenceph Clin Neurophysiol 40:43–58

22. Donchin E, Gallaway E, Cooper R, Desmedt JE, Goff WR, Hillyard SA, Sutton S (1977) Publication criteria for studies of evoked potentials in man (report of a committee). In: Desmedt JE (ed) Attention, voluntary contraction and event-related cerebral potentials. Progress in Clinical Neurophysiology, vol 1. Karger, Basel, pp 1–11

23. Drummond JC, Todd MM, Hoi Sang U (1985) The effect of high dose sodium thiopental on brain stem auditory and median nerve somatosensory evoked responses in humans. Anesthesiology 63:249–254

24. Ebner A, Deuschl G (1988) Frontal and parietal components of enhanced somatosensory evoked potentials: a comparison between pathological and pharmacologically induced conditions. Electroenceph Clin Neurophysiol 71:170–179

25. Eccles JC (1951) Interpretation of action potentials evoked in the cerebral cortex. Electroenceph Clin Neurophysiol 3:449–464

26. Freye E, Latasch L, Schmidhammer H (1992) Pharmacodynamic effects of S-(+)-ketamine on EEG, sensory-evoked potentials and respiration. Anaesthesist 41:527–533

27. Freye E, Segeth M, Hartung E (1984) Somatosensory evoked potentials by alfentanil. Anaesthesist 33:103–107

28. Freye E, Dehnen-Seipel H, Rohner D (1985) Somatosensory-evoked potentials (SEP) during isoflurane and enflurane anaesthesia in heart surgery. Anaesthesist 34:670–674

29. Freye E, Hartung E, Schenk GK (1989) Somatosensory-evoked potentials during block of surgical stimulation with propofol. Br J Anaesth 63:357–359

30. Freye E, Neruda B, Falke K (1991) EEG power spectra and evoked potentials during the after alfentanil/midazolam analgosedation in intensive care patients. Anasthesiol Intensivmed Notfallmed Schmerzther 26:384–388

31. Gravenstein MA, Sasse F, Hogan K (1992) Effects of hypocapnia on canine spinal, subcortical, and cortical somatosensory-evoked potentials during isoflurane anesthesia. J Clin Monit 8:126–130

32. Grundy BL, Brown RH, Greenberg PS (1979) Diazepam alters cortical evoked potentials. Anesthesiology 51:S38

33. Grundy BL (1983) Intraoperative monitoring of sensory evoked potentials. Anesthesiology 58:72–87

34. Haghighi SS, Oro JJ (1989) Effects of hypovolemic hypotensive shock on somatosensory and motor evoked potentials. Neurosurgery 24:246–252

35. Heiss WD, Hayakawa T, Waltz AG (1976) Cortical neuronal function during ischaemia. Arch Neurol 33:813–820
36. Hogan K, Gravenstein M, Sasse F (1988) Effects of halothane and stimulus rate on canine spinal, far-field and near-field somatosensory evoked potentials. Electroenceph Clin Neurophysiol 69:277–286
37. Hume AL, Cant BR (1978) Conduction time in central somatosensory pathways in man. Electroencephal Clin Neurophysiol 45:361–375
38. Hume AL, Durkin MA (1986) Central and spinal somatosensory conduction times during hypothermic cardiopulmonary bypass and some observations on the effects of fentanyl and isoflurane anesthesia. Electroenceph Clin Neurophysiol 65:46–58
39. Jewkes D, Wang A, Symon L (1984) Effects of halothane, enflurane and hypotension on central conduction time. Br J Anaesth 56:1302P
40. Jones TH, Morawetz RB, Crowell RM, Marcoux FW, Fitzgibbon, SJ, De Girolami U, Ojemann RG (1979) Thresholds of focal cerebral ischaemia in awake monkeys. J Neurosurg, 54:773–782
41. Jordan WS, Grahn AR, Roberts LS, Wong KC, Dunn H (1983) Nitrous oxide suppression of spinal evoked EEG potentials in surgical patients. Anesth Analg 62:267–270
42. Kalkman CJ, Rheineck Leyssius AT, Bovill JG (1988) Influence of high-dose opioid anesthesia on posterior tibial nerve somatosensory cortical evoked potentials: effects of fentanyl, sufentanil and alfentanil. J Cardiothorac Anesth 2:785–764
43. Kallman CJ, Traast H, Zuurmond WA, Bovill JG (1991) Differential effects of propofol and nitrous oxide on posterior tibial nerve somatosensory cortical evoked potentials during alfentanil anaesthesia. Br J Anaesth 66:483–489
44. Kalkman CJ, Boezeman EH, Ribberink AA, Oosting J, Deen L, Bovill JG (1991) The influence of changes in arterial carbon dioxide tension on the electroencephalogram and posterior tibial nerve somatosensory cortical evoked potentials during alfentanil/nitrous oxide anesthesia. Anesthesiology 75:68–74
45. Kataoka K, Graf R, Rosner G, Heiss WD (1987) Experimental focal ischemia in cats: changes in multimodality evoked potentials as related to local cerebral blood flow and ischemic brain edema. Stroke 18:188–194
46. Kimovec MA, Koht A, Sloan TB (1990) Effects of sufentanil on median nerve somatosensory evoked potentials. Br J Anaesth 65:169–172
47. Kochs E, Schulte am Esch J (1991) Somatosensory evoked responses during and after graded progressive incomplete cerebral ischaemia in goats. Eur J Anaesth 8:257–265
48. Kochs E, Treede RD, Schulte am Esch J (1986) Increase of somatosensory evoked potentials during induction of anaesthesia with etomidate. Anaesthesist 35:359–365
49. Kochs E, Treede RD, Roewer N, Schulte am Esch J (1986) Alterations of somatosensory evoked potentials by etomidate and diprivan. Anesthesiology 65:A353
50. Kochs E, Blanc I, Werner C, Schulte am Esch J (1988) Electroencephalogram and somatosensory evoked potentials following low-dose ketamine (0.5 mg/kg). Anaesthesist 37:625–630
51. Kochs E, Treede RD, Schulte am Esch J, Bromm B (1990) Modulation of pain-related somatosensory evoked potentials by general anesthesia. Anesth Analg 71:225–230

52. Koht A, Schultz W, Schmidt G, Schramm J, Watanabe E (1988) The effects of etomidate, midazolam and thiopental on median nerve somatosensory evoked potentials and the additive effects of fentanyl and nitrous oxide. Anesth Analg 67:435–441
53. Kopf GS, Hume AL, Durkin MA, Hammond GL, Hashim SW, Geha AS (1985) Measurement of central somatosensory conduction time in patients undergoing cardiopulmonary bypass; an index of neurologic function. Am J Surg 149:445–448
54. Lebowitz MH, Blitt CD, Dillon JB (1972) Enflurane-induced central nervous system excitation and its relation to carbon dioxide tension. Anesth Analg 51:355–363
55. Lesser RP, Koehle R, Lueders H (1979) Effect of stimulus intensity of short latency somatosensory evoked potentials. Electroenceph Clin Neurophysiol 47:377–382
56. Loughnan BL, Sebel PS, Thomas D, Rutherford CF, Rogers H (1987) Evoked potentials following diazepam or fentanyl. Anaesthesia 42:195–198
57. Lueders H, Dinner DS, Lesser RF, Klem G (1983) Origin on farfield subcortical potentials evoked by stimulation of the posterior tibial nerve. Electroencephal Clin Neurophysiol 52:336–344
58. Lund C, Hansen OB, Kehlet H (1990) Effect of surgey on sensory threshold and somatosensory evoked potentials after skin stimulation. Br J Anaesth 65:173–176
59. Markand ON, Warren CH, Moorthey SS, Stoelting RK, King RD (1984) Monitoring of multimodality evoked potentials during open heart surgery under hypothermia. Electroenceph Clin Neurophysiol 59:432–439
60. Maurette P, Simeon F, Castagnera L, Esposito J, Macouillard G, Heraut LA (1988) Propofol anaesthesia alters somatosensory evoked cortical potentials. Anaesthesia 43 [Suppl]:44–45
61. McPherson RW, Mahla M, Johnson R; Traystman RJ (1985) Effects of enflurane, isoflurane, and nitorous oxide on somatosensory evoked potentials during fentanyl anesthesia. Anesthesiology 62:626–633
62. McPherson RW, Sell B, Traystman RJ (1986) Effects of thiopental, fentanyl, and etomidate on upper extremity somatosensory evoked potentials in humans. Anesthesiology 65:584–589
63. Meyjes FEP Jr (1969) On some characteristics of early components of the somatosensory evoked response to mechanical stimuli in man. Psychiatr Neurol Neurochir 72:263–268
64. Nishiyama Y, Ito M (1993) Effects of isoflurane, sevoflurane and enfluranc on median nerve somatosensory evoked potentials in humans, Jap J Anesth 42:339–343
65. Nogueira MC, Brunko E, Vandesteen A, DeRood M, Zegers de Beyl D (1989) Differential effects of isoflurane on SEP recorded over parietal and frontal scalp. Neurology 39:1210–1215
66. Nordwall A, Axelgaard J, Harada Y, Valencia P, McNeal DR, Brown JC (1979) Spinal cord monitoring using evoked potentials recorded from feline vertebral bone. Spine 4:486–489
67. Nuwer MR (1986) Evoked potential monitoring in the operating room. Raven, New York
68. Nuwer MR, Dawson E (1984) Intraoperative evoked potential monitoring of the spinal cord: enhanced stability of cortical recordings. Electroenceph Clin Neurophysiol 59:318–327
69. Pathak KS, Brown RH, Cascorbi HF, Nash CL (1984) Effects of fentanyl and morphine on intraoperative somatosensory cortical-evoked potentials. Anesth Analg 63:883–887

70. Pathak KS, Ammadio M, Kalamchi A, Scoles PV, Shaffer JW, Mackay W (1987) Effects of halothane, enflurane, and isoflurane on somatosensory evoked potentials during nitrous oxide anesthesia. Anesthesiology 66:753–757

71. Peterson DO, Drummond JC, Todd MM (1986) Effects of halothane, enflurane, isoflurane and nitrous oxide on somatosensory evoked potentials in humans. Anesthesiology 65:35–40

72. Raudzens PA (1982) Intraoperative monitoring of evoked potentials. Ann NY Acad Sci 388:308–326

73. Russ W, Thiel A, Gerlach H, Hempelmann G (1985) Effects on nitrous oxide and halothane on median nerve somatosensory evoked responses. Anaesthesist 20:186–192

74. Russ W, Thiel A, Schwandt HJ, Hempelmann G (1986) Alternations of somatosensory evoked potentials in response to thiopentone and etomidate. Anaesthesist 35:679–685

75. Russ W, Sticher J, Scheld H, Hempelmann G (1987) Effects of hypothermia on somatosensory evoked responses in man. Br J Anaesth 59:1484–1491

76. Salzman SK, Beckman AL, Marks HG, Naidu R, Bunnell WP, MacEwen GD (1986) Effects of halothane on intraoperative scalp-recorded somatosensory evoked potentials to posterior tibial nerve stimulation in man. Electroenceph Clin Neurophysiol 65:36–45

77. Samara SK, Vanderzant CW, Domer PA, Sackellares J (1987) Differential effects of isoflurane on human median nerve somatosensory evoked potentials. Anesthesiology 66:29–35

78. Scheepstra GL, De Lange JJ, Booji LHD, Ros HH (1989) Median nerve evoked potentials during propofol anaesthesia. Br J Anaesth 62:92–94

79. Schubert A, Drummond JC (1986) The effect of acute hypocapnia on human median nerve somatosensory evoked responses. Anesth Analg 65:240–244

80. Schubert A, Drummond JC, Petersen DO, Saidman LJ (1987) The effect of high-dose fentanyl on human median nerve somatosensory-evoked responses. Can J Anaesth 34:35–39

81. Schubert A, Licina MG, Kineberry PJ (1990) The effect of ketamine on human somatosensory evoked potentials and its modification by nitrous oxide. Anesthesiology 72:33–39

82. Schubert A, Licina MG, Lineberry PJ, Deers MA (1991) The effect of intrathecal morphine on somatosensory evoked potentials in wake humans. Anesthesiology 75:401–405

83. Sebel PS, Flynn PJ, Ingram DA (1984) Effect of nitrous oxide on visual, auditory and somatosensory evoked potentials. Br J Anaesth 56:1403–1407

84. Sebel PS, Heneghan CP, Ingram DA (1985) Evoked responses – a neurophysiological indicator of depth of anaesthesia? (editorial) Br J Anaesth 57:841–842

85. Sebel PS, Erwin CW, Neville WK (1987) Effects of halothane and enflurane on far and near field somatosensory evoked potentials. Br J Anaesth 59:1492–1496

86. Sebel PS, Glass P, Neville WK (1988) Do evoked potentials measure depth of anaesthesia? Int J Clin Monit 5:163–166

87. Debel PS, Withington PS, Rutherford CF, Markham K (1988) The effect of tracheal intubation and surgical stimulation on median nerve somatosensory evoked potentials during anesthesia. Anaesthesia 43:857–860

88. Solan TB, Koht A (1985) Depression of cortical somatosensory evoked potentials by nitrous oxide. Br J Anaesth 57:849–852

89. Sloan TB, Kimovec MA, Serpico LC (1989) Effects of thiopentone on median nerve somatosensory evoked potentials. Br J Anaesth 63:51–55

90. Solan TB, Fugina ML, Toleikis JR (1990) Effects of midazolam on median nerve somatosensory evoked potentials. Br J Anaesth 64:590–593

91. Stone JL, Ghaly RF, Levy WJ, Kartha R, Krinsky L, Roccaforte P (1992) A comparative analysis of enflurane anesthesia on primate motor and somatosensory evoked potentials. Electroenceph Clin Neurophysiol 84:180–187

92. Symon S, Wang AD, Costa da Silva IE, Gentili F (1984) Perioperative use of somatosensory evoked responses in aneurysm surgery. J Neurosurg 60:269–275

93. Tachibana N (1975) Somatosensory evoked potentials and analgesia in man. Int Anesthesiol Clin 13:191–213

94. Taniguchi M, Nadstawek J, Pechstein U, Schramm J (1992) Total intravenous anesthesia for improvement of intraoperative monitoring of somatosensory evoked potentials during aneurysm surgery. Neurosurgery 31:891–897

95. Thiel A, Russ W, Hempelmann G (1988) Evoked potentials and volatile anaesthetics. Klin Wochenschr 66 [Suppl XIV] 11–18

96. Thiel A, Russ W, Kafurke H, Hempelmann G (1987) The effects of enfluane and isoflurane on somatosensory evoked potentials after stimulation of the median nerve. Anasth Intensivther Notfallmed 22:159–165

97. Thornton C (1991) Evoked potentials in anaesthesia. Eur J Anaesth 8:89–107

98. Thurner F, Schramm J, Pasch T (1987) Effects of fentanyl and enflurane on human somatosensory evoked potentials during flunitrazepam/N_2O anesthesia. Anaesthesist 36:548–554

99. Vandesteene A, Nogueira MC, Mavroudakis N, Defevrimont M, Brunko E, Zegers de Beyl D (1991) Synaptic effects of halogenated anesthetics on short-latency SEP. Neurology 41:913–918

100. Vandesteene A, Mavroudakis N, Defevrimont M, Brundo E, Zegers de Beyl D (1993) Topographic analysis of the effects of isoflurane anesthesia on SEP. Electroenceph Clin Neurophysiol 88:77–81

Peri-operative Anesthesiological Monitoring
of Auditory-Evoked Potentials

C. Thornton, P. Creagh-Barry, and D. E. F. Newton

Introduction

This paper discusses the series of steps which have been taken to evaluate auditory evoked potentials (AEP) as a measure of depth of anesthesia. A brief description of the origins of the AEP and the reasons for its choice are given. The following two sections describe the experimental evaluation, where a hypothesis was formulated and statistical tests carried out, and clinical evaluation of the technique during surgery and anesthesia. Finally, progress is reviewed and future work discussed.

Origins of the Auditory Evoked Potential

The AEP has features which makes it worthwhile investigating as the basis for a clinical monitor of depth of anesthesia. It consists of a series of waves (Fig. 1), generated from different levels of the neuraxis (Chatrian et al. 1960; Celesia et al. 1968; Jewett and Williston 1971; Kaga et al. 1980; Woods et al. 1987), which are differentially sensitive to drugs and sensory stimuli (reviewed Thornton 1991). It is not affected by neuromuscular blocking drugs (Harker et al. 1977) and hence will work in paralysed patients, where the need for a clinical monitor of anesthetic depth is greatest.

Experimental Evaluation

What Aspect of Anesthesia does the Auditory Evoked Potential Measure?

Anesthesia is difficult to define and can be divided into at least two measurable components, analgesia and hypnosis. In two studies, we compared the effects of isoflurane, a strong hypnotic, with nitrous oxide, a strong analgesic, but weak hypnotic drug. There was greater depression of the AEP waves P_a and N_b by

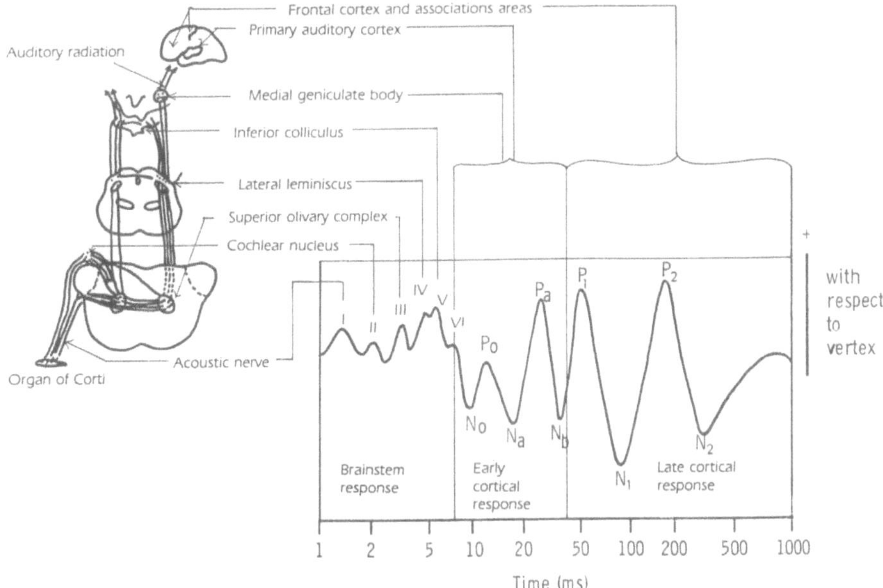

Fig. 1. The auditory evoked response waveform and the proposed generators in the brain as electrical activity passes from cochlea to cortex (reproduced from Thornton and Newton 1989)

isoflurane compared to nitrous oxide. Therefore, we concluded that these waves, which are generated from the medial geniculate and primary auditory cortex and hence named the early cortical waves, measure the hypnotic component of anesthesia (Newton et al. 1989; Thornton et al. 1992).

Validation of the Auditory Evoked Potential as a Measure of Depth of Anesthesia

In the absence of a gold standard with which to compare the AEP, there is a problem in validating it as a measure of depth of anesthesia. There are two possibilities: (1) to compare the AEP technique with other measurements of depth of anesthesia and (2) to define criteria which the technique should fulfill if it is a valid measure of depth of anesthesia.

We have had limited success with the first approach. Although changes in the AEP correlate with autonomic signs around the time of induction and intubation, when taken over the entire period of anesthesia and surgery the correlation is poor. The AEP changes also showed a poor correlation with oesophageal contractility, which Evans and Davies (1984) claim is a measure of anesthetic depth. However, we (Thornton et al. 1989b) and others (Cullen et al. 1972) have found both these methods to be unreliable, as they show inconsistencies at similar depth between patients and at different times within patients.

The second approach has been more successful. To measure depth of anesthesia, a technique should fulfill the following criteria. It should show:

1. Graded changes with anesthetic concentration; these should be similar for different general anesthetics
2. Appropriate changes with surgical stimulation
3. Changes with loss of consciousness

Graded Changes with Anesthetic Concentration, Similar for Different Anesthetics

Six general anesthetics belonging to different chemical groups were tested, at equipotent concentrations over the clinical concentration range. The early cortical AEP waves P_a and N_b showed similar graded changes, i.e. increases in latency and decreases in amplitude, with all the general anesthetics studied (Thornton et al. 1989). An example of these changes and the fact that they are similar for different classes of general anesthetic drugs is given in Fig. 2, where

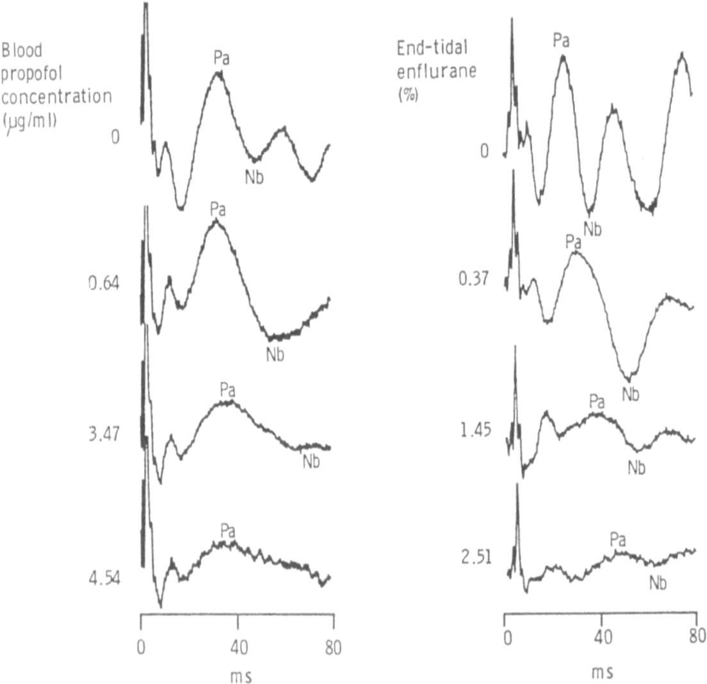

Fig. 2. Early cortical auditory evoked responses in patients given propofol (*left*) and enflurane (*right*) in increasing concentrations. In addition to the concentrations shown, the patients received 70% nitrous oxide in oxygen (Thornton and Newton 1989)

the AEP for a patient given propofol and one given enflurane are shown. These patients were not being surgically stimulated at the time.

Appropriate Changes with Surgical Stimulation

Depth of anesthesia can be viewed as a balance between the depression of the central nervous system (CNS) by anesthetic drugs and the stimulation by sensory stimulation such as surgery. In a study where the halothane concentration was kept constant at 0.3% end-tidal halothane against a background of 70% nitrous oxide, changes in the AEP brought about by the general anesthetic agent were reversed by surgical stimulation (Thornton et al. 1988). The AEP appeared similar to that of a lower concentration of general anesthetic (Fig. 3). This is important because it shows that this is not simply a monitor of concentration, but that these changes in the AEP reflect true depth of anesthesia.

Changes with Loss of Consciousness

Characteristics of the AEP have been identified which indicate potential awareness. A short N_b latency such that three waves fitted into the 100-ms time interval was associated with a positive response to the isolated forearm in

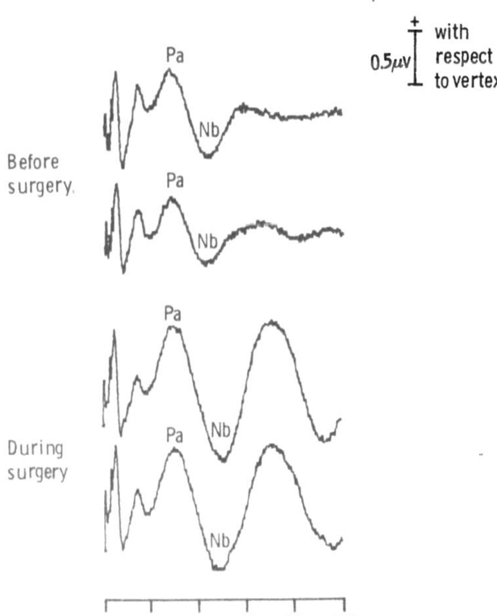

Fig. 3. Early cortical auditory evoked response in a patient before and during surgery (reproduced from Thornton 1991)

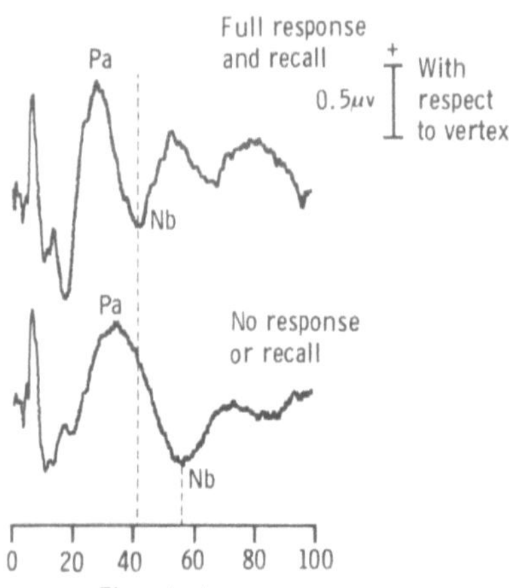

Fig. 4. The auditory evoked response in an individual subject at two anesthetic concentrations: the first showing a "three-wave" pattern and associated with full response and recall, and the second a "two-wave" response, when neither response nor recall was elicited. Note the increase in N_b latency which characterizes this change (reproduced from Thornton and Newton 1989)

general surgery patients studied before their surgery commenced (Thornton et al. 1989) and with a positive response to lists of commands and words and their subsequent recall in volunteer anesthetists receiving four sub-MAC (minimum alveolar concentration to prevent movement at incision in 50% of patients) levels of isoflurane (Newton et al. 1992). Figure 4 shows the AEP of a volunteer when he was responding and then not responding to command. The N_b latency lengthens and the three AEP waves reduce to two or less as the response to command is lost.

Clinical Evaluation

Methods

For clinical monitoring, the derivation of the index which reflects depth of anesthesia has to be automated. Our present system calculates the double differential of the waveform between a pre-defined latency window (Thornton and Newton 1989). Our library of data suggests that the AEP shown in Fig. 5 correspond to the particular levels of anesthesia indicated. Using a pre-set latency window, gain and filtering (all these factors affect the index), the double differential derived is given on the left of the figure.

To evaluate the AEP as a clinical measure of depth of anesthesia, we recorded this index along with automatic variables, assessed surgical stimulation and measured anesthetic concentration. The autonomic variables were used to construct a PRST score (Evans and Davies 1984; Table 1) and the severity of surgical stimulation was scored from 0 to 3 (0, hands off; 3, very severe stimulus).

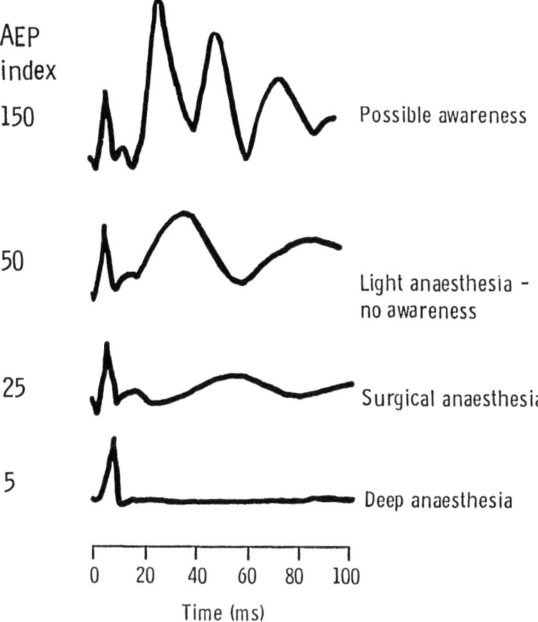

Fig. 5. Early cortical auditory evoked response of a patient in relation to adequacy of anesthesia with automatically derived AEP index (reproduced from Thornton 1991)

Table 1. Calculation of PRST score

Index	Condition	Score
Systolic pressure (mmHg)	< Control + 15	0
	< Control + 30	1
	> Control + 30	2
Heart rate (beats/min)	< Control + 15	0
	< Control + 30	1
	> Control + 30	2
Sweating	Nil	0
	Skin moist to touch	1
	Visible beads to sweat	2
Tears	No excess of tears in open eye	0
	Excess of tears in open eye	1
	Tear overflow from closed eye	2

The individual scores are summed to give a total score.

The following anesthetic protocol was followed: 20 patients consented to participate in the studies approved by the Harrow ethical committee. They were premedicated with 10 mg temazepam and induced with $2\,mg\,kg^{-1}$ propofol. Tracheal intubation was with 6 mg vecuronium and they were mechanically ventilated to maintain the CO_2 concentration at 5 kPa. They were then randomly allocated to receive isoflurane to give an end-tidal concentration 0.4 MAC or propofol by standard infusion regime to give a blood concentration of approximately $3\,\mu g\,ml^{-1}$. Both groups were given nitrous oxide to give an end-tidal of 67% nitrous oxide.

Results

An example of the data is given in Fig. 6 from a patient who had an uneventful propofol anesthetic.

At the top of the data plot, you see the AEP index. It decreases following induction, changes very little at intubation and remains relatively constant and

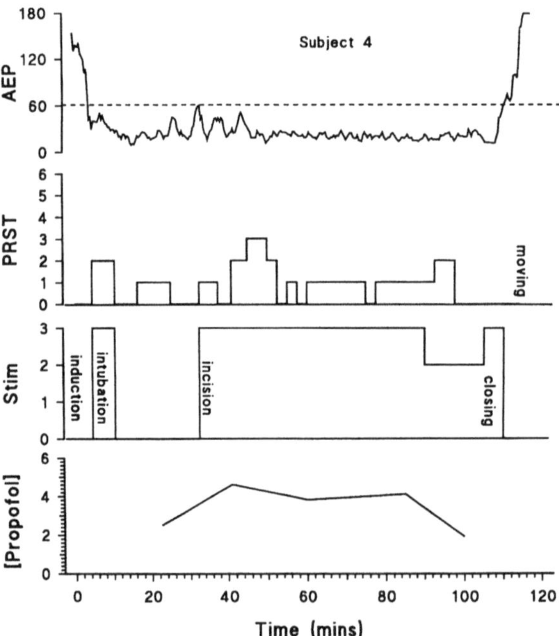

Fig. 6. Changes in the auditory evoked response (*AEP*) index and clinical assessment of a patient during propofol anesthesia (and 70% nitrous oxide) undergoing abdominal hysterectomy. The *PRST* scores are calculated from the autonomic variables (see Table 1). Severity of surgical stimulus (*Stim*) is scored on a scale of 0–3, where 0 represents "hands off" and 3 severe surgical stimulus, surgical events are marked. Blood propofol concentration is also shown

below the dotted line, which indicates adequate clinical anesthesia, for most of the operation. It returns to the high level when anesthesia is reversed. Individual AEP traces at intervals through the operation are shown in Fig. 7.

The propofol levels in this patient averaged 3–4 μg ml^{-1} for throughout most of anesthesia and surgery. The PRST score averaged 1.5. This is in contrast to another patient who received propofol and whose data are plotted in Fig. 8. This patient had lower propofol blood levels during anesthesia and surgery. They averaged 2 μg ml^{-1}. For lower blood concentration, the PRST scores were in general higher (average 2.5) and the AEP index was above the dotted line for a substantial amount of the time. There was a response in the AEP index (and also an autonomic response) to intubation and to other surgical events. At one point in the operation, the patient moved. The patients allocated to the isoflurane group showed a similar range of differences as those given for propofol.

Discussion and Conclusions

The AEP changes seen were compatible with valid depth of anesthesia measurement. Decreases in the AEP index could be seen with both lowering the anesthetic blood level and surgical stimulation in a way that might be expected in routine anesthetic practice. It is interesting to compare the two patients whose data are plotted in Figs. 7 and 8. At a point approximately 25–26 min from

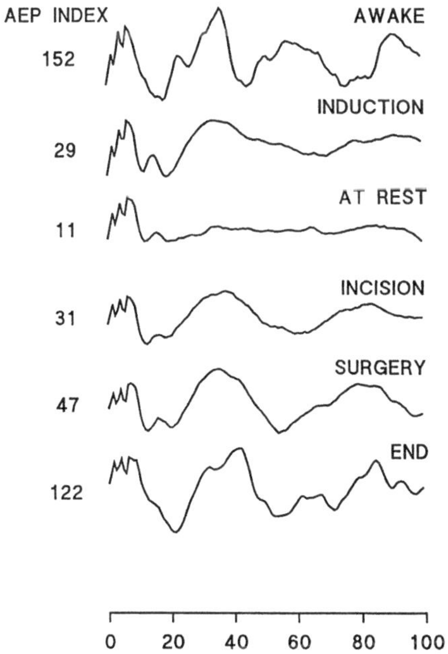

Fig. 7. Early cortical auditory evoked responses from the patient whose data are plotted in Fig. 6 sampled at intervals throughout anesthesia and surgery

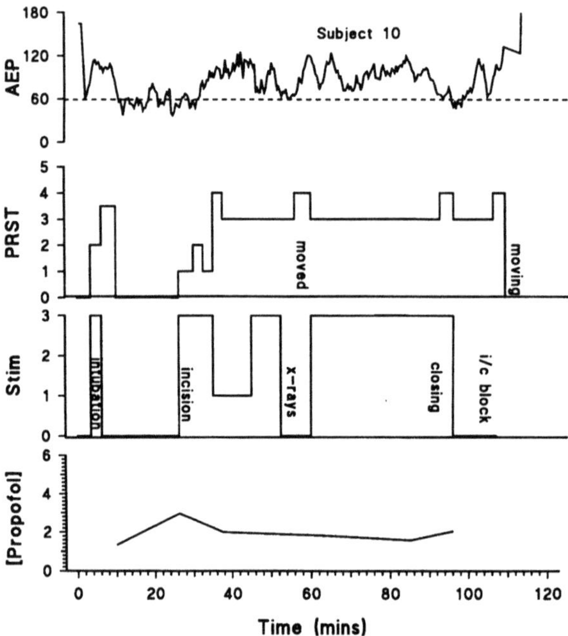

Fig. 8. Changes in the auditory evoked response (*AEP*) index and clinical assessment of a patient during propofol anesthesia (and 70% nitrous oxide) undergoing abdominal cholecystectomy. The *PRST* scores are calculated from the autonomic variables (see Table 1). Severity of surgical stimulus (*Stim*) is scored on a scale of 0–3, where 0 represents "hands off" and 3 severe surgical stimulus, surgical events are marked. Blood propofol concentration is also shown

induction where the blood propofol concentrations were similar at $3 \mu g\,ml^{-1}$ and neither patient was being surgically stimulated, the AEP indices were virtually the same, i.e. 44 for the patient in Fig. 7 and 47 for the patient in Fig. 8. This is an important point, because for a measure of depth of anesthesia to be reliable, the same depth in different patients should give the same number.

Overview of Progress, Future Work

So far, the technique shows promise as a clinical monitor. It only requires three adhesive electrodes (behind each ear and on the forehead) and small ear inserts to be attached to the patient. The display needs to be designed to give the anesthetist quick access to the information required to make a decision. We are examining different ways of analysing the data to produce an index and reduce the time required to produce a reliable response, taking into consideration the fact that the response needs to be sufficiently rapid to track the changes which occur during anesthesia and surgery.

Other university sites are now involved in a collaborative study to evaluate the AEP technique in a wide range of patients, different types of surgery and anesthesia. It is important to ensure that factors such as neurological disease, low blood pressure, oxygen saturation and temperature do not invalidate the results and the technique performs satisfactorily in harsh electrical environments such in the presence of diathermy.

References

Celesia GC, Broughton RJ, Rasmussen T, Branch C (1968) Auditory evoked responses from the exposed human cortex. Electroencephalogr Clin Neurophysiol 24:458–466

Chatrain GE, Petersen MC, Lazarte JA (1960) Responses to clicks from the human brain: some depth electrographic observations. Electroencephalogr Clin Neurophysiol 12:479–489

Cullen DJ, Eger EI. II, Stevens WC, Smith NT, Cromwell TH, Cullen BF, Gregory GA, Balman SH, Dolan WM, Stoelting RK, Fourcade HE (1972) Clinical signs of anesthesia. Anesthesiology 36:21–36

Evans JM, Davies WL (1984) Monitoring anaesthesia. Clin Anaesthesiol 2:243–262

Harker LA, Hosick EC, Voots RJ, Mendel MI (1977) Influence of succinylcholine on middle component auditory evoked potentials. Arch Otolaryngol 103:133–137

Jewett DL, Williston JS (1971) Auditory-evoked far fields averaged from scalp of humans. Brain 94:681–696

Kaga K, Hink RF, Shinoda Y, Suzuki J (1980) Evidence for a primary cortical origin of a middle latency auditory evoked potential in cats. Electroencephalogr Clin Neurophysiol 50:254–266

Newton DEF, Thornton C, Creagh-Barry P, Doré CJ (1989) The early cortical auditory evoked response in anaesthesia: comparison of the effects of nitrous oxide and isoflurane. Br J Anaesth 62:61–65

Newton DEF, Thornton C, Konieczko KM, Jordan C, Webster NR, Luff NP, Frith CD, Doré CJ (1992) The auditory evoked response and awareness: a study in volunteers at sub-Mac concentrations of isoflurane. Br J Anaesth 69:122–129

Thornton C (1991) Evoked potentials in anaesthesia. Eur J Anaesth 8:89–107

Thornton C, Newton DEF (1989) The auditory evoked responses. In: Jones JG, (ed) Ballière's clinical anaesthesiology – depth of anaesthesia, vol 3 (No. 3) Harcourt Brace Jovanovich, London

Thornton C, Konieczko K, Jones JG, Jordan C, Doré CJ, Heneghan CPH (1988) Effect of surgical stimulation on the auditory evoked response. Br J Anaesth 60:372–378

Thornton C, Barrowcliffe MP, Konieczko KM, Ventham P, Doré CJ, Newton DEF, Jones JG (1989a) The auditory evoked response as an indicator of awareness. Br J Anaesth 63:113–115

Thornton C, Konieczko KM, Knight AB, Kaul B, Jones JG, Doré CJ, White DC (1989b) The effect of propofol on the auditory evoked response and on oesophageal contractility. Br J Anaesth 63:411–417

Thornton C, Craegh-Barry P, Jordan C, Luff NP, Doré CJ, Henley M, Newton DEF (1992) Somatosensory and auditory evoked responses recorded simultaneously: differential effects of nitrous oxide and isoflurane. Br J Anaesth 68:504–514
Woods DL, Clayworth CC, Knight RT, Simpson GV, Naeser MA (1987) Generators of middle- and long latency auditory evoked potentials: implications from studies of patients with bitemporal lesions. Electroencephalogr Clin Neurophysiol 68:132–148

The 40-Hz Auditory Steady State Response
for Monitoring Level of Consciousness:
Methodological Considerations

C. Villemure, G. Plourde, and P. April

Introduction

Delivering auditory stimuli at a rate of approximately 40 per s produces an electrical cerebral response which can be recorded from the scalp. This response, called the 40-Hz auditory steady state response (40-Hz ASSR), consists of a sinusoidal waveform that has the same frequency as that of stimulus delivery [7, 11, 16, 22] (Fig. 1). Recent evidence suggests that the 40-Hz ASSR may offer a way to monitor the level of consciousness during anesthesia. The amplitude of the 40-Hz ASSR was reduced to noise level during anesthesia with isoflurane 1% end-tidal in oxygen [15]. The amplitude remained maximally reduced after decreasing the concentration of the isoflurane to 0.5% end-tidal in oxygen and allowing 10 min for equilibration [15]. The amplitude of the 40-Hz ASSR was also maximally reduced by enflurane 0.5%, 0.8%, or 1.1% and 60% N_2O (end-tidal concentrations) [25]. The return of the ability to open the eyes on command after termination of the anesthetic seemed reliably associated with a clear, stepwise increase in the amplitude of the 40-Hz ASSR, which until then had remained markedly reduced [15, 25]. This suggests that the profound attenuation of the 40-Hz ASSR by enflurane–N_2O or isoflurane may reflect unconsciousness (unresponsiveness to verbal command). Additional evidence that the attenuation of the 40-Hz ASSR reflects unconsciousness was obtained during induction of anesthesia with thiopental [15], propofol (unpublished observations), or sufentanil [14]. With these agents, profound attenuation of the 40-Hz ASSR and unresponsivenes to verbal command occurred at the same time.

The presence of the 40-Hz ASSR, however, does not in general prove consciousness, because the 40-Hz ASSR sometimes persists in comatose patients [6]. Nevertheless, in the context of anesthesia, it appears that maximal suppression of the 40-Hz ASSR by anesthetic agents reflects unconsciousness and that a sudden increase of the amplitude of the 40-Hz ASSR signals the regaining of the ability to follow simple commands [13].

One advantage of the 40-Hz ASSR over the better-known transient auditory middle latency response [24] is the simplicity of the 40-Hz ASSR waveform [19]. Because the 40-Hz ASSR can be approximated as a sinusoid, it is completely determined by three parameters: (1) frequency (which is equal to

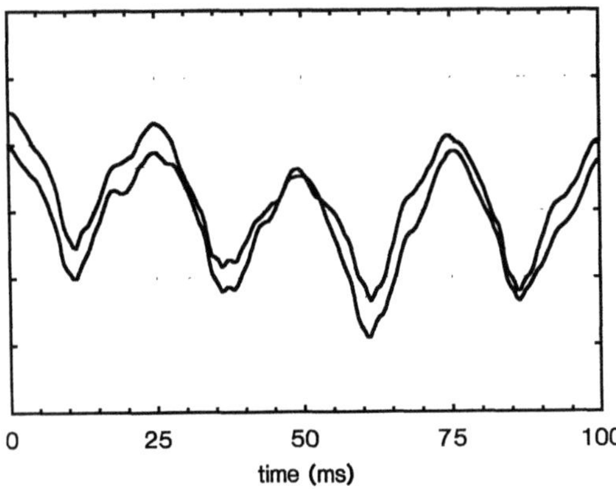

Fig. 1. 40-Hz auditory steady state response from one subject. The sinusoidal shape of the waveforms is readily apparent. Cz is referenced to right mastoid–0.3- to 100-Hz bandpass–500-Hz tonebursts were presented at the rate of 40 per s. A total of 2000 responses were averaged for each trace. Relative negativity at Cz is plotted upward; 0.5 μV per vertical tick

the rate of stimulus delivery); (2) amplitude (the height of the waveform); and (3) phase (position of the waveform corresponding to the onset of stimulus). Phase is the frequency domain equivalent of latency.

The aim of this paper is to discuss the methodology required for recording the 40-Hz ASSR using an IBM compatible microcomputer (Intel 80386) for stimulus control and signal acquisition. It is assumed that the reader is familiar with the techniques for recording auditory evoked potentials [3, 12, 23].

Stimuli

Clicks or pure tones (500 to 4000-Hz tonal frequency) are commonly used. For monitoring purposes in anesthesiology, a large response is desirable.

Stimulus Parameters

Three factors determine the amplitude of the 40-Hz ASSR: tonal frequency, stimulus intensity, and rate of stimulus delivery. These three parameters also influence the phase of the response.

Tonal Frequency

Tonal frequency denotes the pitch of the stimulus and is expressed in Hz. It must not be confused with the rate of stimulus delivery, which may also be expressed in Hz (e.g., 40 Hz). The amplitude of the 40-Hz ASSR is larger for low (approximately 500 Hz) tonal frequencies [20, 22]. It is therefore preferable to use low-frequency (500 Hz) pure tones, rather than high-pitched tones (greater than 2000 Hz) or clicks, which have a wide frequency content). Another advantage of low tonal frequency is that such tones are in general less affected than high tonal frequencies in most type of presbycusis [1]. The phase of the 40-Hz ASSR is smaller for low tonal frequencies [20].

Stimulus Intensity

The amplitude of the 40-Hz ASSR increases linearly with stimulus intensity over the range 20–80 dB HL (hearing level). Phase is reduced by increasing stimulus intensity [20, 22]. We have found that binaural stimuli (500 Hz tonebursts, delivered at an intensity of 80–90 dB peak equivalent SPL, i.e., approximately 68.5–78.5 dB HL; SPL, sound pressure level) provide an adequate response (baseline to peak amplitudes equal to or greater than 0.4 μV) [15]. The stimuli are not presented continuously for more than 5 min. There is a rest period of 1 min for every 5-min recording period

Rate of Stimulus Delivery

Despite the "40-Hz" label, it is not necessary to use a rate of stimulus delivery of exactly 40 per s. The amplitude of the response remains about the same for any stimulus rate between 35–45 per s. Phase increases linearly with the rate of stimulus delivery (from 30 to 60 Hz) [22].

Presentation of the Stimuli

The waveforms for the stimuli are generated by the microcomputer (Intel 80386) and fed into a digital-to-analog (DA) converter (Model 2821-Data Translation Inc., 100 Locke Drive, Marlboro MA 01752-USA). The output from the converter is directly fed into a passive, adjustable, resistive circuit connected to two insert earphones (E-A-R TONE 3A, Cabot Corporation, Indianapolis, IN 46268, USA). The intensity of the stimuli is determined by the interposed resistance and periodically verified by calibration with a sound-level meter and an oscilloscope. Before testing, it is essential to confirm by otoscopy that external auditory meatus is not obstructed and that the tympanic membrane is intact. It is also desirable, particularly for experimental work, to perform a pure tone audiometric screening (with a portable audiometer) prior to testing.

Signal Acquisition

Electrodes

Gold-plated cup electrodes with a hole on top (Grass Instruments Company, Quincy, MA, USA) provide adequate recordings and are easy to use and comfortable for the subjects. After gentle abrasion of the scalp sites with an abrasive gel (Omni Prep, D. O. Weaver and Co., 565-C Nucla Way, Aurora, CO 80011, USA), the electrodes are filled with saline gel (Teca Corporation, Pleasantville, NY, USA) and attached with gauze impregnated with collodion glue. A hair-dryer (set to "cool") is used to dry the collodion. Collodion is highly flammable both in liquid and vapor form; the electrodes must be applied in a well-ventilated area outside the operating room and away from possible ignition sources. Interelectrode impedances (10-Hz A.C. ohmmeter) of less than 5 kΩ with interelectrode differences of 1.0 kΩ or less are easily achieved with no or only minimal, transient discomfort for the subjects. If the impedance of an electrode is too high, a sterile needle is used to gently scratch the skin and add conductive gel via the hole on top of the electrode. Proper cleaning and disinfection of the electrodes according to current recommendations [18] is mandatory.

The 40-Hz ASSR is largest at the vertex [10] (Cz according to the 10–20 system), which is therefore the preferred site for recording. We use the right mastoid as reference and the posterior cervical sagittal midline as ground. The forehead would be a more convenient site than Cz for routine monitoring, because it allows the use of self-adhesive electrodes. Preliminary, unpublished data indicate however that the 40-Hz ASSR recorded from the forehead on the sagittal midline near the hairline (midway between Fz and Fpz) is reduced by 20% or more compared with the Cz recordings.

Amplification and Filtering

Selection of the amplifier bandpass must take into account the roll-off rate of the filters to ensure that the frequencies of interest (say, 0.5–50 Hz, to include the range of relevant electroencephalogram (EEG) frequencies and the 40-Hz response) are not attenuated by the bandpass filter settings. The information can usually be found in the documentation provided with most commercial EEG amplifiers. High- and low-pass nominal cutoffs of 0.5 Hz and 100 Hz are generally adequate for recording the 40-Hz ASSR and the EEG. The use of "notch" filters to attenuate interference from power lines (60 Hz in North America) must be avoided, because frequencies in the 40-Hz range will be markedly attenuated.

Signal Analysis

The amplified signal can be analyzed with analog or digital methods.

Signal Analysis–Analog Methods

Regan [19 (p. 90)] has described an analog Fourier analyzer which can be easily assembled from inexpensive electronic components for analysis of steady state visual evoked potentials. This device multiplies the amplified EEG signal by the sine and cosine of the stimulus repetition frequency. These multiplications convert the components of the EEG that have the same frequency as that of the stimulus repetition (including, of course, the 40-Hz ASSR) into a steady D.C. output. All other frequencies are converted to A.C. output. The D.C. output can be used to provide a continuous read-out of the amplitude and phase of the 40-Hz ASSR. Stapells et al. [22] found that this device is adequate for the 40-Hz ASSR. Its main advantage is that the analysis is done in real time and that no averaging is required. Furthermore, this technique can be used concurrently with the digital analysis methods described below.

Signal Analysis–Digital Methods

This approach has two main advantages: permanent data storage in digital format on disk and a wide choice of on-line and off-line procedures for data analysis. Digital signal processing [2, 8, 19 (pp.20–25)] requires that the signal first be transformed from an analog (continuous signal) to a digital (a series of numerical values measured at a fixed time interval) format (Fig. 2). For this procedure, we use a Model 2821-F-SE analog-to-digital (AD) converter (Data Translation Inc.) controlled by the microcomputer (Intel 80386).

Rate of Analog-to-Digital Conversion

An important consideration in AD operations is how often the conversion must be made. Shannon's sampling theorem [21] states that the rate of AD conversion must be at least twice the highest frequency present in the signal. The sampling frequency which just fulfils this requirement is called the Nyquist rate. Failure to fulfil this requirement causes frequency components faster than half the sampling rate to mimic slower components. This phenomenon is known as aliasing and must be avoided. There are no known procedures that can correct aliasing once it has occurred. The highest frequency present is determined by the low-pass filter settings (nominally 100 Hz). Therefore, the rate of AD conversion must be at least 200 Hz per channel, meaning that the analog signal must be measured 200 times per s or every 5 ms. In practice, it is preferable to use 2.5 or three times the low-pass settings, because frequencies slightly above the nominal cut-off may not be sufficiently eliminated. Therefore, 300 Hz is a practical sampling rate. Oversampling should in general be avoided, because it increases the amount of data to be processed.

Averaging

One must decide whether or not to use signal averaging to increase the signal-to-noise ratio before performing the fast Fourier transformation (FFT).

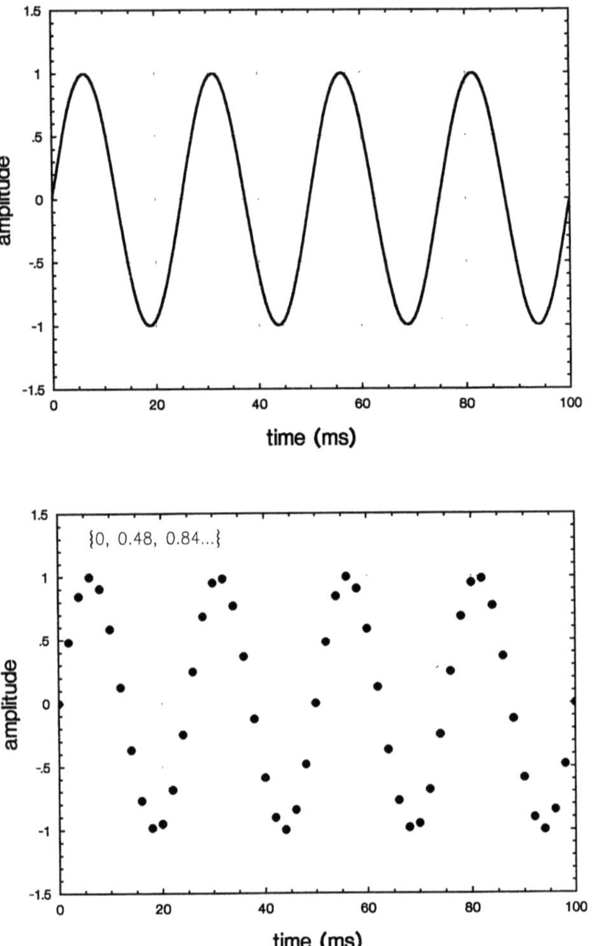

Fig. 2. Conversion of an arbitrary analog signal (*top*) to digital representation (*bottom*). Measuring the voltage at regular time intervals (every 2 ms) yields a series of discrete numerical values represented by the *dots*

The 40-Hz ASSR can be seen in amplitude spectra based on short (3.4 s), nonaveraged EEG segments, but it is not present on every trace (Fig. 3). Whether this observation reflects physiologic fluctuations of the 40-Hz ASSR or artifacts of signal processing is not known. Averaging ten EEG segments (3.4 s) followed by FFT of the averaged waveform allows rapid and reliable identification of the 40-Hz ASSR (Fig. 3). It is also possible to average in the frequency domain, i.e., to do an FFT on every EEG segment and average the spectra. The disadvantage of this approach is that the multiple FFT required increase the time needed for analysis.

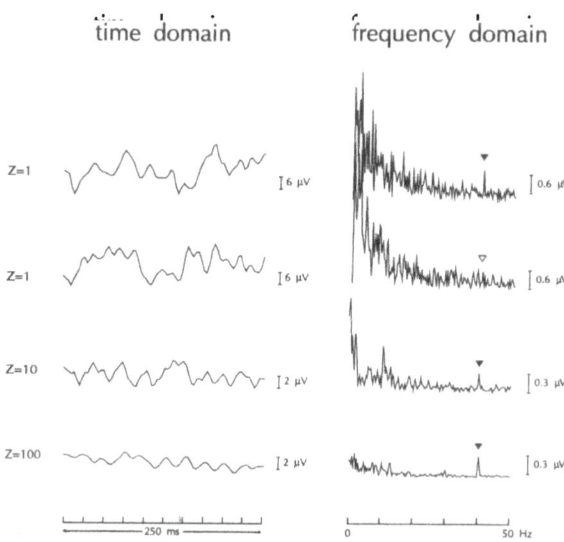

Fig. 3. Effect of signal averaging on response identification. Stimuli, 500-Hz tonebursts presented at the rate of 40 per s. Cz is referenced to right mastoid; bandpass, 0.3-100Hz. Analog-to-digital (AD) conversion rate, 298.9781 Hz; 1024 data points; epoch duration, 3.425 s. Z is the number of electroencephalogram (EEG) segments averaged; $Z = 1$ means no averaging used because there is only one epoch. *Top tracings*, on the left, the EEG (time domain) is shown; for clarity, only a portion (250 ms) of the epoch is included. On the *right*, the corresponding amplitude spectrum based on the entire (3.425 s) epoch is shown. The 40-Hz auditory steady state response (ASSR) is visible on the spectrum (*filled triangle*). *Second row tracings*, recording obtained moments after the above recording and in a similar manner. Adequate delivery of the stimuli was confirmed by the subject. The 40-Hz ASSR is not visible on the spectrum (*open triangle*). When such epochs are successively recorded, the 40-Hz ASSR is visible on about 25% of the tracings. *Third row tracings*, as above, except that ten EEG segments were averaged in the time domain. The amplitude spectrum of the averaged waveform shows the 40-Hz ASSR. This is consistent from trial to trial. *Bottom tracings*, as above, except that 100 EEG segments were averaged. Background noise is further reduced

Relative Timing Between the Stimuli and Electroencephalogram Recording

It is important that the relative timing between the stimuli and EEG recording be precisely controlled to ensure that the beginning of any EEG epoch coincides with the onset of a stimulus. Otherwise it will not be possible to demonstrate that the 40-Hz activity present in the recording is time-locked to the stimuli. Averaging would no longer increase the signal to noise ratio and phase information would become meaningless.

Because the 40-Hz ASSR likely arises from the superimposition of transient responses evoked by each stimulus [7, 9, 16], it is preferable to start delivering the stimuli shortly (0.5 s) before starting the EEG recording to allow time for

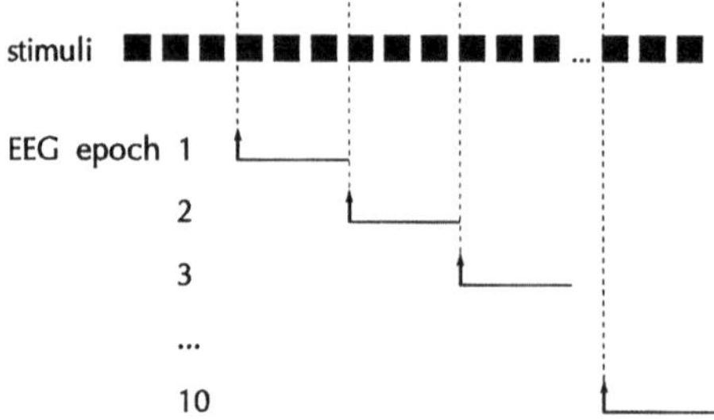

Fig. 4. The relative timing between stimuli and electroencephalogram (EEG) recording must be constant. The beginning of an EEG epoch always coincides with the onset of a stimulus (*filled square*). For clarity, we included only three stimuli per epoch. The stimuli preceding the first epoch are to build up the response

building up the 40-Hz ASSR (Fig. 4). If averaging of successive epochs is used, the above is only required before the first epoch.

Spectral Analysis

Whether or not averaging is used, the FFT should only be performed on EEG segments that are free from artifacts [2]. We minimize artifacts by rejecting EEG segments that include values outside a specified range. In practice, a segment is rejected if 10% or more data points fall outside the -100 to $+100\ \mu V$ range. FFT also requires that the data be stationary (i.e., possess statistical properties that do not fluctuate with time). This is impossible to verify during on-line recording. Brief artifact-free EEG segments can be considered stationary [4].

Spectral analysis changes the data format from *voltage versus time* (time domain representation) to voltage *versus* frequency (frequency domain representation). This greatly facilitates the individual evaluation of the rhythmic components present in the waveforms. Fourier transformation is based on well-established mathematical concepts and is the most popular method for spectral analysis. Cooley and Tukey developed in 1965 an efficient algorithm for computing the Fourier transform [5]. This method, FFT, is very versatile and widely used. It requires, however, that the number (n) of data points in the time domain recording be a power of 2.

The FFT of a trace consisting of n data points will yield a spectrum with $n/2$ frequencies and provide the amplitude and the phase for each frequency. The FFT frequency range is from 0 (D.C.) to about half the AD conversion frequency. For example, using $n = 1024$, EEG data points and an AD conversion frequency of 300 Hz will yield 512 ($n/2$) equidistant frequencies ranging from 0 to 149.7070 Hz, e.g., 0.0000, 0.2930, 0.5859, 0.8789, 1.1719 . . . Hz. The 149.7070 value is obtained by dividing the AD conversion frequency by two and subtracting one frequency step (0.2930) (e.g., [300/2] $-$ 0.2930 = 149.7070). If $n = 128$ and the AD conversion rate is still 300 Hz the FFT will yield 64 equidistant frequencies ranging from 0 to 147.6563 Hz e.g. 0.0000, 2.3438, 4.6875, 7.0313, 9.3750, . . . Hz. Reducing by a factor of 8 the number of data points (from 1024 to 128) increased the frequency steps from 0.2930 Hz to

2.3438 Hz, an eightfold decrease in frequency resolution. It therefore follows that, for a given AD conversion rate, the frequency resolution of the FFT is proportional to the number of data points included in the analysis and consequently to the duration of the recording. In practice, we use 1024 data points with a sampling frequency of approximately 300 Hz, yielding an epoch of approximately 3.4 s. Averaging ten epochs requires 34 s. Less than 2 s are needed to compute the FFT of the averaged trace and display the results. An updated waveform can therefore be easily obtained every minute for two EEG channels.

It is crucial to ensure that the rate of stimulus delivery corresponds to a frequency term of the FFT. Otherwise, the amplitude of the response will be falsely reduced; the attenuation will increase with the difference between the rate of stimulus delivery and its closest frequency term on the FFT. An effective approach for avoiding this problem is to adjust the AD conversion frequency such that there will be a frequency term exactly at the rate of stimulus delivery. Using $n = 1024$ and 300 Hz AD conversion frequency will not yield a term at 40 Hz. The frequencies closest to 40 Hz will be 39.8438 and 40.1367 Hz. By changing the AD conversion frequency to 298.9781 Hz, one will obtain a data point at exactly 40 Hz. The steps required to calculate the exact AD sampling frequency are as follows, assuming a rate of stimulus delivery of 40 Hz.

1. $ST = REQ/n$
2. $SEQ = 40/ST$
3. If SEQ is an integer, then $ACT = REQ$
4. If SEQ is not an integer, then round off to nearest integer (SEQr)
5. $ACT = (n \times 40)/SEQr$

(n, number of data points in the time domain trace; ST, frequency step obtained with REQ; REQ, requested AD conversion frequency; ACT actual AD conversion rate; SEQ, sequential position of the frequency closest to 40 Hz).

Another advantage of having a frequency term corresponding to the rate of stimulus delivery is that there will be no spectral leakage [17 (pp. 439–447)] from the 40-Hz peak.

Conclusion

We cannot deny that the methods required to record the 40-Hz ASSR are more complex than those required to record conventional transient auditory evoked potentials. We are convinced that the effort invested in setting up an adequate recording system is well compensated for by the ease and rapidity with which the response can subsequently be recorded and interpreted.

Acknowledgements. We thank Ms. Suzan Caney for typing the manuscript.

References

1. Arnst DJ (1985) Presbycusis. In: Katz J (ed) Handbook of clinical audiology, 3rd edn. Williams and Wilkins, Baltimore, pp 707–720
2. Beauchamp KG, Yuen CK (1992) Digital methods for signal analysis. Allen and Unwin, London
3. Chiappa KH (1990) Evoked potentials in clinical medicine, 2nd edn. Raven, New York
4. Cohen BA, Sances A (1977) Stationarity of the human electroencephalogram. Med Biol Eng Comput 15:513–518
5. Cooley JW, Tukey JW (1965) An algorithm for the machine calculation of complex Fourier series. Math Comp pp 297–301
6. Firsching R, Luther J, Eidelberg E, Brown WE Jr, Story JL, Boop FA (1987) 40 Hz - middle latency auditory evoked response in comatose patients. Electroencephalogr Clin Neurophysiol 67:213–216
7. Galambos R, Makeig S, Talmachoff PJ (1981) A 40-Hz auditory potential recorded from the human scalp. Prac Natl Acad Sci USA 78:2643–2647
8. Gotman J (1992) The use of computers in analysis and display of EEG and evoked potentials. In: Daly DD, Peddley TA (eds) Current practice of clinical electroencephalography, 2nd edn. Raven, New York, pp 51–83
9. Hari R, Hamalainen M, Joutsiniemi SR (1989) Neuromagnetic steady state response to auditory stimuli. J Acoust Soc Am 86:1033–1039
10. Makela JP, Hari R (1987) Evidence for cortical origin of the 40 Hz auditory evoked response in man. Electroencephalogr Clin Neurophysiol 66:539–546
11. Picton TW (1987) Human auditory steady-state responses. In: Barber C, Blum T (eds) Evoked potentials III. Butterworths, Toronto, pp 117–124
12. Picton TW (1990) Auditory evoked potentials. In: Daly DD, Pedley TA (eds) Current practice of clinical electroencephalography, 2nd edn. Raven, New York pp 625–678
13. Plourde G (1991) Depth of anaesthesia. Can J Anaesth 38:270–274
14. Plourde G, Boylan JF (1991) The auditory steady state response during sufentanil anaesthesia. Br J Anaesth 66:683–691
15. Plourde G, Picton TW (1990) Human auditory steady-state response during general anesthesia. Anesth Analg 71:460–468
16. Plourde G, Stapells DR, Picton TW (1991) The human auditory steady-state evoked potentials. Acta Otolaryngol Suppl (Stockh) 491:153–160
17. Press WH, Flannery BP, Teukolsky SA, Vetterling WT (1988) Numerical recipes in C. The art of scientific computing. Cambridge University Press, Cambridge
18. Putnam LE, Johnson R, Roth WT (1992) Guidelines for reducing the risk of disease transmission in the psychophysiology laboratory. Psychophysiology 29:127–141
19. Regan D (1989) Human brain electrophysiology. Evoked potentials and evoked magnetic fields in science and medicine. Elsevier, New York
20. Rodriguez R, Picton TW, Linden D, Hamel G, Laframboise G (1986) Human auditory steady state responses: effects of intensity and frequency. Ear Hear 7:300–313
21. Shannon CE (1948) A mathematical theory of communication. Bell System Tech J 27:623–656

22. Stapells DR, Linden D, Suffield JB, Hamel G, Picton TW (1984) Human auditory steady state potentials. Ear Hear 5:105–113
23. Starr A, Don M (1992) Brain potentials evoked by acoustic stimuli. In: Picton TW (ed) Human event-related potentials. Elsevier, New York, pp 97–155
24. Thornton C (1991) Evoked potentials in anaesthesia. Eur J Anaesthesiol 8:89–107
25. Villemure C, Plourde G, Castillo O (1991) The 40 Hz auditory steady-state response for monitoring the level of consciousness. Anesth Analg 74:S336 (abstract)

Motor-Evoked Potentials

C. J. Kalkman

Introduction

During the last decade, somatosensory evoked potentials (SSEP) have become established as a practical method for monitoring the spinal cord during various surgical procedures where there is a risk of paraplegia, e.g., scoliosis surgery, thoracic aortic surgery, and neurosurgical procedures upon the spinal cord. However, it has also become apparent that SSEP have limitations concerning their ability to monitor the entire spinal cord. SSEP travel exclusively in ascending sensory pathways (dorsal columns and posterolateral tracts). Accordingly, selective injury to the more anteriorly located motor tracts and motor neuronal systems in the central gray matter and anterior horn may go undetected. A number of case reports have described false negative results with SSEP monitoring, i.e., postoperative paraplegia despite unaltered intraoperative SSEP [4, 27, 38]. A recent survey by the Scoliosis Research Society among physicians performing intraoperative SSEP monitoring during spinal surgery revealed that five out of 27 major neurological complications (17%) that occurred with monitoring in place were not diagnosed by changes in SSEP [8]. Even if technical errors or lack of experience are taken into account that may have hampered the acquisition of reliable SSEP waveforms in some of these cases, this figure suggests that injury to the spinal cord is sometimes limited to the motor pathways. Given the differences in blood supply to the anterior and posterior spinal cord, there are several clinical situations where selective ischemia of the anterior part of the cord may ensue. This is particularly true for the thoracic spinal cord, where in some patients the anatomical variation of the anterior spinal artery may be such that interruption of a single intercostal or lumbar feeder vessel will result in spinal cord ischemia.

Methods of Monitoring the Motor Pathways

Various investigators have attempted to monitor the motor pathways intraoperatively. Levy employed direct spinal cord stimulation and recorded motor evoked potentials (MEP) from the caudal spinal cord [28, 29]. Using this technique, it was possible to record a descending spinal MEP in humans during surgery. However, for several reasons this spinal monitoring system has never been widely employed. The technique is highly invasive, since the stimulating electrodes are placed directly on the surgically exposed spinal cord and recording electrodes are in the epidural space. This in itself might result in injury to the cord. Moreover, Machida et al. have subsequently demonstrated that peripheral nerve or muscle MEP are more sensitive to spinal cord injury than MEP that are recorded directly from the spinal cord [32, 34, 35].

There are various systems for monitoring MEP during surgery that might be used clinically (Fig. 1). Electrical or magnetic transcranial stimulation (TCS) of the motor cortex can be accomplished with special stimulators, while the response may be recorded either from an electrode positioned in the epidural space or as a compound muscle action potential (CMAP). Alternatively, it is possible to use electrical spinal cord stimulation at a cervical or high thoracic level and record a response from the sciatic nerve ("neurogenic" MEP, NMEP).

Transcranial Motor Evoked Potentials

When conventional constant current stimuli are applied to the scalp, it is extremely difficult to produce sufficient depolarization of pyramidal cells to result in muscle contraction. This is due to the high resistivity of the scalp and dispersion of the stimulus current in the skin. In an attempt to overcome this

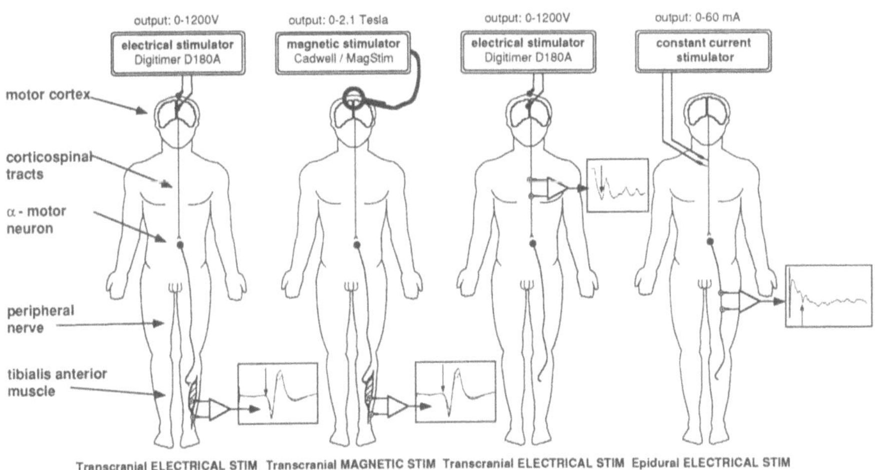

Fig. 1. Systems for intraoperative monitoring of motor evoked potentials: electrical or magnetic transcranial stimulation (*STIM*) with recording of compound muscle action potentials, epidural recording of responses to transcranial stimulation, and neurogenic response to electrical thoracic spinal cord stimulation

problem, Levy employed transcranial stimulation via an electrode on the vertex and an electrode placed against the palatum [30]. Nevertheless, signal averaging of several hundred trials at a rate of 5–20 Hz was needed in order to obtain a reproducible MEP signal. In 1980, Merton and Morton demonstrated that it was possible to produce limb movement and record a reproducible CMAP after application of a single exponentially decaying high-voltage transcranial stimulus applied to the scalp via conventional electroencephalogram (EEG) electrodes [36]. Five years later, Barker et al. demonstrated that a transient magnetic field generated by passing a large current through a coil applied over the scalp was able to produce similar myogenic motor evoked responses [3]. Since the arrival of commerically available electrical and magnetic transcranial stimulators, several groups have investigated the conducting pathways of transcranial MEP (tcMEP) and their possible utility in clinical neurophysiology [9, 11, 12, 43, 44, 45]. More recently, these techniques have been brought into the operating room.

Electrical Transcranial Stimulation

Stimulators designed for electrical TCS consist of a capacitor charged to 400–1200 V that is discharged through a thyristor and delivered to the patient via an isolated low-output impedance transformer (Digitimer D180, Digitimer Ltd., Welwyn Garden City, UK). Time constant of the exponentially decaying impulse typically is 50 or 100 μs. Anodal stimuli over the cortical region of interest are the most effective in generating myogenic responses [5, 11].

A single electrical transcranial stimulus applied to the scalp will produce direct activation of pyramidal cell bodies and/or axons that can be recorded as an early direct (D) wave from electrodes in the epidural space (Fig. 2). This early D wave is followed several milliseconds later by a train of smaller, indirect (I) waves. I waves are thought to be the result of transsynaptic activation of the pyramidal cell by cortical interneurons. The descending volley arrives at the spinal interneuronal system (only a small percentage of corticospinal fibers directly synapse on the α-motor neuron). Both D and I waves produce excitatory postsynaptic potentials (EPSP) at the spinal motoneuronal system. When the threshold for firing of the α-motor neuron is exceeded, it will fire, resulting in activation of a number of muscle fibers depending on the size of the motor unit.

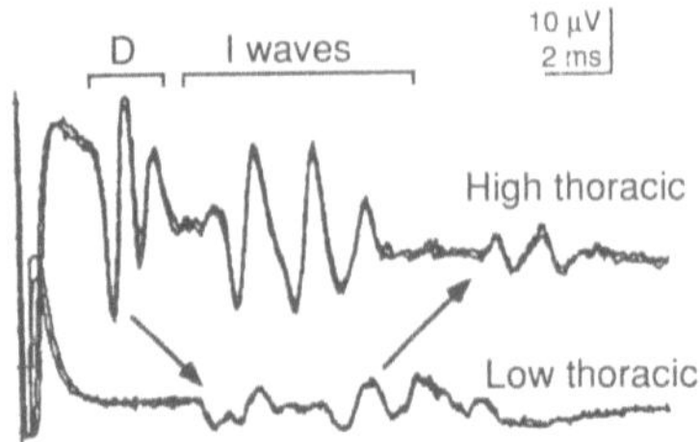

Fig. 2. Epidurally recorded motor evoked potentials to transcranial electrical stimulation. Note initial *D wave* followed by several *I waves*. (Reproduced with permission from [18])

Magnetic Transcranial Stimulation

Magnetic transcranial stimulators produce a transient, time-varying magnetic field (2–2.5 T) by discharging a bank of capacitors through an isolated coil with a diameter of 6–12 cm. Very large currents (up to 5000 A) need to be switched, which makes these stimulators large and relatively expensive. The repetition rate is currently limited to once every 2–3 s, because the capacitors have to recharge. Magnetic stimulators differ in their maximum output power and pulse type (biphasic vs. monophasic), and several types and sizes of stimulating coils are available. Larger coils tend to produce activation of deeper structures, which is critically important when the leg area needs to be stimulated for intraoperative spinal cord monitoring. On the other hand, less powerful, but more focal, cortical stimulation can be achieved with smaller coils and double ("butterfly") coils. Stimulus intensity is expressed as a percentage of maximal stimulator output. The time-varying magnetic field stimulates the cortex by generating a current in brain tissue below the coil. Coil orientation is important: with the coil placed in the midline over the vertex, clockwise current flow in the coil (as seen from above) preferentially activates the left hemisphere and vice versa. Magnetic stimulation with the coil placed tangentially over the scalp produces predominantly I waves, suggesting that it activates corticospinal axons trans-synaptically, as opposed to the direct depolarization by electrical stimulation. Since magnetic stimulation usually does not produce D waves, the latency of magnetic tcMEP is 1–3 ms longer than the latency of electrical tcMEP [2, 10]. The major advantage of magnetic TCS is that it is well tolerated by awake patients. In contrast, electrical TCS is moderately painful. However, in the anesthetized patient it is of more importance which stimulator produces the largest responses in the presence of anesthetic-induced MEP amplitude depression.

Motor Evoked Potentials to Spinal Cord Stimulation

Direct spinal cord stimulation stimulates both sensory and motor tracts, and recordings from the distal spinal cord are therefore not specific for the motor tracts, because they reflect both the orthodromic and antidromic axonal conduction. Machida et al. stimulated the spinal cord from electrodes placed in the epidural space at the T5 or T6 level and recorded CMAP from the soleus muscle. They found this signal to be more sensitive to ischemic and distractive spinal cord injury than MEP recorded from the lower spinal cord in experimental animals and humans [33, 34]. Owen et al. introduced the technique of NMEP, in which high thoracic spinal cord stimulation is achieved via needles positioned in the base of the spinous process, with recordings made from the sciatic nerve. Experimental work by this group showed that NMEP were insensitive to dorsal root rhizotomies, but highly sensitive to selective motor tract lesioning [40]. However, in a recent study the sciatic nerve response to spinal cord stimulation was abolished after dorsal column transaction in dogs, suggesting that a neurogenic response may also occur as the result of antidromic sensory pathway conduction [48]. The authors recommended that if spinal cord stimulation is to be employed, the myogenic rather than neurogenic response should be monitored.

Effects of Anesthetic Drugs on Motor Evoked Potentials

A major obstacle to the clinical application of MEP for intraoperative spinal cord monitoring is the fact that they are extremely sensitive to depression by anesthetic drugs. Most standard anesthetic regimens preclude the recording of myogenic MEP or produce unacceptable amplitude depression. Table 1 shows the depressant effect of anesthetic drugs on tcMEP in decreasing order.

Volatile Anesthetics

While SSEP are degraded by volatile anesthetics in a dose-dependent manner [41], the myogenic transcranial motor response is completely abolished when isoflurane is introduced during the N_2O–opioid anesthetic technique [6, 22, 46]. Figure 3 shows that the response is abolished at an expired isoflurane concentration of 0.3% [22]. After discontinuation of isoflurane, the MEP gradually

Table 1. Depressant effect of anesthetic drugs on myogenic transcranial motor evoked potentials (MEP) in order of decreasing influence on MEP amplitude

Drug	Effect
Isoflurane	↓ ↓ ↓
Enflurane/halothane	↓ ↓
Propofol	↓ ↓
Thiopental	↓ ↓
Midazolam	↓
Nitrous oxide	↓
Droperidol	↓
Etomidate	↓
Fentanyl	=
Ketamine	= ↑

↓ ↓ ↓, large depressant effect; ↓ ↓, medium depressant effect; ↓, small depressant effect; =, no depressant effect; ↑, stimulant effect.

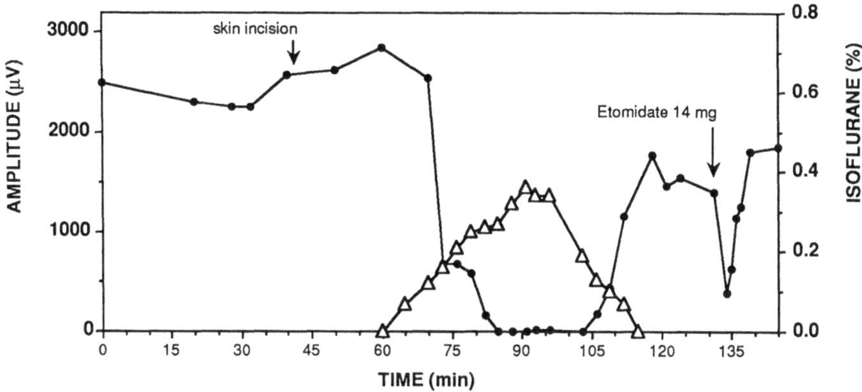

Fig. 3. Amplitude (µV; *filled circles*) of motor evoked responses to transcranial electrical stimulation and end-tidal isoflurane concentration (*open triangles*) versus time in one patient during nitrous oxide–sufentanil anesthesia. (Reproduced with permission from [22])

returns as the expired isoflurane concentration declines, although some hysteresis is present. Isoflurane probably exerts its effect both at the cortical and spinal level. Hicks et al. recorded MEP to electrical TCS in the epidural space and observed multiple I waves following the D wave during N_2O–opioid anesthesia. However, when isoflurane was introduced, the number of I waves decreased and individual I waves became smaller in amplitude the greater the isoflurane concentration, while there were only minor changes in the D wave. The maximum depressant effect on I waves was reached at an end-tidal isoflurane concentration of 0.5% [18]. These data support the concept of I waves reflecting transsynaptic activation of the pyramidal cells by cortical interneurons, since anesthetics primarily depress synaptic processes.

Nitrous Oxide

N_2O appears to be a powerful depressant of amplitude. Zentner et al. showed that inhalation of 66% N_2O in volunteers decreased the amplitude of myogenic MEP to electrical TCS to less than 10% of baseline. The same authors demonstrated in rats that this effect is caused–at least in part–at the spinal level, because the effect of N_2O was also present when spinal cord stimulation was employed [50]. Ghaly et al. found that in ketamine-anesthetized monkeys, N_2O caused small increases in magnetic tcMEP amplitude with concentrations up to 50%, but a 60% amplitude decrease was observed with 75% N_2O [16].

Schmid et al. recorded magnetic tcMEP from the hand muscles in surgical patients. They found that inhalation of N_2O in concentrations up to 79% depressed amplitude to only 61% of baseline [46]. A possible explanation for the apparent discrepancy in sensitivity to N_2O could be that Schmid et al.

recorded from small hand muscles, which are more easily stimulated than the leg muscles, since the cortical representation of the hand area is on the convex side of the motor cortex. Addition of N_2O to a propofol anesthetic increased latencies and decreased amplitudes [21]. With 50% N_2O, amplitude was decreased to 50%, but sharply decreased to 11% of control with 70% N_2O. In conclusion, it appears that N_2O in concentrations up to 50% may be safely administered, but that higher concentrations sharply diminish tcMEP amplitude.

Intravenous Anesthetics

Although intravenous anesthetics have relatively little effect on the amplitude of SSEP, their effects on tcMEP vary from mild enhancement to pronounced amplitude depression. Figure 4 shows the effects of single bolus doses of propofol, etomidate, midazolam, or fentanyl tcMEP in human volunteers.

Barbiturates. Thiopental and thiamylal, produce substantial amplitude depression of tcMEP and NMEP to less than 10% of awake baseline values [31, 46].

Propofol. This is also a powerful depressant of tcMEP [20, 25, 31, 46] and NMEP [42]. An interesting observation is that the depressant effect of an induction dose of 2 mg propofol kg^{-1} on tcMEP amplitude persisted for more than 20 min after the subjects regained consciousness [25].

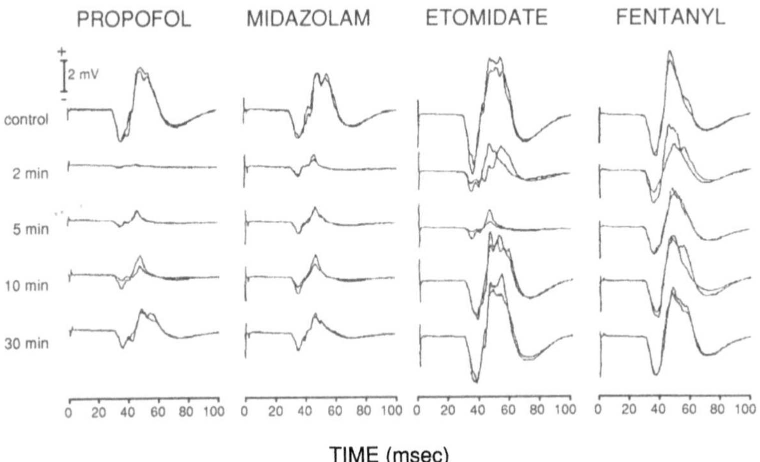

Fig. 4. Motor evoked response waveforms recorded from tibialis anterior muscle to transcranial electrical or magnetic stimulation after injection of 2 mg propofol kg^{-1}, 0.3 mg etomidate kg^{-1}, 0.05 mg midazolam kg^{-1}, or 3 μg fentanyl kg^{-1} in one subject (Reproduced with permission from [25])

Midazolam. This has a strong depressant effect on myogenic tcMEP, even when administered in sedative doses [25, 46, 47].

Etomidate. This has a more variable effect on tcMEP. In monkeys, it decreases the scalp area where magnetic stimulation is effective, and repeated doses increase the threshold for a detectable response by 15%–30%. Amplitude depression was 15%–50% for the lower limb muscles [15]. In humans, a bolus dose of 0.3 mg etomidate/kg caused initial amplitude depression, followed by a return to baseline within 5–10 min [25].

Fentanyl. In doses between 1.5 and 8 μg/kg, fentanyl had only minimal effects on tcMEP amplitude [25, 46].

Ketamine. Administered in incremental doses to monkeys, ketamine produced no amplitude depression until a cumulative dose of 15–20 mg/kg had been reached. Thereafter, only small decreases in amplitude were observed. In human volunteers, administration of 1 mg ketamine/kg produced no effect or, in some subjects, enhancement of the magnetic tcMEP [24].

It has been suggested that responses to magnetic TCS might be more susceptible to depression by anesthetic drugs, since magnetic stimulation predominantly activates motor pathways transsynaptically [6]. However, in the study by Kalkman et al., both modes of TCS produced about the same degree of amplitude depression, although the amplitude of the magnetic tcMEP was always somewhat smaller than that of the electrical tcMEP [25]. In some volunteers, the response to magnetic stimulation was abolished after propofol, while a small, but detectable, response was present after electrical stimulation.

Muscle Relaxants

Some authors have stated that recording myogenic MEP precludes the administration of muscle relaxants. However, there are several reasons why it may be impractical to perform surgery and tcMEP monitoring without muscle relaxation. There is a risk that the patient will move, either spontaneously or in response to stimulation. This could be dangerous during some operations, especially when the surgeon is working in the vicinity of the spinal cord or nerve roots. Fortunately, it is possible to titrate the administration of muscle relaxants to a level at which contraction of muscles is greatly diminished, while still being able to record a CMAP from the leg muscles. When one mechanical twitch response of the thumb was present to train-of-four ulnar nerve stimulation (10% twitch height) during neuromuscular blockade with vecuronium, tcMEP amplitude was reduced by about 60% [23]. Maintenance of a constant level of neuromuscular blockade, for example using a computer-controlled servo mechanism, for the administration of relaxant is mandatory. The CMAP to peripheral nerve stimulation of the target muscle may be used to verify a constant level of relaxation.

Methods of Facilitating Motor Evoked Potentials

It has been shown numerous times that in awake human volunteers, a slight voluntary contraction of the muscle studied produces significant facilitation of the transcranial motor evoked response, effectively doubling or tripling CMAP amplitude or revealing a response that was absent without voluntary contraction [1, 37]. A possible explanation for this phenomenon is that afferent input to the dorsal horn from muscle afferents results in spatial and temporal summation of EPSP at the α-motor neuron. Unfortunately, this type of facilitation cannot be employed in anesthetized patients, but it may be mimicked by applying electrical stimulation to the dermatome corresponding to the muscle [13, 26]. Conversely, Cowan et al. showed that the soleus muscle H-reflex could be facilitated by applying a subliminal transcranial stimulus (one not resulting in a discernible CMAP) [7]. Based on this principle, experiments using paired stimuli have shown that with interstimulus intervals of between 1.5 and 2 ms there was facilitation, while the tcMEP was inhibited with an interstimulus interval of between 10 and 100 ms [19]. Facilities for application of double pulse stimulation are currently being developed by manufacturers of magnetic and electrical transcranial stimulators. Zentner recently reported significant improvement of tcMEP amplitude using five consecutive stimuli of a 500-Hz pulse train (2 ms inter-stimulus interval; J. Zentner 1992, personal communication).

Clinical Experience with Intraoperative Use
of Motor Evoked Potentials

Clinical experience with MEP during surgery is limited. Levy was the first to employ transcranial MEP techniques intraoperatively in humans. He used constant current stimulation (vertex–palatum) and recorded averaged nerve and muscle evoked responses [30]. During this type of continuous TCS at a rate of 5-20 Hz, there were significant increases in heart rate and blood pressure. This method has now been abandoned in favor of single pulse, high-voltage, low-output impedance devices. Zentner et al. recorded myogenic responses to single electrical transcranial stimuli during neurosurgical operations [49, 51]. They used a commercially available electrical transcranial stimulator (Digitimer D180) that uses capacitor discharges to deliver single exponentially decaying high-voltage stimuli of upto a maximum of 1200 V. Responses were recorded from the anterior tibial muscles. The averages of five to 15 signals were evaluated. Although potentials were obtained preoperatively in all 50 patients, during neuroleptanesthesia intraoperative recording from the anterior tibial muscles was possible in 43 patients (86%) and from the thenar muscles in 21 patients (88%). The authors considered amplitudes superior to latencies as evaluation criteria for intraoperative changes in potentials. Using a 50% amplitude decrease at the end of the operation as a criterion for abnormal tcMEP response, they found a good correlation between abnormal MEP and postoperative neurological status. There were false positive results in about 20% of patients, but no false negative results were encountered.

Jellinek et al. recently presented their experiences with tcMEP monitoring during propofol anesthesia. The responses were extremely small (generally less than 100 μV). Nevertheless, they were able to record CMAP responses, even in patients with preexisting neurological deficits.

Hicks et al. recorded SSEP and tcMEP simultaneously from electrodes placed in the epidural space in 40 adolescent patients undergoing Cotrel–Dubousset instrumentation for scoliosis. Since no new neurological deficits occurred, the authors were unable to comment on the relative sensitivity of SSEP and tcMEP [17]. Edmonds et al. were able to record myogenic responses to magnetic TCS in nine out of 11 patients during scoliosis surgery with N_2O–opioid anesthesia [14].

Owen et al. stimulated the spinal cord above the level of potential injury by having the surgeon place electrodes in the base of two adjacent spinous processes [39]. They recorded SSEP and the NMEP from the sciatic nerve over the ischial tuberosity in 300 patients, but later changed this to sciatic nerve in the popliteal fossa. They reported that variability of SSEP was higher than that of NMEP and attributed this to anesthesia and "unknown factors." There were 17 false positive NMEP results and 54 occurrences of false positive SSEP. Since there were no iatrogenic spinal cord injuries in the 240 cases involving elective surgery for spinal deformity, the authors were unable to comment on the relative sensitivity and specificity of NMEP versus SSEP. However, following intrathecal or intramedullary tumor surgery, there were five patients who awoke with motor deficits, each of which had been predicted by intraoperative loss of NMEP, but with preservation of SSEP.

Conclusions

Intraoperative MEP monitoring of spinal cord function is a promising technique. Addition of this modality to existing techniques of spinal cord monitoring may decrease the incidence of false negative results reported with SSEP monitoring. However, there are several issues that need to be addressed before routine application of MEP techniques can be advocated:

— What is the relative sensitivity and specificity of the various MEP techniques for the detection of spinal cord injury?
— Which anesthetic techniques are optimal for MEP monitoring?
— How can the amplitude of intraoperative myogenic MEP be increased, and variability in an individual patient be reduced?
— Which type of TCS is preferable during surgery, electrical or magnetic?

This author's bias is towards recording CMAP in response to electrical TCS, because this technique is both highly sensitive to spinal cord injury and selective for the motor tracts. However, myogenic responses show a large inherent variability and are probably the most sensitive to anesthetic-induced amplitude depression, because they rely on synaptic processes in the motor cortex, the spinal cord, and at the neuromuscular junction.

Spinal cord stimulation is a more invasive technique. The neurogenic responses are small ($\pm 1\ \mu V$), but are possibly somewhat more resistant to anesthetic drugs. NMEP may be less specific for the motor tracts, because electrical stimulation of the spinal cord excites both sensory and motor elements in the spinal cord, resulting in antidromic conduction in sensory fibers.

No formal comparisons have yet been made between the various anesthetic techniques that might be useful for MEP monitoring. Acceptable responses have been recorded during N_2O–opioid anesthesia and during propofol anesthesia, but amplitudes were severely depressed with both techniques. Future studies should focus on finding anesthetic regimens that minimally depress MEP, while maintaining good operating conditions. The possible role of ketamine as an adjunct during anesthesia for MEP monitoring should be evaluated.

The problem of anesthetic-induced amplitude depression of tcMEP may be partly overcome by the introduction of multiple pulse electrical or magnetic TCS techniques in combination with appropriate anesthetic regimens. By using SSEP and MEP as complementary techniques, the functional status of the spinal cord can be reliably assessed during surgery to the benefit of patients who are at risk of suffering iatrogenic spinal cord injury.

References

1. Ackermann H, Scholz E, Koehler W, Dichgans J (1991) Influence of posture and voluntary background contraction upon compound muscle action potentials from anterior tibial and soleus muscle following transcranial magnetic stimulation. Electroencephalogr Clin Neurophysiol 81:71–80
2. Amassian VE, Cracco RQ, Maccabee PJ (1989) Focal stimulation of human cerebral cortex with the magnetic coil: a comparison with electrical stimulation. Electroencephalogr Clin Neurophysiol 74:401–416
3. Barker AT, Jalinous R, Freeston IL (1985) Noninvasive magnetic stimulation of human motor cortex. Lancet II:1106–1107
4. Ben-David B, Haller G, Taylor P (1987) Anterior spinal fusion complicated by paraplegia. A case report of a false-negative somatosensory-evoked potential. Spine 12:536–539
5. Benecke R, Meyer BU, Gohmann M, Conrad B (1988) Analysis of muscle responses elicited by transcranial stimulation of the corticospinal system in man. Electroencephalogr Clin Neurophysiol 69:412–422
6. Calancie B, Klose KJ, Baier S, Green BA (1991) Isoflurane-induced attenuation of motor evoked potentials caused by electrical motor cortex stimulation during surgery. J Neurosurg 74:897–904

7. Cowan JM, Day BL, Marsden C, Rothwell JC (1986) The effect of percutaneous motor cortex stimulation on H reflexes in muscles of the arm and leg in intact man. J. Physiol (Lond) 377:333–347

8. Dawson EG, Sherman JE, Kanim LE, Nuwer MR (1991) Spinal cord monitoring: results of the Scoliosis Research Society and the European Spinal Deformity Society Survey. Spine 16:S361–S364

9. Day BL, Dick JPR, Marsden CD, Thompson PD (1986) Differences between electrical and magnetic stimulation of the human brain. J Physiol (Lond) 378:36P

10. Day BL, Dressler D, Maertens de Noordhout A, Marsden CD, Nakashima K, Rothwell JC, Thompson PD (1989) Electric and magnetic stimulation of human motor cortex: surface EMG and single motor unit responses. J Physiol (Lond) 412:449–473

11. Day BL, Maertens de Noordhout A, Marsden CD, Nakashima K, Rothwell JC, Thompson PD (1987) A comparison of the effects of anodal and cathodal stimulation of the human motor cortex through the intact scalp. J Physiol (Lond) 394:118P

12. Day BL, Rothwell JC, Thompson PD, Dick JP, Cowan JM, Berardelli A, Marsden CD (1987) Motor cortex stimulation in intact man. II. Multiple descending volleys. Brain 110:1191–1209

13. Deletis V, Dimitrijevic MR, Sherwood AM (1987) Effects of electrically induced afferent input from limb nerves on the excitability of human cortex. Neurosurgery 20:195–197

14. Edmonds HL Jr, Paloheimo MP, Backman MH, Johnson JR, Holt RT, Shields CB (1989) Transcranial magnetic motor evoked potentials (tcMMEP) for functional monitoring of motor pathways during scoliosis surgery. Spine 14:683–686

15. Ghaly RF, Stone JL, Levy WJ, Roccaforte P, Brunner EB (1990) The effect of etomidate on motor evoked potentials induced by transcranial magnetic stimulation in the monkey. Neurosurgery 27:936–942

16. Ghaly RF, Stone JL, Levy WL, Kartha R, Aldrete JA (1990) The effect of nitrous oxide on transcranial magnetic-induced electromyographic responses in the monkey. J Neurosurg Anesth 2:175–181

17. Hicks RG, Burke DJ, Stephen JP (1991) Monitoring spinal cord function during scoliosis surgery with Cotrel-Dubousset instrumentation. Med J Aust 154:82–86

18. Hicks RG, Woodforth IJ, Crawford MR, Stephen JPH, Burke DJ (1992) Some effects of isoflurane on I waves of the motor evoked potential. Br J Anaesth 69:130–136

19. Inghilleri M, Berardellli A, Cruccu G, Priori A, Manfredi M (1990) Motor potentials evoked by paired cortical stimuli. Electroencephalogr Clin Neurophysiol 77:382–389

20. Jellinek D, Jewkes D, Symon L (1991) Noninvasive intraoperative monitoring of motor evoked potentials under propofol anesthesia: effects of spinal surgery on the amplitude and latency of motor evoked potentials. Neurosurgery 29:551–557

21. Jellinek D, Platt M, Jewkes D, Symon L (1991) Effects of nitrous oxide on motor evoked potentials recorded from skeletal muscle in patients under total anesthesia with intravenously administered propofol. Neurosurgery 29:558–562

22. Kalkman C, Drummond J, Ribberink A (1991) Low concentrations of isoflurane abolish motor evoked responses to transcranial electrical stimulation during nitrous oxide/opioid anesthesia in humans. Anesth Analg 73:410–415

23. Kalkman CJ, Drummond JC, Kennelly NA, Patel PM, Partridge BL (1992) Intraoperative monitoring of tibialis anterior muscle motor evoked responses to transcranial electrical stimulation during partial neuromuscular blockade. Anesth Analg 75:584–589

24. Kalkman CJ, Drummond JC, Patel PM, Sano T, Chesnut RM (1992) Effects of droperidol, pentobarbital and ketamine on myogenic transcranial motor evoked responses in humans. Anesthesiology 77:A163 (abstr)
25. Kalkman CJ, Drummond JC, Ribberink AA, Patel PM, Sano T, Bickford RG (1992) Effects of propofol, etomidate, midazolam and fentanyl on motor evoked responses to transcranial electrical or magnetic stimulation in humans. Anesthesiology 76:502–509
26. Kasai T, Hayes KC, Wolfe DL, Allatt RD (1992) Afferent conditioning of motor evoked potentials following transcranial magnetic stimulation of motor cortex in normal subjects. Electroencephalogr Clin Neurophysiol 85:95–101
27. Lesser RP, Raudzens P, Lüders H, Nuwer MR, Goldie WD, Morris HH, Dinner DS, Klem G, Hahn JF, Shetter AG, Ginsburg HH, Gurd AR (1986) Post-operative neurological deficits may occur despite unchanged intraoperative somatosensory evoked potentials. Ann Neurol 19:22–25
28. Levy WJ (1983) Spinal evoked potentials from the motor tracts. J Neurosurgery 58:38–44
29. Levy WJ, York DH (1983) Evoked potentials from the motor tracts in humans. Neurosurgery 4:422–429
30. Levy WJ, York DH, McCaffrey M, Tanzer F (1984) Motor evoked potential from transcranial stimulation of the motor cortex in humans. Neurosurgery 15:287–302
31. Losasso T, Boudreaux J, Muzzi D, Cucchiara R, Daube J (1991) The effect of anesthetic agents on magnetic motor evoked potentials (TMEP) in neurosurgical patients. Anesthesiology 75:A1032 (abstr)
32. Machida M, Weinstein SL, Imamura Y, Usui T, Yamada T, Kimura J, Toriyama S (1989) Compound muscle action potentials and spinal evoked potentials in experimental spine maneuver. Spine 14:687–691
33. Machida M, Weinstein SL, Yamada T, Kimura J (1985) Spinal cord monitoring: electrophysiological measures of sensory and motor function during spinal surgery. Spine 10:407–413
34. Machida M, Weinstein SL, Yamada T, Kimura J, Itagaki T, Usui T (1988) Monitoring of motor action potentials after stimulation of the spinal cord. J Bone Joint Surg [Am] 70:911–918
35. Machida M, Weinstein SL, Yamada T, Kimura J, Toriyama S (1988) Dissociation of muscle action potentials and spinal somatosensory evoked potentials after ischemic damage of spinal cord. Spine 13:1119–1124
36. Merton PA, Morton HB (1980) Stimulation of the cerebral cortex in the intact human subject. Nature 285:227
37. Mills KR, Murray NM, Hess CW (1987) Magnetic and electrical transcranial brain stimulation: physiological mechanisms and clinical applications. Neurosurgery 20:164–168
38. Mustain W, Kendig R (1991) Dissociation of neurogenic motor and somatosensory evoked potentials. A case report. Spine 16:851–853
39. Owen JH, Bridwell KH, Grubb R, Jenny A, Allen B, Padberg AM, Shimon SM (1991) The clinical application of neurogenic motor evoked potentials to monitor spinal cord function during surgery. Spine 16:S385–S390
40. Owen JH, Jenny AB, Naito M, Weber K, Bridwell KH, McGhee R (1989) Effects of spinal cord lesioning on somatosensory and neurogenic-motor evoked potentials. Spine 14:673–682

41. Peterson DO, Drummond JC, Todd MM (1986) Effects of halothane, enflurane, isoflurane and nitrous oxide on somatosensory evoked potentials in humans. Anesthesiology 65:35–40
42. Peterson R, Mongan P (1991) Effect of intravenous anesthetics on neurogenic motor evoked potentials recorded at the spinal and sciatic level. Anesthesiology 75:A179
43. Rossini P (1990) Methodological and physiological aspects of motor evoked potentials (Supplement 41 to Electroencephalogr Clin Neurophysiol). New trends and advanced techniques in clinical neurophysiology. Elsevier Science, Amsterdam
44. Rossini PM, Caramia MD, Zarola F (1987) Mechanisms of nervous propagation along central motor pathways: noninvasive evaluation in healthy subjects and in patients with neurological disease. Neurosurgery 20:183–191
45. Rothwell JC, Thompson PD, Day BL, Dick JP, Kachi T, Cowan JM, Marsden CD (1987) Motor cortex stimulation in intact man. 1. General characteristics of EMG responses in different muscles. Brain 110:1173–1190
46. Schmid UD, Boll J, Liechti S, Schmid J, Hess CW (1992) Influence of some anesthetic agents on muscle responses to transcranial magnetic cortex stimulation–a pilot study in humans. Neurosurgery 30:85–92
47. Schönle PW, Isenberg C, Crozier TA, Dressler D, Machetanz J, Conrad B (1989) Changes of transcranially evoked motor responses in man by midazolam, a short acting benzodiazepine. Neurosci Lett 101:321–324
48. Su C, Haghighi S, Oro J, Gaines R (1992) "Backfiring" in spinal cord monitoring: high thoracic spinal cord stimulation evokes sciatic response by antidromic sensory pathway conduction, not motor tract conduction. Spine 17:504–508
49. Zentner J (1989) Noninvasive motor evoked potential monitoring during neurosurgical operations on the spinal cord. Neurosurgery 24:709–712
50. Zentner J, Ebner A (1989) Nitrous oxide suppresses the electromyographic response evoked by electrical stimulation of the motor cortex. Neurosurgery 24:60–62
51. Zentner J, Schumacher M, Bien S (1988) Motor evoked potentials during interventional neuroradiology. Neuroradiology 30:252–255

Part IV

Evoked Response: Special Applications

Auditory Evoked Potentials to Monitor Intraoperative Awareness

D. Schwender, S. Klasing, C. Madler, E. Pöppel, and K. Peter

Introduction

Awareness During Anesthesia

General anesthesia provides unconsciousness, analgesia, muscle relaxation, and stabilization of vital, autonomic functions during a surgical procedure. During combined anesthetics, each of these components can be achieved nearly independently from each other by combining different anesthetic drugs or procedures. Thus, undesirable, dose-dependent side effects of each single anesthetic can be minimized. However, common to all combined anesthetics is the fact that monitoring unconsciousness based on autonomic clinical signs is difficult, especially when adequate surgical analgesia is provided by high-dose opioids or conductance anesthesia. This fact favours the incidence of unwanted intraoperative episodes of wakefulness and awareness; the literature reports that the incidence of intraoperative awareness which can be actively recalled by the patients is about 0.5%–2% [11, 14–16].

Awareness and Auditory Perception

Various case reports and many clinical studies about intraoperative awareness deviate according to the methods applied, the anesthetic agents used, and the patient groups studied. Nevertheless, all reports showed that auditory information in particular can be perceived intraoperatively and recalled postoperatively. Therefore, the auditory modality seems to be the most important sensory channel for sensory information processing during general anesthesia [1, 3–8, 11, 14, 20]. As yet it is far from clear which anesthetic agents suppress auditory perceptions during general anesthesia most reliably. Therefore, it is particularly interesting to study the effects of different anesthetics on auditory stimulus processing.

The Auditory Pathway and Auditory Evoked Potentials

One possibility of investigating auditory information processing during general anesthesia is to record auditory evoked potentials (AEP). They reflect the response of the central nervous system to auditory stimulation at different stages of the brain.

Figure 1 shows the conduction of auditory stimuli from the cochlea to the primary auditory cortex and the frontal cortex. Postsynaptic neurons in the cochlea transmit auditory information via the cochlear nerve to the cochlear nucleus in the brain stem. From there, the lateral lemniscus ascends to the

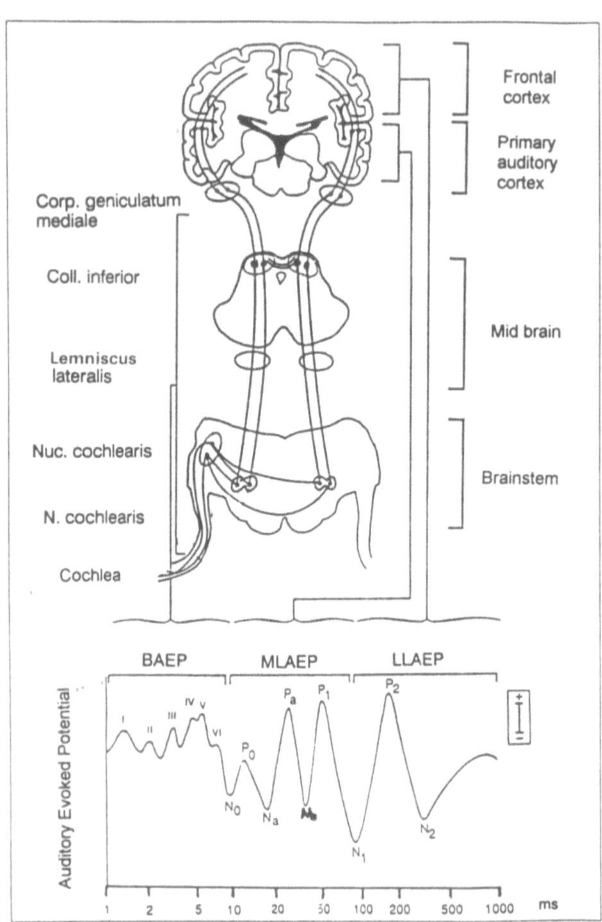

Fig. 1. The auditory pathway and auditory evoked potentials (AEP). *Corp. geniculatum mediale*, corpus geniculatum mediale; *coll. inferior*, colliculus inferior; *Nuc. cochlearis*, nucleus cochlearis; *N. cochlearis*, nervus cochlearis; *BAEP*, brain stem AEP; *MLAEP*, midlatency AEP; *LLAEP*, late-latency AEP

inferior colliculus and medial geniculate body of the mid brain. This finally gives input to the primary auditory cortex in the temporal lobe. The primary auditory cortex is strongly connected with the frontal cortex [23].

AEP consist of a series of positive and negative waves. They are generated by the successive activation of structures in the auditory pathway. They represent processes of transduction, transmission, and processing of auditory information from the cochlea to the brain stem, the primary auditory cortex, and the frontal cortex.

Early peaks of the potentials, brain stem AEP (BAEP), are generated in the cochlear nerve and relays of the brain stem. They reflect successful stimulus transduction and primary stimulus transmission [25]. Midlatency AEP (MLAEP) consist of overlapping activation in different structures of the primary auditory cortex [10, 17, 25, 29, 31]. Finally, late-latency AEP (LLAEP) depict the neuronal activity of association cortices in the frontal lobe. They reflect the process of emotional stimulus evaluation as well as cognitive analysis of the auditory information [2, 19, 22, 26].

BAEP remain nearly unchanged during general anesthesia [21, 30]. LLAEP are highly variable in awake subjects and rely strongly on processes of attention and orientation to the stimulus [24]; in contrast, MLAEP are intra- and interindividually stable. Recordings of MLAEP therefore offer the opportunity to monitor auditory information processing in the primary auditory cortex during general anesthesia.

Auditory Evoked Potential Recordings in Surgical Patients

After institutional approval and informed consent, 175 patients undergoing elective intra-abdominal, gynecological, urological, or cardiac surgery were studied. In all patients, AEP were recorded in the awake state and during general anesthesia.

AEP were recorded from vertex (positive) and mastoids on both sides (negative) against the forehead as common electrode. Auditory clicks were presented binaurally at 70 dB above the normal hearing level with a stimulation frequency of 9.3 Hz. Using the electrodiagnostic system Pathfinder I (Nicolet), 1000 successive stimulus responses were averaged over a 100-ms poststimulus period and analyzed off-line.

Latencies of the peaks V, N_a, P_a, N_b, and P_1 were measured. By fast Fourier transformation power spectra were calculated to analyze the energy of the different AEP frequencies. Figure 2 shows an original tracing of an AEP (left side) and its power spectrum (right side) of an awake patient. Peak V belongs to the brain stem-generated potentials, which demonstrates that auditory stimuli were correctly transduced. N_a, P_a, N_b, and P1 are generated in the primary auditory cortex of the temporal lobe. They are the electrophysiological correlate of the primary cortical processing of the auditory stimuli [10, 17, 25, 29, 31]. The MLAEP has a characteristic, periodic waveform, and the power spectrum has its maximal energy in the 30–40 Hz frequency range.

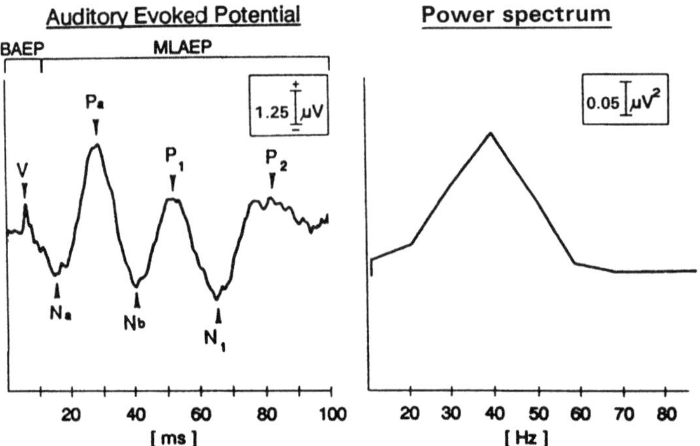

Fig. 2. Auditory evoked potential (AEP) of an awake patient. *BAEP*, brain stem AEP; *MLAEP*, midlatency AEP

Auditory Evoked Potentials and General Anesthetics

As the first step of our study, we investigated the effects of different anesthetic agents on MLAEP. The anesthetics can be roughly divided in two subgroups. The first group are general anesthetics such as the volatile anesthetics isoflurane and enflurane, the barbiturate thiopentone, and the intravenously administered anesthetics etomidate and propofol. These anesthetics act in a rather nonspecific way, i.e., they affect nearly every excitable biological membrane in the human brain. In contrast, receptor-specific agents such as the benzodiazepines midazolam, diazepam, and flunitrazepam, the opioid fentanyl, and the phencyclidine derivative ketamine interact specifically with certain receptors or defined brain regions. AEP were recorded before and during general anesthesia with these agents.

Isoflurane, Enflurane, Thiopentone, Etomidate, and Propofol

AEP during general anesthesia with isoflurane, enflurane, thiopentone, etomidate, and propofol are presented in Fig. 3. The upper part of each trace shows the control AEP of awake patients. The brain stem-generated peak V can be identified easily in each potential. The MLAEP of awake patients are characterized by high peak-to-peak amplitudes and periodic waveforms. During anesthesia with these agents, the peak V remains unchanged, whereas the MLAEP show marked increases of latencies and decreases of amplitudes or are even completely suppressed. This indicates a successful stimulus transmission up to the level of the brain stem and midbrain. However, stimulus processing in the

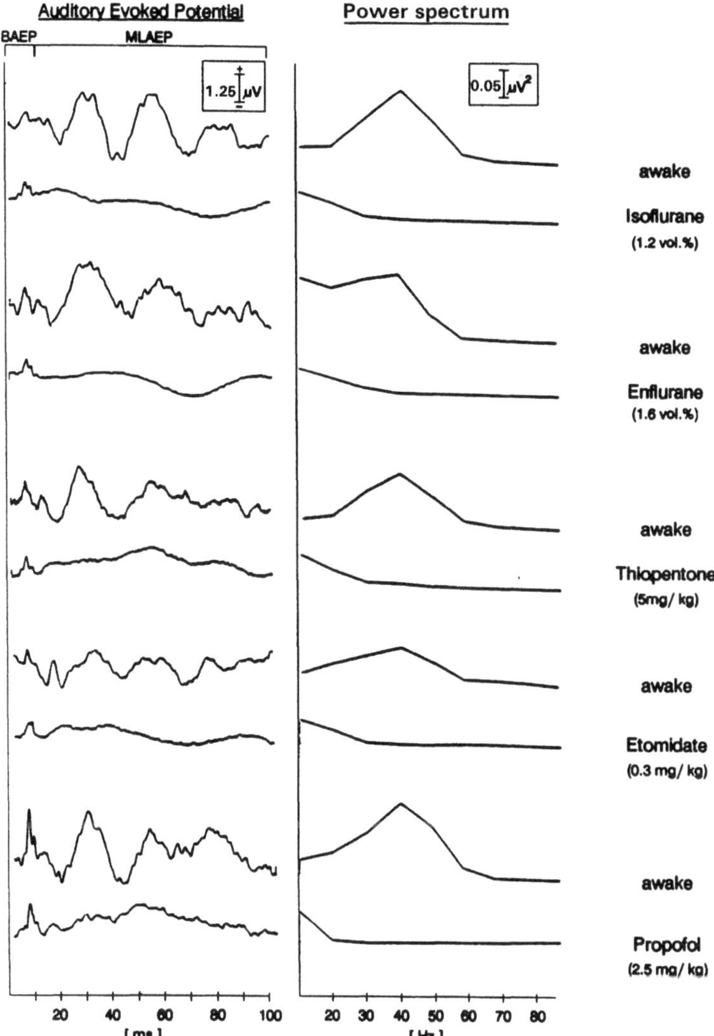

Fig. 3. Auditory evoked potentials (AEP) during general anesthesia with *isoflurane, enflurane, thiopentone, etomidate,* and *propofol. BAEP,* brain stem AEP; *MLAEP,* mid-latency AEP

primary auditory cortex is blocked. This result is also reflected in the power spectra, which show a significant decrease in the dominating 30–40 Hz activity to the low-frequency range.

To demonstrate that the effects of general anesthesia on MLAEP are not an "all or nothing" phenomenon, we recorded MLAEP under increasing concentrations of isoflurane. In Fig. 4, the upper trace shows the AEP of the awake

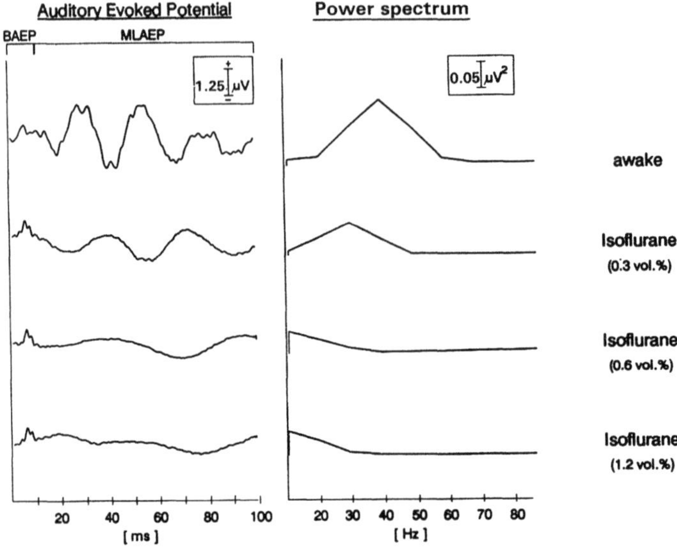

Fig. 4. Auditory evoked potentials (AEP) under increasing end-expiratory concentrations of isoflurane. *BAEP*, brain stem AEP; *MLAEP*, midlatency AEP

patients with its characteristic, periodic waveform. Under increasing end-expiratory concentrations of isoflurane, the BAEP do not change in latency or amplitude and can be recorded as in the awake state. In contrast, MLAEP show a dose-dependent increase in latencies and decrease in amplitudes. Under surgical anesthesia with 1.2 vol.% in the last tracing, MLAEP are nearly completely suppressed.

Midazolam, Flunitrazepam, Diazepam, Fentanyl, and Ketamine

A different picture can be seen when MLAEP were recorded during anesthesia with receptor-specific anesthetics. In Fig. 5, the top traces in each of the five sections show AEP of awake patients. The brain stem-generated potential V in particular can be easily identified in each recording. The MLAEP show high peak-to-peak amplitudes and periodic waveforms. During anesthesia with receptor-specific anesthetics, the brain stem peak V and the midlatency components remain nearly unchanged compared with AEP from awake patients. This indicates that auditory stimuli reach the primary auditory cortex and are processed at a primary cortical level.

To demonstrate that this effect does not depend on the administered doses of receptor-specific anesthetics, we recorded MLAEP under increasing doses of the opioid fentanyl. In Fig. 6, the top trace shows the characteristic AEP of awake patients; it is superimposed with high-frequency muscle artifacts. With

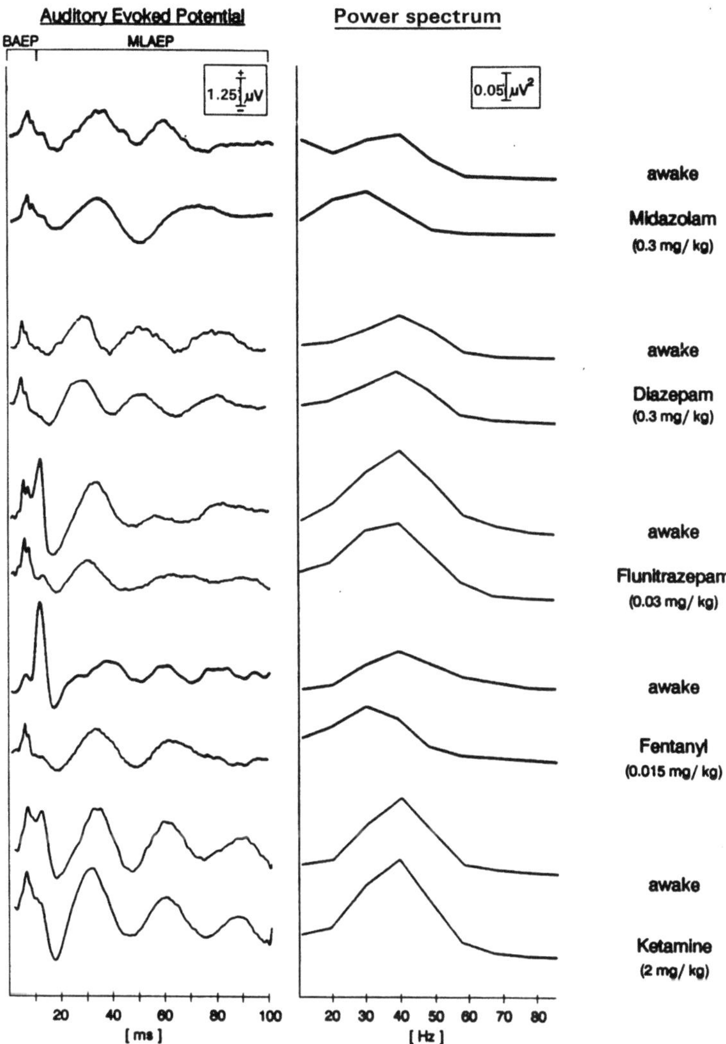

Fig. 5. Auditory evoked potentials (AEP) during general anesthesia with *midazolam,*
flunitrazepam, diazepam, fentanyl, and *ketamine. BAEP,* brain stem AEP; *MLAEP,*
midlatency AEP

increasing doses of fentanyl only a significant decrease in amplitudes for the late
component P_1 can be observed. This effect does not depend on the given
dosages and can be seen after the first fentanyl bolus injection. In contrast,
BAEP and the early cortical responses N_a, P_a, and N_b do not change significantly compared with the awake state. Primary cortical processing of auditory
stimuli seems to be preserved.

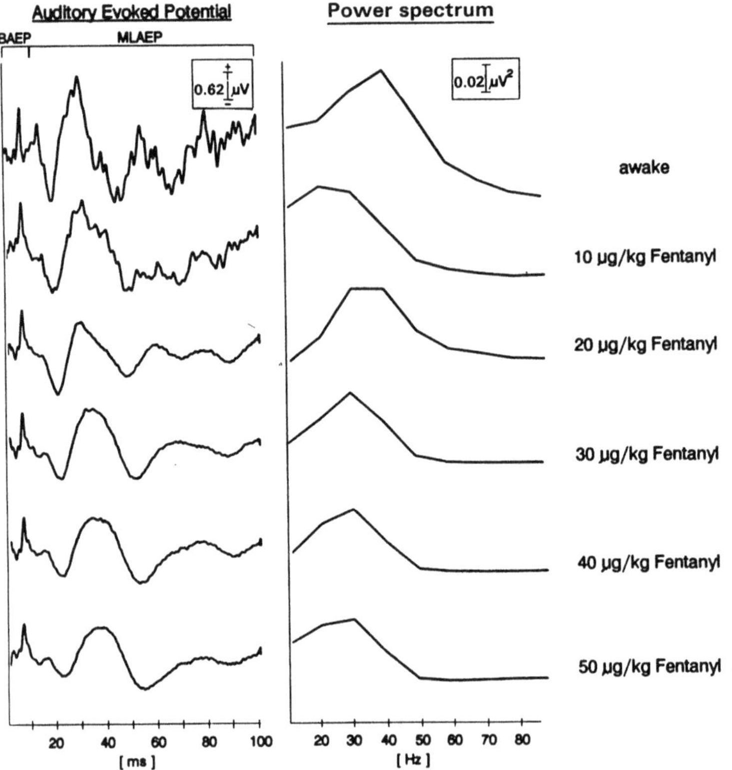

Fig. 6. Auditory evoked potentials (AEP) under increasing doses of fentanyl. *BAEP*, brain stem AEP; *MLAEP*, midlatency AEP

The main questions that arise from these data are: Is there any relation between AEP and intraoperative awareness? Does intact primary information processing – indicated by AEP – reflect the prerequisite for postoperative recall of intraoperative events? If so, a sensitive method to detect recalls of intraoperative events postoperatively has to be developed. A prior condition for the registration, retention, and retrieval of auditory information is an intact auditory stimulus processing and a partially intact memory system.

Explicit and Implicit Memory systems

A distinction that is often made in cognitive psychology is the one between explicit and implicit memory. Explicit memory is the deliberate active, and conscious recollection of an experience in time and space, e.g., of episodes in a human life. In contrast, implicit memory remembers passively and unconsciously in an associative or semantic context without being related to time and space, e.g., knowledge about language and the world [12, 18, 27].

Functional differences between explicit and implicit memory can be demonstrated by performing a memory task. If I ask you which was the main political event in February 1991, which newspapers and news broadcasts all over the world reported about, I am testing your explicit memory. You will remember what you read in the newspapers or watched on television in February 1991. It was of course, the Gulf War.

A different way of testing your memory would be to ask question: "What is the first word that comes into your mind in association with the word "gulf"?" It might, for example be a stream in the Atlantic ocean or a war. This question is testing implicit knowledge which will be remembered independently of an active, conscious, and explicit recollection of a biographic episode such as reading a newspaper or watching the news.

Another useful example to explain the difference between explicit and implicit memory is an experiment performed by Claparede 90 years ago [9]. He welcomed his amnesic patients holding a needle between his fingers. This was naturally a very unpleasant experience for his patients. The following day the patients did not remember ever having met Claparede before. Nevertheless, they refused to shake hands with him, stating that hands sometimes contain pins. This experiment clearly demonstrates the existence of the two distinct memory systems. Claparede's patients were able to recollect their previous experience implicitly without any explicit memory of the circumstances in which they had gained the information [28].

Explicit and Implicit Memory and Auditory Evoked Potentials During Cardiac Surgery

We tried to transfer these findings to general anesthesia. Recollection of intraoperative events most often occurs after cardiac surgery [13]. We therefore investigated postoperative memory in 45 patients undergoing elective cardiac surgery. After oral premedication with a benzodiazepine, the patients were randomly assigned to one of four groups. Anesthesia was induced in group 1 (10 patients) with flunitrazepam and fentanyl (0.01 mg/kg), in group 2 (10 patients) and group 3 (10 patients) with etomidate (0.25 mg/kg) and fentanyl (0.005 mg/kg). For maintenance of anesthesia, all patients received high-dose opioid analgesia using fentanyl (1.2 mg/h). Additionally, the patients in group 1 received the benzodiazepine flunitrazepam (1.2 mg/h), in group 2 isoflurane (0.6–1.2 vol.%), and in group 3 propofol (4–8 mg/kg per h). A total of 15 patients (group 4) served as a control, and patients were randomly anesthetized as in group 1, 2, or 3. After sternotomy, an audiotape was played to the patients in groups 1–3 which included an implicit memory task. As well as positive suggestions about the course of the operation, a short version of the story of

Robinson Crusoe was told and it was suggested to the patients that they should remember Robinson Crusoe when they were asked about what they associated with the word "Friday". Three to five days after the operation, the patients were asked about their recollections of intraoperative events and about what they associated with "Friday". All experimental evaluations were conducted under double-blind conditions, i.e., neither the patients nor the interviewer knew which anesthetic had been employed or whether an audiotape had been played or not. AEP were recorded in the awake state and during general anesthesia before and after the audiotape had been played.

In the postoperative interview, none of the patients had explicit memories of intraoperative events. Some patients stated hearing voices or noise, but these perceptions could not definitely be related to the intraoperative situation. Typical associations with the word "Friday" were "start of the weekend," "last working day of the week," or "fish." These statements were also given by patients who then spontaneously associated "Robinson Crusoe" with "Friday." Typical reproductions of the story of Robinson Crusoe were as follows: "When you say "Friday," I have to think of an island and the story of Robinson Crusoe, but I don't think this has anything to do with your question." Another typical answer was: "When you say "Friday," I remember that when I was a child we used to play on a little island. We called it 'Robinson Island'." It is interesting to note that all patients with implicit memory strongly denied that their associations had anything to do with information heard during anesthesia.

The incidence of implicit memories of the intraoperatively played story are presented in Fig 7. There were no implicit memories in patients belonging to the control group. One patient in the propofol group and one patient in the isoflurane group made the expected association between "Friday" and

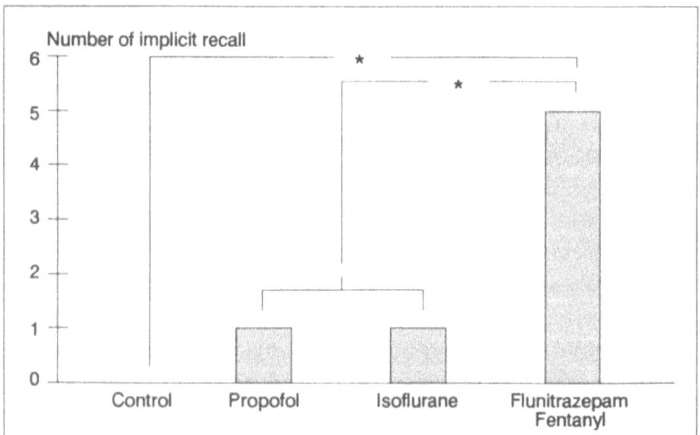

Fig. 7. Incidence of implicit memories of the intraoperatively presented story in the four groups (control, propofol, isoflurane, and flunitrazepam/fentanyl)

"Robinson Crusoe," whereas five out of ten patients in the flunitrazepam/fentanyl group remembered the keyword "Robinson Crusoe" when asked what they associated with the word "Friday."

There were marked differences between AEP recorded from patients showing implicit memories and patients who did not recall information from the story. Figure 8 shows the AEP of patients without implicit memory: in the upper part the propofol group, and in the lower part, the isoflurane group. The upper traces show the AEP of the awake patients. BAEP can be identified easily. MLAEP show high peak-to-peak amplitudes and a characteristic, periodic wave-form. During general anesthesia in these patients, BAEP were found to be similar to those in the awake state, whereas MLAEP show marked increases in latencies and decreases in amplitudes or are even completely suppressed. This

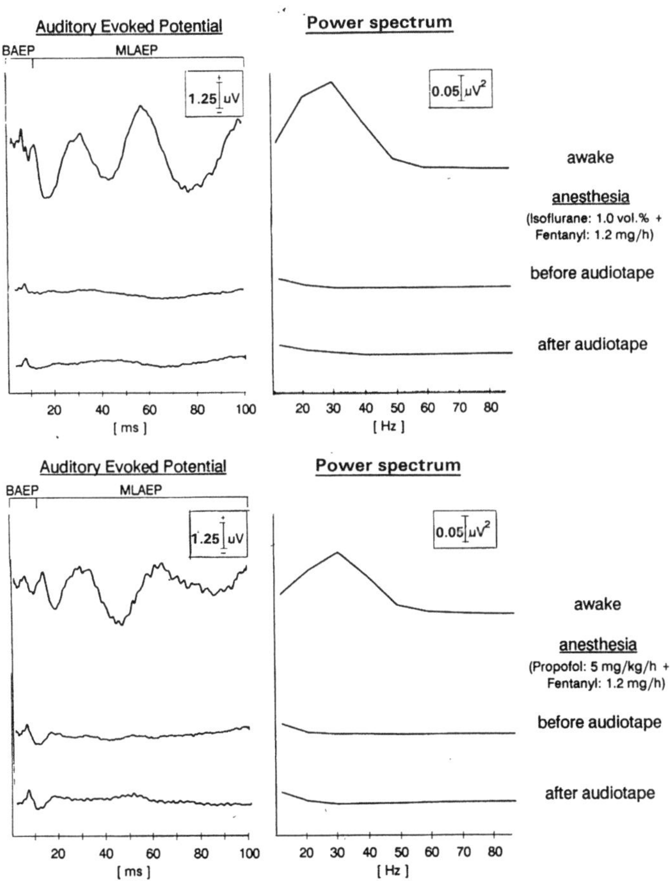

Fig. 8. Auditory evoked potentials (AEP) of patients of the isoiiurane (*top*) and propofol (*bottom*) group without implicit memory of the intraoperative imformation. *BAEP*, brain stem AEP; *MLAEP*, midlatency AEP

means that during general anesthesia in these patients, auditory stimuli are regularly transduced and processed upto brain stem or midbrain level, whereas the primary cortical processing of the auditory stimuli in the primary auditory cortex of the temporal lobe is blocked.

A different picture can be seen in the MLAEP of the patients in the flunitrazepam/fentanyl group, as presented in Fig. 9. These patients could remember Robinson Crusoe implicitly after the operation. The upper trace shows the AEP of the awake patients. They are characterized by high peak-to-peak amplitudes and a periodic waveform. During general anesthesia before

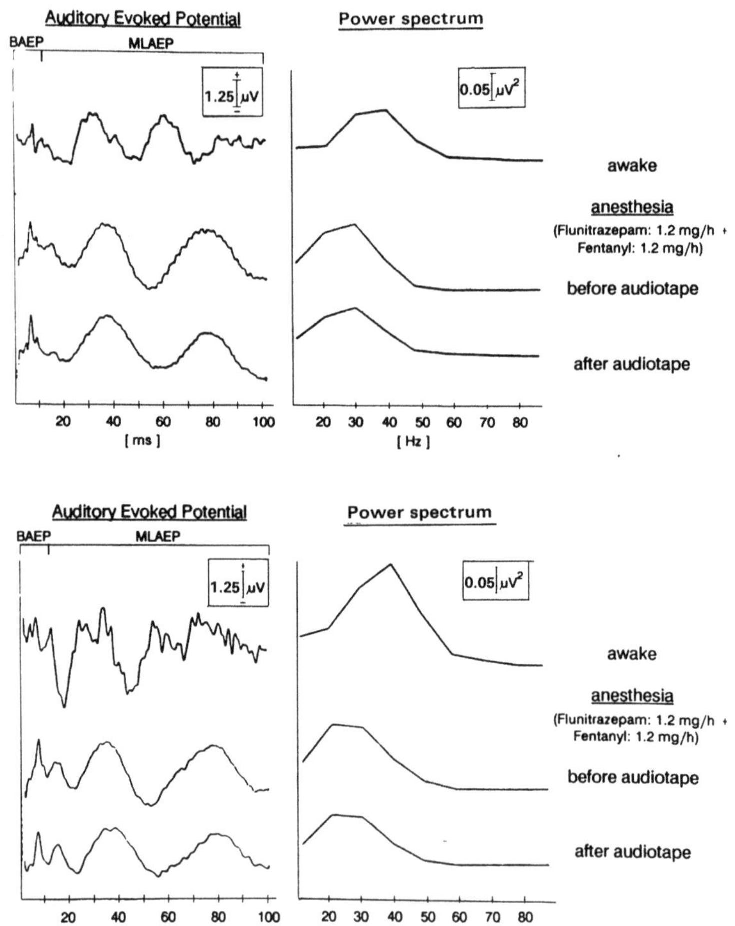

Fig. 9. Auditory evoked potentials (AEP) of patients of the flunitrazepam/fentanyl group with implicit memory of the intraoperative information. *BAEP*, brain stem AEP, *MLAEP*, midlatency AEP

and after the audiotape was played, MLAEP similar to those found in the awake state were recorded. The high amplitudes and the periodic waveform were preserved, which means that the electrophysiological conditions of primary cortical processing of auditory stimuli are at least partly preserved.

Conclusions

During general anesthesia with isoflurane, enflurane, thiopentone, etomidate, and propofol, MLAEP are suppressed in a dose-dependent way. This indicates that the primary sensory information processing in the brain is blocked at the level of the primary sensory cortex. In contrast, receptor-specific anesthetics such as midazolam, diazepam, flunitrazepam, fentanyl, and ketamine do not suppress MLAEP. This supports the assumption that under the influence of these agents, auditory information processing remains intact to some extent.

Our results demonstrate that auditory information can be processed during general anesthesia and remembered postoperatively by implicit memory function. We were able to find a close correlation between implicit memory and MLAEP. Implicit memory was observed when MLAEP and primary cortical processing of auditory stimuli were at least partly preserved. No implicit memory could be detected when MLAEP were suppressed.

MLAEP are widely preserved and implicit memory is observed more often when high-dose opioid analgesia is combined with receptor-specific agents such as benzodiazepines than under nonspecific anesthetics such as isoflurane or propofol. Nonspecific anesthetics seem to provide more effective suppression of auditory stimuli processing and unconscious perceptions during general anesthesia than receptor-specific agents.

References

1. Anon (1968) Is your anesthetized patient listening? JAMA 206:1004–1005
2. Bailey PL, Stanley TH (1990) Narcotic intravenous anesthetics. In: Miller RD (ed) Anesthesia. Churchill Livingstone, New York, pp 281–366
3. Bennett HL (1986) Response to intraoperative conversation. Br J Anaesth 58:134–135
4. Bennett HL, Davis HS, Giannini JA (1985) Non-verbal response to intraoperative conversation. Br J Anaesth 57:174–179
5. Breckenridge JL, Aitkenhead AR (1983) Awareness during anaesthesia: a review. Ann R Coll Surg 65:93–96
6. Cheek DB (1962) The anesthetized patient can hear and can remember. Am J Proctol 13:287–290

7. Cheek DB (1964) Further evidence of persistence of hearing under chemo-anesthesia: detailed case report. Am J Clin Hypn 7:55–59
8. Cheek DB (1980) Awareness of meaningful sounds under general anesthesia: considerations and a review of the literature 1959–1979. In: Wain HJ (ed) Theoretical and clinical aspects of hypnosis. Symposia Specialists, Miami, pp 87–106
9. Claparede E (1911) Recognition et moiite. Arch Psychol 11:79–90 (Reprinted as: Recogniton and 'me-ness'. In: Rappaport D (ed) Organizations and pathology of thought. Columbia University Press, New York, 1951).
10. Deiber MP, Ibanez V, Fischer C, Perrin F, Maugiere F (1988) Sequential mapping favours the hypothesis of distinct generators for Na and Pa middle latency auditory evoked potentials. Electroencephalogr Clin Neurophysiol 71:187–197
11. Goldmann L (1988) Information processing under general anaesthesia: a review. J R Soc Med 81:224–227
12. Goldmann L, Ogg TW, Levey AB (1988) Hypnosis and daycase anaesthesia. A study to reduce pre-operative anxiety and intra-operative anaesthetic requirements. Anesthesia 43:466–469
13. Goldmann L, Shah MV, Hebden MW (1987) Memory and cardiac anaesthesia. Anesthesia 42:596–603
14. Jones JG (1986) Hearing and memory in anaesthetised patients. Br Med J 292:1291–1293
15. Jones JG (1988) Awareness under anaesthesia. Anaesthesia Rounds 21:1–28
16. Jones JG (1989) Depth of anaesthesia and awareness. In: Nunn JF, Utting JE, Brown BR Jr (eds) General anaesthesia. Butterworths, London, pp 419–427
17. Kaga K, Hink RF, Shinoda Y, Suzuki J (1980) Evidence for a primary cortical origin of a middle latency auditory evoked potential in cats. Electroencephalogr Clin Neurophysiol 50:254–266
18. Kihlstrom JF (1987) The cognitive unconscious. Science 237:1445–1452
19. Kutas M (1988) Review of event-related potential studies of memory. In: Gazzaniga MS (ed) Perspectives in memory research. MIT Press, Cambridge, pp 181–217
20. Levinson BW (1965) States of awareness during general anaesthesia. Br J Anaesth 37:544–546
21. Madler C, Keller I, Schwender D, Pöppel E (1991) Sensory information processing during general anaesthesia: effect of isoflurane on auditory evoked neuronal oscillations. Br J Anaesth 66:81–87
22. Näätänen R, Picton TW (1987) The N1 wave of the human electric and magnetic response to sounds: a review and analysis of the component structure. Psychophysiology 25(4):375–425
23. Netter FH (1987) In: Krämer G (ed) Farbatlanten der Medizin, vol 5: Nervensystem I. Thieme, Stuttgart, pp 177–178
24. Picton TW, Hillyard SA (1974) Human auditory evoked potentials. II. Effects of attention. Electroencephalogr Clin Neurophysiol 36:191–199
25. Picton TW, Hillyard SA, Krausz HI, Galambos R (1974) Human auditory evoked potentials. I. Evaluation of components. Electroencephalogr Clin Neurophysiol 36:179–190
26. Pockberger H, Rappelsberger P, Petsche H (1988) Cognitive processing in the EEG. In: Basar E (ed) Dynamics of sensory and cognitive processing by the brain. Springer, Berlin Heidelberg New York, pp 266–274
27. Schacter DL (1987) Implicit memory: history and current status. J Exp Psychol 13:501–518

28. Schacter DL, McAndrews MP, Moscovitch M (1988) Access to consciousness: Dissociations between implicit and explicit knowledge in neuropsychological syndromes. In: Weiskrantz L (ed) Thought without language. Oxford University Press, Oxford, pp 242–278
29. Scherg M, Volk SA (1983) Frequency specificity of simultaneously recorded early and middle latency auditory evoked potentials. Electroencephalogr Clin Neurophysiol 56:443–452
30. Thornton C, Heneghan CP, James MFM, Jones JG (1984) Effects of halothane or enflurane with controlled ventilation on auditory evoked potentials. Br J Anaesth 56:315–323
31. Woods DL, Clayworth CC, Simpson GV, Naeser MA (1987) Generators of middle- and long-latency auditory evoked potentials: implications from studies of patients with bitemporal lesions. Electroencephalogr Clin Neurophysiol 68:132–148

Evoked Potential Monitoring for Vascular Surgery

M. Dinkel, H. Lörler, H. Langer, H. Schweiger, and E. Rügheimer

Introduction

Reconstructive surgery on vessels supplying the brain and spinal cord frequently requires temporary interruption of blood flow, which entails cerebral hypoperfusion of a varying degree. The consequences of temporary cross-clamping range from almost unchanged to fully abolished perfusion to the corresponding region as blood flow is contingent on the availability and sufficiency of collateral circulation. Moreover, due to the minimal ischemic tolerance of the central nervous system (CNS), neurological deficits are dreaded complications in vascular surgery [15, 17, 41].

On average, 2%–3% of all patients undergoing carotid endarterectomy sustain a stroke in the perioperative period. Following thoracoabdominal aortic surgery, 0.2%–24% of patients suffer paraplegia or paraparesis.

Etiology of Spinal Cord Ischemia During Aortic Surgery

In addition to the etiology and extent of the underlying disease, the incidence of paraplegia following aortic surgery is mainly dependent on the level and duration of aortic cross-clamping during the operation [17, 31, 41]. Due to the lack of collateral vessels, patients with acute dissecting or ruptured aneurysms show a higher rate of complications than those with well-collaterated aortic coarctation: the more extensive the aortic aneurysm and the greater the number of intercostal and lumbar arteries excluded from aortic blood supply by the graft, the higher the probability of severe spinal cord ischemia. Most at risk are patients who experience complete interruption of blood flow via the great radicular artery (artery of Adamkiewicz), as this vessel is critical in maintaining perfusion to the motor pathways of the terminal spinal cord which are badly collaterated and thus particularly prone to ischemic damage. In 75% of people, the artery of Adamkiewicz originates between T9 and T12. This explains why the risk of spinal hypoperfusion is particularly high during cross-clamping of the thoracic part of the aorta. Last but not least, the incidence of irreversible paralysis increases overproportionally if the clamping period exceeds 30 min.

Prevention of Ischemic Spinal Cord Lesions

As preoperative evaluation of spinal collateral circulation and of the significance of different aortic segmental arteries for spinal cord perfusion is not possible, a variety of measures are taken to prevent ischemic spinal cord damage, especially if a longer clamping period is expected and if aortic cross-clamping at a level higher than T12 will be necessary (Table 1) [28, 36, 44]. This approach is, however, in most cases not only superfluous and dangerous, but also not very effective, as even these measures cannot prevent devastating neurological deficits. In order to avoid ischemic complications, early intraoperative detection of critical spinal hypoperfusion, purposeful shunt placement, drainage of spinal fluid, reimplantation of segmental arteries, and repeated checking of the efficacy of these measures in restoring and maintaining adequate spinal perfusion seem more important [17, 18, 31, 41].

Spinal Cord Monitoring

Tibial Nerve Somatosensory Evoked Potentials

For this purpose, somatosensory evoked potentials (SEP) generated by uni- or bilateral peripheral tibial nerve stimulation have been used for several years to monitor the functional integrity of the spinal cord during thoracoabdominal

Table 1. Methods of preventing paraplegia during aortic surgery

---

Shunt and bypass techniques
-Heparin-coated shunts
-Left heart bypass (with a pump)
-Cardiopulmonary bypass (with hypothermia)

Enhancement of the spinal perfusion pressure
-Papaverine intravenous or intrathecal
-Cerebrospinal fluid drainage

Pharmacologic agents
-Prostaglandin El
-Steroids
-Naloxone
-Barbiturates
-Calcium channel blockers

---

aortic surgery. The principle of this monitoring technique is based on the expectation that critical spinal cord ischemia after aortic cross-clamping adversely affects spinal impulse propagation, causing prolonged latencies and reduced amplitudes of the cervical or cortical potentials. Apparently, according to the experience of most investigators, critical hypoperfusion of the spinal cord, sufficient to cause paraplegia, is reliably detected by loss of the cortical SEP response, as the low rate of false negative findings indicates (Table 2). The high rate of false positive findings, however, implies that SEP recording is a very unspecific monitor. In other words, a lot of patients do not demonstrate a neurological deficit in the postoperative period even when intraoperative loss of SEP occurred [7, 8, 11, 14, 29, 32, 33].

This is in accordance with our experience, as the following example of a 54-year-old patient who underwent elective thoracoabdominal aortic aneurysm repair shows (Fig. 1). After aortic cross-clamping, we observed a protracted decrease in the SEP amplitude resulting in complete flattening of the cortical SEP waveform after 15 min. The potential was absent for the remainder of the clamping period, but it recovered quite rapidly after declamping. This patient was neurologically intact in the postoperative period, although no measures had been taken to maintain spinal cord perfusion.

This case reflects typical findings in thoracoabdominal aortic surgery: on the one hand, neurological deficits are rarely encountered if the SEP loss lasts less than 30 min and, on the other hand, almost all patients not receiving a shunt for maintenance of distal aortic perfusion experience SEP loss after aortic cross-clamping. This is not necessarily a sign of critical spinal cord ischemia, but may instead be attributable to disturbed peripheral impulse propagation due to hypoperfusion, hypothermia, or anesthesia. The resultant low specificity is the limiting factor of the usefulness of this technique in establishing the necessity of preventive measures [8, 11, 20].

Table 2. Assessing spinal cord ischemia with the use of peripheral evoked potentials

Author	Year	Patients	False negative results	False positive results	
		(n)	(n)	(n)	%
Cunningham et al. [8]	1987	33	0	11	33.9
Crawford et al. [7]	1988	99	3	41	41.4
Fava et al. [14]	1988	16	0	15	93.8
Maeda et al. [32]	1989	19	0	2	10.5
McNulty et al. [33]	1991	3	0	1	33.3
Drenger et al. [11]	1992	18	0	8	44.4
Own results	Unpublished	13	0	3	23.1

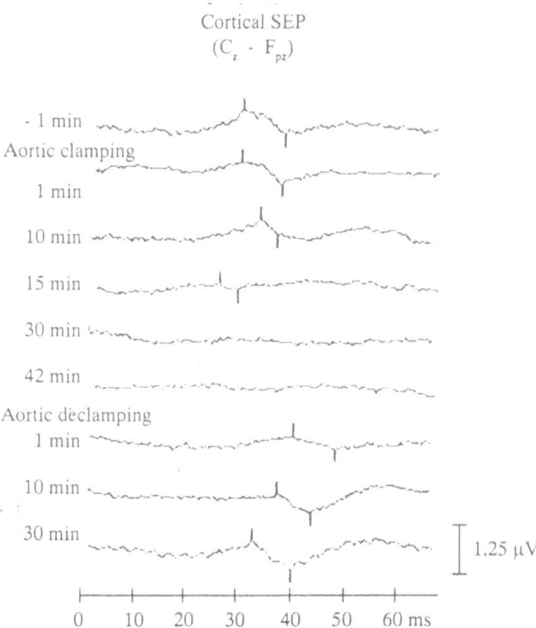

Fig. 1. Tibialis somatosensory evoked potential (*SEP*) tracings of a 54-year-old man during elective thoracoabdominal aneurysmectomy without use of protective measures. The cortical SEP flattened during the aortic occlusion and reappeared after declamping. Postoperative neurologic examination was negative

Direct Spinal Cord Stimulation

This disadvantage of tibial nerve SEP monitoring can be overcome by direct stimulation of the spinal cord via an epidurally placed electrode. As a matter of fact, by using simultaneous stimulation of the tibial nerve and spinal cord, Drenger et al. [11], in their study of 18 thoracoabdominal aneurysmectomies, were able to show that only those two patients suffering postoperative neurological deficits also experienced cortical electrical silence after spinal cord stimulation, whereas a further eight patients showed electrical silence after tibial nerve stimulation (Table 2). In spite of these convincing results, which have been corroborated by other investigators, and despite the fact that under anesthetic conditions spinal SEP are more stable and of a higher amplitude than tibial nerve SEP, there are some reservations about this method for several reasons [8, 19, 26]. Firstly, placing the spinal stimulating electrode is often difficult and the use of two epidural catheters in patients under anticoagulant medication is usually contraindicated. Secondly, by spinal stimulation only the sensory pathways in the dorsal column can be monitored directly, whereas only indirect assessment of the integrity of motor pathways, which are especially prone to ischemic damage, is possible. For this reason, false negative findings have to be expected even after spinal stimulation.

Motor Evoked Response

Taking into consideration the pathophysiological background, an improvement in spinal neurological monitoring in vascular surgery is to be expected by recording the spinal neuronal response or the peripheral muscular response after transcranial stimulation of the motor cortex. These motor evoked responses (MER) facilitate the control of motor pathways particularly prone to ischemia. Animal studies, however, revealed conflicting results as to the sensitivity and specificity of MER [12, 17, 30]. Moreover, MER are difficult to record under general anesthesia, and methodological problems, such as the issue of magnetic or electrical stimulation, single or double stimulation and simultaneous sensory facilitation have not been fully solved [23, 24, 26]. Therefore, there has only been minimal experience with this monitoring technique, and further studies are necessary to elucidate the utility and limitations of motor evoked potentials.

Conclusion

At present, not only the question of the appropriate monitoring technique, but also the question of the significance of spinal monitoring in vascular surgery remains unanswered. Whether recording of evoked potentials can reduce the overall rate of ischemic spinal cord complications is difficult to establish as a lot of factors influence the outcome. A lower rate of complications is, however, indicated if neurophysiological monitoring is performed [18]. Therefore, at least during particularly riskful thoracoabdominal procedures, accompanying neuro-monitoring is required.

Problems and Pathogenesis of Cerebral Ischemia During Supra-aortic Surgery

Whereas there is generally no doubt about the indication for aortic surgery because of the threat to the patient's life, the indication for supra-aortic procedures is, due to the risk of irreversible neurological deficits, still disputed 40 years after the first carotid endarterectomy was performed.

One reason to generally question carotid surgery is the fact that there is so far no convincing evidence to prove the advantage of surgery over conservative therapy. In 1991, however, two prospective comparative studies had to be discontinued for ethical reasons because it became obvious that the incidence of strokes in patients with symptomatic, severe carotid stenosis can be lowered significantly by performing carotid endarterectomy as opposed to therapy solely with platelet aggregation inhibitors [13, 35]

These studies also show that patients will particularly benefit from surgery if the initial perioperative mortality and morbidity is as low as possible. In order to reduce the rate of perioperative strokes, which on average is 2%–3%, we have to know and forestall the essential pathomechanisms in perioperative strokes [15].

In addition to periods of postoperative hyperperfusion, which can effectively be prevented by strict antihypertensive treatment, neurological deficits are caused by embolization of arteriosclerotic and thrombotic debris. This explains why, especially in the case of ulcerative carotid lesions, one should refrain from all measures which could cause the dislodgement of emboli. With regard to the increased risk of embolization, unnecessary manipulation of the carotid artery and routine shunt placement should be avoided. A further pathomechanism involved in the development of perioperative neurological deficits is cerebral hypoperfusion caused by insufficient collateral blood flow after carotid cross-clamping.

Cerebroprotective Measures

To avoid ischemic damage during the clamping period, several cerebroprotective measures are recommended, such as pharmacologic agents, hyperbaric oxygenation, a short clamping period, hypothermia, general anesthesia, induced hypertension, and shunt placement. Most of these, however, have proved inadequate in the clinical setting because of low efficacy and severe side effects.

Surgery under general anesthesia generally produces better effects than operations under local anesthesia: on the one hand, cerebral oxygen consumption is reduced and on the other hand, oxygenation can be improved in critical situations if the need arises.

During the clamping period, blood pressure is frequently raised 10%–20% above ward level, although better perfusion to ischemic areas of the brain by means of induced hypertension has never been substantiated. Moreover, with this technique, a higher risk of myocardial ischemia is caused by increased myocardial oxygen consumption as the coronary reserve is often already reduced.

The most effective means of maintaining sufficient perfusion to the ipsilateral hemisphere is the use of a shunt. Placement of a shunt, however, not only increases the direct risk of cerebral embolization; intimal damage also favors formation of adhesive thrombi and recurrent stenoses. The higher rate of embolization due to shunt placement elucidates the reason why extensive statistical analysis did not reveal any difference in neurological outcome between centers never using a shunt and others who routinely place shunts [15].

Cerebral Neuromonitoring

Significance

A further reduction of neurological complications is thus only to be expected if one succeeds in taking these potentially hazardous cerebroprotective measures (e. g. shunt placement) when they are really necessary. For this purpose, objective criteria for the detection of critical cerebral hypoperfusion by means of suitable neuromonitoring are vital.

Requirements

CNS monitoring will only contribute to the guidance of the surgical and anesthetic procedures and to the improvement of neurological outcome in carotid surgery if it meets essential requirements. The most important criterion is high sensitivity. A suitable monitoring technique must identify all patients who will not tolerate carotid cross-clamping because of insufficient collateral circulation and who will thus benefit from cerebroprotective measures. For routine monitoring further requirements must be met, such as high specificity, ease of application, continuous monitoring, reliable interpretation of measurements, low rate of technical failure, no monitor-induced hazards.

Cerebral Monitoring with Somatosensory Evoked Potentials

Fundamental Principles

SEP seemed a particularly useful monitor during carotid surgery. Following contralateral median nerve stimulation, electrical activity is evoked in the postcentral region on the side of surgery. Normal functioning of this cerebral region is put particularly at risk after carotid cross-clamping, as it is contingent on sufficient blood flow via the middle cerebral artery. Branston and Symon [5] have been able to show in animal studies that a reduction in cerebral blood flow to below 16 ml/100 g per min causes an increasing reduction of amplitudes and that blood flow below 12 ml/100 g per min causes loss of cortical SEP amplitudes.

Validation

With these facts and the clinical situation in mind, we assessed the suitability of SEP with regard to the requirements of appropriate clinical neuromonitoring in a prospective study of 665 patients undergoing 753 carotid operations.

In order to be able to unequivocally verify the validity of SEP monitoring in carotid surgery and to directly compare SEP findings with postoperative neurological outcome, SEP were recorded in the first phase of the study without information being given to the surgeon about SEP changes and without shunt placement. All patients were extubated immediately after surgery to facilitate detection even of short-lived neurological deficits. During this validation period, it was shown that critical clamp-related cerebral ischemia is reliably indicated and easily recognized by loss of the cortical SEP waveform. In the further course of the study, we therefore checked whether the selective use of cerebroprotective measures can prevent neurological deficits once SEP loss has occurred.

Results

In our data of 753 carotid operations, about 7% of patients revealed insufficient collateral blood flow after carotid cross-clamping (Table 3). All these patients were identified by complete loss of the cortical evoked response. Apart from the patients with SEP loss, only five patients without intraoperative SEP changes showed neurological deficits in the postoperative period. However, as could be proved by surgical revision, these were not caused by clamp-related ischemia, but by perioperative thromboembolic events. Early and sufficient shunt placement after the occurrence of SEP loss can effectively prevent neurological deficits (Table 3). Only three of 12 patients without shunt placement remained neurologically intact after the SEP loss, compared to 28 of 38 patients with a bypass shunt. Two patients suffered a stroke despite the use of a shunt. In one of them, the cortical evoked response did not recover, owing to inadequate shunt blood flow. The other patient had thrombotic occlusion of the carotid artery postoperatively.

Sensitivity and Specificity

In contrast to their positive influence on cerebral hypoperfusion, cerebroprotective measures do not prevent, but rather increase, neurological sequelae due to

Table 3. Somatosensory evoked potential (SEP) findings and neurologic outcome of 753 patients undergoing carotid endarterectomy

SEP intraoperatively	Shunt placement	neurologic outcome		
		Unchanged	Stroke	TIA
Lost ($n = 50$)	Without shunt ($n = 12$)	3	1	8
	With shunt ($n = 38$)	28	2	8
Identifiable ($n = 703$)	Without shunt ($n = 703$)	698	3	2

TIA, transient ischemic attack.

Table 4. Somatosensory evoked potential monitoring during carotid surgery: sensitivity and specificity

Author	Year	Patients (n)	Sensitivity %	Specificity %
Ruß et al. [48]	1985	106	83	99
Gigli et al. [16]	1987	40	100	83
Amantini et al. [1]	1987	58	100	98
Lam et al. [25]	1991	64	100	94
Dinkel et al. [9]	1991	482	100	99

embolization. For this reason, the crucial test criterion for a suitable monitor in carotid surgery is fail-safe detection of clinically relevant clamp-related hypoperfusion. As illustrated by our own results and those of other teams, recording of SEP reliably indicates critical regional hypoperfusion during the clamping period (Table 4) [1, 9, 16, 25, 40].

Somatosensory Evoked Potentials in Comparison with Other Central Nervous System Monitors

Electroencephalogram

As shown above, SEP are not only a reliable monitor regarding the indication of the need for cerebroprotective measures, but they can also be used as a method of reference against which other CNS monitors can be tested.

From this perspective, a comparison with electroencephalogram (EEG) monitoring seems reasonable (Table 5), as 16-channel raw EEG recordings have been the gold standard of neuromonitoring in carotid surgery because of their high sensitivity with regard to cerebral hypoperfusion and the possibility of detecting even circumscribed regional ischemia in any part of the cortex [15, 34, 43]. However, conventional EEG recording in the operating theatre is technically demanding and in up to 40% of operations it cannot be utilized to assess cerebral well-being because of superimposed artifacts or other interferences [2]. Moreover, interpretation of conventional EEG recordings requires an experienced neurophysiologist. Considering the large amount of data generated by 16-channel raw EEG recording, even these experts sometimes find it impossible to detect critical ischemic changes early and to reliably distinguish between such changes and those due to anesthetic or other factors.

However, it is not only difficult to assign certain EEG changes to possible causes; it also seems impossible to quantify the extent of cerebral ischemia by means of EEG changes and to unequivocally detect truly critical hypoperfusion

[4, 22]. It is true that cerebral blood flow below 20 ml/100 g per min causes slowing of the EEG, but, on the other hand, this perfusion is way above the critical level of 6 ml/100 g per min, at which interference with cellular metabolism takes place [34, 43]. This accounts for the relatively high percentage of shunt use (20%–30%) in operations with EEG monitoring [15].

So far, only a few studies comparing EEG and SEP monitoring in carotid surgery, which yielded similar results for both monitoring techniques, are available [15, 25, 26]. In his study of 64 carotid endarterectomies, Lam [25] observed intraoperative SEP changes in both patients with postoperative neurological deficits, whereas only one of them exhibited significant changes in the conventional 16-channel EEG. This is in accordance with our own experience. Using quantitative EEG analysis (CATEEM Medisyst) during 26 carotid procedures, we were able to detect typical and, because of unequivocal SEP loss, foreseeable EEG changes during the operation in only two cases. In a further patient, postoperative off-line EEG analysis was necessary, which yielded isolated reduction in alpha activity as the correlate of the SEP loss observed during the operation. Even if the number of comparative studies is too small to substantiate the superiority of one electrophysiological monitor over another, there are crucial advantages of the SEP over the EEG, including high sensitivity and specificity, insignificant interference of anesthesia, low rate of failure, and existence of easily interpretable and reliable criteria for ischemia (total loss of the SEP amplitude) [37].

Carotid Stump Pressure

Internal carotid artery stump pressure (CSP) is a hemodynamic parameter frequently used to assess sufficient collateral blood flow and impending ischemia and is defined as the remaining pressure in the distal segment of the carotid artery produced by retrograde collateral blood flow after carotid cross-clamping. Several studies describe a higher incidence of ischemic EEG changes and compromised cerebral blood flow at decreased CSP values [6, 10]

In our study of 125 carotid endarterectomies, it became obvious that, as an indication of compromised cerebral perfusion, CSP levels of patients with a loss of cortical SEP after carotid cross-clamping are on average only half as high as those of patients with an intact SEP response (Table 6). No pathological SEP recording was seen in patients with stump pressures greater than 50 mm Hg. However, this limit, which was hypothesized in various studies, is only a very unspecific parameter. This is because 61 out of 125 of our patients had CSP values of less than 50 mm Hg without experiencing changes in the SEP or neurological deficits. The low specificity of CSP monitoring, which is verified in a lot of studies, can be explained by the fact that CSP, which is measured at the skull base, is dependent on a variety of factors, such as altered autoregulation, intracerebral arteriosclerotic vascular lesions, cerebral metabolic rate, and choice of anesthetic technique. Therefore, CSP measurement does not allow

Table 6. Carotid stump pressure (CSP) compared to somatosensory evoked potential findings after carotid artery cross-clamping ($n = 125$)

CSP	SEP	
	Complete loss	Identifiable
< 50 mm Hg	12	$n = 61$
> 50 mm Hg	0	$n = 52$
Mean	25.8 mm Hg	49.8 mm Hg
Range	15–41 mm Hg	17–109 mm Hg

conclusions about perfusion to the brain area supplied by the clamped artery [10]. Furthermore, for surgical reasons only, discontinuous registration of CSP is feasible. Up to 30% of cerebral ischemic events, however, are seen during the course of cross-clamping and may hence not be detected by CSP measurements taken immediately after cross-clamping.

The most important reason why CSP monitoring should no longer be used as a criterion for selective shunting and induced hypertension, however, is that it subjects patients to the hazards of unnecessary cerebroprotection, because it cannot reliably identify patients not at risk from critical ischemia.

Transcranial Doppler Sonography

A further hemodynamic parameter helpful in detecting cerebral hypoperfusion is blood flow velocity in the ipsilateral middle cerebral artery, which can be measured continuously by means of transcranial Doppler sonography (TCD). Early studies aiming at its validation indicate that blood flow velocity in the middle cerebral artery correlates well with CSP, EEG, and SEP changes and, under certain conditions, even with regional cerebral blood flow [3, 21, 38, 42, 45]. At the moment there is, however, no general accord about the extent of reduction in mean blood velocity which is tolerated without ischemic deficit after carotid cross-clamping. The reason for this is that mean blood flow velocity is compared with different monitors in various studies. Halsey, for example, by comparing it with EEG monitoring, found that mean blood flow velocity of 15 cm/s was a critical lower limit [21, 38].

We were able to perform simultaneous SEP and TCD monitoring in 48 patients (Table 7). Of these, only those four patients with completely erased Doppler signals demonstrated SEP loss.

Although TCD apparently permits the reliable distinction between patients with and without sufficient collateral circulation, there are limitations to its usefulness as an intraoperative monitor. In more than 30% of patients, technically satisfactory recordings cannot be obtained.

Table 7. Mean blood flow velocity of the middle cerebral artery (v-mean MCA) and Somatosensory evoked potential (SEP) findings during 79 carotid procedures

v-mean MCA	SEP	
	Complete loss	Identifiable
Not feasible	$n = 6$	$n = 25$
Feasible	$n = 5$	$n = 44$
Median	0 cm/s	26 cm/s
Range	0–0 cm/s	10–56 cm/s

Jugular Venous Oxygen Saturation

The possibility of continuous registration of jugular venous oxygen saturation by means of fiberoptic catheters breathed new life into cerebrovenous oximetry. Basically, measurement of jugular venous oxygen saturation is a suitable monitor of cerebral ischemia, as cerebral hypoperfusion can decrease jugular venous oxygen saturation owing to increased cerebral oxygen extraction [39]. However, this relationship is apparently only true for global cerebral ischemia. We took jugular venous blood samples during cross-clamping in 24 carotid operations and found no difference in oxygen saturation between patients with and without loss of SEP. In agreement with Larson, we arrived at the conclusion that jugular venous oxygen saturation is an unsuitable monitor of regional cerebral ischemia in carotid surgery [27].

Regional Cerebral Blood Flow

Among the other cerebral monitors in carotid surgery, direct measurement of regional cerebral blood flow by means of analysis of the washout pattern of intra-arterially or intravenously injected radioactive substances is the most exact method for determination of the extent of cerebral hypoperfusion, which explains its predominant role in the clarification of scientific issues. As this type of monitoring is very demanding and not continuous, it is not a suitable routine monitor in the clinical setting.

Neurological Examination

Surgery with local anesthesia allows the assessment of multiple neurological functions, which facilitates easy detection of cerebral ischemia after carotid cross-clamping without the need for technical equipment. Operations with local anesthesia, however, preclude the use of possible cerebroprotective effects of general anesthesia; in addition, the patient may not tolerate surgery without a significant amount of sedatives. When the patient is sedated, it is no longer possible to reliably ascertain whether changes in the neurostatus are drug-related or due to ischemia. The specificity of this monitor is thus restricted.

Advantages of Somatosensory Evoked Potential Monitoring

Among the different monitors, SEP monitoring holds an outstanding position for a variety of reasons, i.e., its easy application, unambiguous interpretation, resistance to interference, continuous character, and its high sensitivity and specificity (Table 5).

The high sensitivity facilitates the selective use of cerebroprotection in patients who really profit from it. Early shunt placement after loss of SEP can effectively prevent neurological deficits, as is indicated by recovering potentials which, at the same time, can be used to ensure proper functioning of the shunt (Fig. 2). Due to the high specificity of SEP monitoring, side effects of unnecessary cerebral protection, i.e. increased cardiac workload following induced hypertension and cerebral embolization caused by shunt placement, can a priori be prevented in most patients. As long as the cortical evoked response remains clearly identifiable, anesthetic management can be adjusted to the needs of reduced oxygen consumption without compromising cerebral integrity. This conflict of aims in carotid surgery, concisely described by Wade as protection of heart versus protection of brain, can, in most cases, be solved in favor of protection of heart and brain with the use of SEP monitoring [46].

An important advantage of SEP monitoring is that stable SEP amplitudes allow the surgeon to perform carotid desobliteration without undue haste. This may contribute to reduced technical complications and improved long-term outcome of carotid surgery.

Monitoring of SEP can even unmask hidden pathophysiological mechanisms. A neurological deficit without pathological SEP findings in the postoperative period, for example, points towards a thromboembolic origin. Thus, the need for surgical revision can be more easily established and the burden of additional diagnostic examinations is no longer necessary.

Conclusion

Whereas the usefulness of neuromonitoring during aortic surgery is still the object of controversial discussion, the profit derived from cerebral observation

Table 5. Cerebral monitoring during carotid surgery

Monitor	Parameter	Application	Continuous	Interpretation	Sensitivity	Specificity
EEG	Spontaneous electrical activity	Requires extensive training	Yes	Requires experience	High	Mediocre
Stump pressure	Collateral pressure	Easy	No	Easy, misleading	High	Low
TCD	Blood flow velocity	Prone to failure	Yes	Easy, misleading	Low	Mediocre
SjO_2	O_2 saturation	Easy	Yes	Easy	Low	Low
rCBF	Regional blood flow	Requires training	No	Difficult, safe	High	High
Awake patient	Neurologic examination	Easy	Yes	Easy, safe	High	High
SEP	Evoked electrical activity	Easy	Yes	Easy, safe	High	High

EEG, electroencephalogram; TCD, transcranial Doppler ultrasonography; SjO_2, jugular venous oxygen saturation; rCBF, regional cerebral blood flow; SEP, somatosensory evoked potentials.

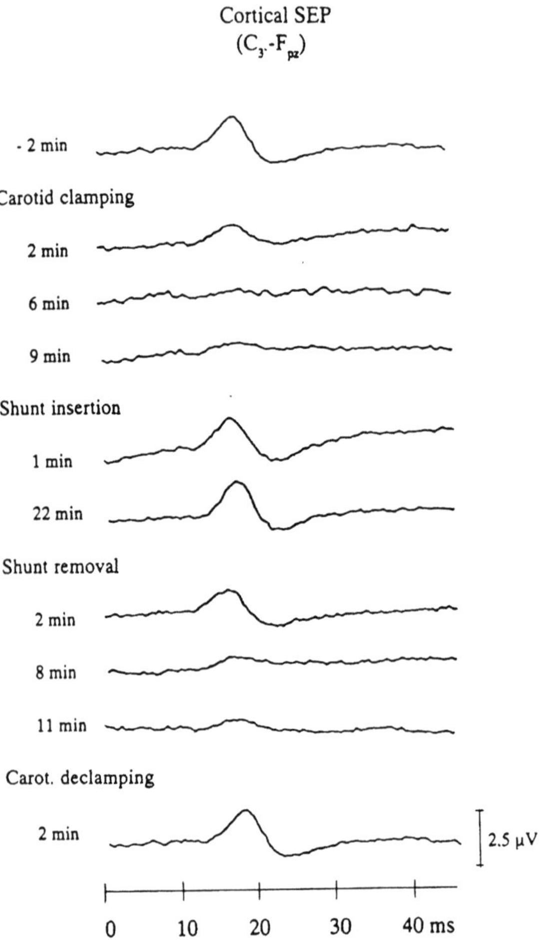

Fig. 2. Scalp-recorded somatosensory evoked responses (*SEP*) during left carotid endarterectomy. After carotid cross-clamping, the cortical waveform disappeared. SEP was totally restored by shunt placement. After shunt removal, SEP flattened, but complete reversal occurred after carotid declamping. On emergence from anesthesia, the patient demonstrated a right hemiparesis, from which he recovered within 12 h

during reconstructive surgery of supra-aortic vessels is obvious. Neurophysiological monitoring of SEP in particular facilitates better surgical and anesthetic management with the possibility of adjusting it to the individual needs of the patients. This contributes to the prevention of neurological sequelae, particularly in carotid surgery. Therefore, SEP can be recommended as routine monitoring during all carotid procedures.

References

1. Amantini A, De Scisciolo G, Bartelli M, Lori S, Ronchi O, Pratesi C, Bertini D, Pinto F (1987) Selective shunting based on somatosensory evoked potential monitoring during carotid endarterectomy. Int Angiol 6:387–390
2. Bashein G, Nessly M, Bledsoe S, Townes B, Davis K, Coppel D, Hornbein T (1992) Electroencephalography during surgery with cardiopulmonary bypass and hypothermia. Anesthesiology 76:878–891
3. Benichou H, Bergeron P, Ferdani M, Jausseran JM, Reggi M, Courbier R (1991) Pre-and intraoperative transcranial Doppler:prediction and surveillance of tolerance to carotid clamping. Ann Vasc Surg 1:21–25
4. Blume WT, Ferguson GG, McNeill DK (1986) Significance of EEG changes at carotid endarterectomy. Stroke 17:891–897
5. Branston NM, Symon L, Crockard HA, Pasztor E (1974) Relationship between the cortical evoked potential and local cortical blood flow following acute middle cerebral artery occlusion in the baboon. Exp Neurol 45:195–208
6. Cherry KJ, Roland CF, Hallett JW, Gloviczki P, Bower TC, Toomey BJ, Pairolero PC (1991) Stump pressure, the contralateral carotid artery, and electroencephalographic changes. Am J Surg 162:185–189
7. Crawford ES, Mizrahi EM, Hess KR, Coselli JS, Safi HJ, Patel VM (1988) Thoracic and cardiovascular surgery. J Thorac Cardiovasc Surg 95:357–367
8. Cunningham JN, Laschinger JC, Spencer FC (1987) Monitoring of somatosensory evoked potentials during surgical procedures on the thoracoabdominal aorta. J Thorac Cardiovasc Surg 94:275–285
9. Dinkel M, Kamp HD, Schweiger H (1991) Somatosensorisch evozierte Potentiale in der Karotischirurgie. Anaesthesist 40:72–78.
10. Dinkel M, Schweiger H, Goerlitz P (1992) Monitoring during carotid surgery: somatosensory evoked potentials vs. carotid stump pressure. J Neurosurg Anesthesiol 4:167–175
11. Drenger B, Parker StD, McPherson RW, North RB, Williams GM, Reitz BA, Beattle C (1992) Spinal cord stimulation evoked potentials during thoracoabdominal aortic aneurysm surgery. Anesthesiology 76:689–695
12. Elmore JR, Gloviczki P, Harper M, Pairolero PC, Murray MJ, Bourchicr RG, Bower TC, Daube JR (1991) Failure of motor evoked potentials to predict neurologic outcome in experimental thoracic aortic occlusion. J Vasc Surg 14:131–139
13. European Carotid Surgery Trialists Collaborative Groupe (1991) MRC European carotid surgery trial:interim results for symptomatic patients with severe (70–99%) or with mild (0–29%) carotid stenosis. Lancet 337:1235–1243
14. Fava E, Bortolani EM, Ducati A, Ruberti U (1988) Evaluation of spinal cord function by means of lower limb somatosensory evoked potentials in reparative aortic surgery. J Cardiovasc Surg 29:421–427
15. Gewertz BL, McCaffrey MT (1987) Intraoperative monitoring during carotid endarterectomy. Curr Probl Surg 24:475–53
16. Gigli Gl, Caramia M, Marciani MG, Zarola F, Lavaroni F, Rossini PM (1987) Monitoring of subcortical and cortical somatosensory evoked potentials during carotid endarterectomy:comparison with stump pressure levels. Electroencephalogr Clin Neurophysiol 68:424–432

17. Goto T, Crosby G (1992) Anesthesia and the spinal cord. In:Benumof JL, Bissonette B. Cerebral protection, resuscitation, and monitoring. A look into the future of neuroanesthesia. Anesthesiol Clin N Am 10:493–519

18. Grabitz K, Freye E, Prior R,Sandmann W (1990) Protection of the spinal cord with prostaglandin (PGE 1) and prostacyclin (Iloprost) during aortic cross-clamping. Thorac Cardiovasc Surg 38:116

19. Grossi EA, Laschinger JC, Krieger KH, Nathan IM, Colvin SB, Weiss MR, Baumann FG (1988) Epidural-evoked potentials: a more specific indicator of spinal cord ischemia. J Surg Res 44:224–228

20. Gugino LD, Kraus KH, Heino R, Aglio LS, Levy WJ, Cohn L, Maddi R (1992) Peripheral ischemia as a complicating factor during somatosensory and motor evoked potential monitoring of aortic surgery. J Cardiothorac Vasc Anesthes 6:715–719

21. Halsey JH, McDowell HA, Gelmon S, Morawetz RB (1989) Blood velocity in the middle cerebral artery and regional cerebral blood flow during carotid endarterectomy. Stroke 20:53–58

22. Hanowell LH, Soriano S, Bennett HL (1992) EEG power changes are more sensitive than spectral edge frequency variation for detection of cerebral ischemia during carotid artery surgery:a prospective assessment of processed EEG monitoring. J Cardiothorac Vasc Anesthes 6:292–294

23. Kalkman CJ, Drummond JC, Ribberink AA, Patel PM, Sano T, Bickford RG (1992) Effects of propofol, etomidate, midazolam, and fentanyl on motor evoked responses to transcranial electrical or magnetic stimulation in humans. Anesthesiology 76:502–509

24. Kalkman CJ, Drummond JC, Kenelly NA, Patel PM, Partridge BL (1992) Intraoperative monitoring of tibialis anterior muscle motor evoked responses to transcranial electrical stimulation during partial neuromuscular blockade. Anesth Analg 75:584–589

25. Lam AM, Manninen PH, Ferguson GG, Nantau W (1991) Monitoring electro physiologic function during carotid endarterectomy:a comparison of somatosensory evoked potentials and conventional electroencephalogram. Anesthesiology 75:15–21

26. Lam AM (1992) Do evoked potentials have any value in anesthesia? In:Benumof JL, Bissonette B. Cerebral protection, resuscitation, and monitoring. A look into the future of neuroanesthesia. Anesthesiol Clin N Am 10:657–681

27. Larson CP, Ehrenfeld WK, Wade JG, Wylie EJ (1967) Jugular venous oxygen saturation as an index of adequacy of cerebral oxygenation. Surgery 62:31–39

28. Laschinger JC, Cunningham JN, Cooper MM, Krieger K, Nathan IM, Spencer FC (1984) Prevention of ischemic spinal cord injury following aortic cross-clamping: use of corticosteroids. Ann Thorac Surg 38:500–507

29. Laschinger JC, Cunningham JN, Cooper MM, Baumann FG, Spencer FC (1987) Monitoring of somatosensory evoked potentials during surgical procedures on the thoracoabdominal aorta. J Thorac Cardiovasc Surg 94:260–265

30. Laschinger JC, Owen J, Rosenbloom M, Cox JL, Kouchoukos NT (1988) Direct non-invasive monitoring of spinal cord motor function during thoracic aortic occlusion: use of motor evoked potentials. J Vasc Surg 7:161–171

31. Livesay JJ, Cooley DA, Ventemiglia RA, Montero CG, Warrian RK, Brown DM, Duncan JM (1985) Surgical experience in descending thoracic aneurysmectomy with and without adjuncts to avoid ischemia. Ann Thorac Surg 39:37–46

32. Maeda S, Miyamoto T, Murata H, Yamashita K (1989) Prevention of spinal cord ischemia by monitoring spinal cord perfusion pressure and somatosensory evoked potentials. J Cardiovasc Surg 30:565–570

33. McNulty St, Arkoosh V, Goldberg M (1991) The relevance of somatosensory evoked potentials during thoracic aorta aneurysm repair. J Cardiothorac Vasc Anesthes 5:262–265
34. Messick JM, Casement B, Sharbrough FW, Milde LN, Michenfelder JD, Sundt TM (1987) Correlation of regional cerebral blood flow (rCBF) with EEG changes during isoflurance anesthesia for carotid endarterectomy: critical rCBF. Anesthesiology 66:344–349
35. North American Symptomatic Carotid Endarterectomy Trial Collaborators (1991) Beneficial effect of carotid endarterectomy in symptomatic patients with high-grade carotid stenosis. N Engl J Med 325:445–453
36. Nugent M (1992) Pro:cerebrospinal fluid drainage prevents paraplegia. J Cardiothorac Vasc Anesth 6:366–368
37. Nuwer MR (1988) Use of somatosensory evoked potentials for intraoperative monitoring of cerebral and spinal cord function. Neurol Clin 6:881–897
38. Padayachee TS, Goslin RG, Lewis RR, Bishop CC, Browse NL (1987) Transcranial doppler assessment of cerebral collateral during carotid endarterectomy. Br J Surg 74:260–262
39. Robertson C, Narayan RK, Gokaslan ZL, Pahwa R, Grossmann RG, Caram P, Allen E (1989) Cerebral arteriovenous oxygen difference as an estimate of cerebral blood flow in comatose patients. J Neurosurg 70:222–230
40. Ruß W, Fraedrich G, Hehrlein FW, Hempelmann G (1985) Intraoperative somatosensory evoked potentials as a prognostic factor of neurologic state after carotid endarterectomy. Thorac Cardiovasc Surg 33:392–396
41. Shenaq SA, Svensson LG (1992) Con:cerebrospinal fluid drainage does not afford spinal cord protection during resection of thoracic aneurysms. J Cardiothorac Vasc Anesthes 6:369–372
42. Spencer MP, Thomas GI, Moehring MA (1992) Relation between middle cerebral artery blood flow velocity and stump pressure during carotid endarterectomy. Stroke 23:1439–1445
43. Sundt TM, Sharbrough FW, Piepgra DG, Kearns TP, Messick JM, O' Fallon WM (1981) Correlation of cerebral blood flow and electroencephalographic changes during carotid endarterectomy. Mayo Clin Proc 56:533–543
44. Svensson LG, Von Ritter CM, Groeneveld HT, Rickards ES, Hunter SJS, Robinson MF, Hinder RA (1986) Cross-clamping of the thoracic aorta. Ann Surg 204:38–47
45. Thiel A, Russ W, Nestle HW, Hempelmann G (1989) Early detection transcranial dopplersonography and somatosensory evoked potentials. Thorac Cardiovasc Surg 37:115–118
46. Wade JG (1979) Anesthesia for carotid endarterectomy:protection of brain vs protection of heart. ASA 30th Annual Refresher Course Lectures B 234

Assessment of Analgesic Drug Treatment

E. Scharein

An Experimental Pain Model in Healthy Humans

In the last decade, attempts were increasingly made to "objectify" the subjective pain experience and the degree of the pharmacologically induced pain relief by electrophysiologically measurable variables in standardized and experimentally controlled pain models [1, 13, 14, 15]. Like any other sensory sensation, pain is the result of changes in the neuronal activity of a highly specialized sensory system which can be measured with the established neurophysiological methods ([16, 18, 23] for review, see [2]). In the following, some results of our attempts to determine the efficacy of a variety of differently acting analgesic treatments in a standardized non-inflammatory pain model using healthy human subjects are presented. In this model, the nociceptive system is activated by short, standardized, intracutaneously applied electrical currents, and the analgesic potency is objectified by cerebral reactions to the pain-inducing stimuli.

Standardized Activation of the Nociceptive System

To elicit an unequivocal pain reaction, it is necessary to stimulate the nociceptive system in a standardized way. The nociceptive system can be activated by activating the peripheral nociceptive afferents. These are thin, myelinated A-delta fibres with conduction velocities of approximately 15 m/s and unmyelinated C-fibers, which conduct the information at less than 1 m/s toward the brain. Examples of experimental pain stimuli are mechanical and electrical skin stimuli and radiated heat pulses. All of these stimuli activate not only nociceptive afferents, but also the sensitive A-beta mechanoreceptors responsible for correctly localizing the event.

One method that we use to activate predominantly nociceptive fibers and to induce a clearly defined pain sensation in a standardized way is the intracutaneous electrical stimulation technique [7]. After drilling a small hole into the superficial keratinized layers of the epidermis of a finger tip, a special electrode is inserted and fixed. This technique is illustrated in Fig. 1.

In this way the electrical currents are applied in the immediate vicinity of the most superficial afferent skin nerves: A-delta or C-fibers, mostly belonging to

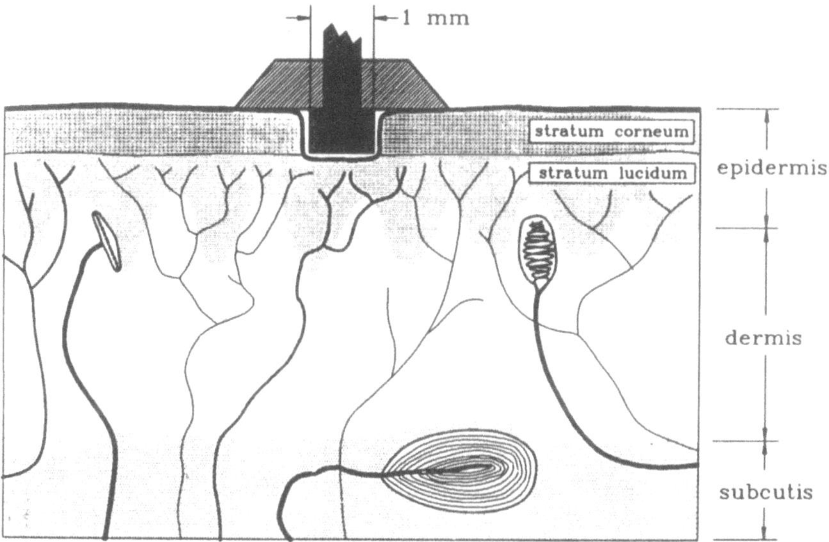

Fig. 1. The intracutaneous pain stimulus; types of receptors in the hairless skin. The nerve endings of unmyelinated C and thin myelinated A-delta fibres reach the most superficial layers of the *stratum lucidum*, whereas the other encapsulated endings terminate in the *dermis* or *subcutis*. By drilling a hole through the *stratum corneum*, the electrode is directly placed in the vicinity of A-delta and C-fibres, mostly belonging to the nociceptive system. Consequently, the brief electric shock, intracutaneously applied, induces a sharp, stabbing, hot, burning pain sensation. Adapted from [7]

the nociceptive system. As a result, the intracutaneous stimulus causes a clear and well-localized pain sensation, characterized as a stabbing, hot, and sharp sensation, very similar to the pain induced by tooth pulp stimulation [2, 12]. This is in contrast to the unpleasant paraesthesia induced by the conventionally applied skin stimulation.

Cerebral Potentials as Correlates of Experimental Pain

On the long path along the spinal cord to the brain, the standardized experimental pain stimuli elicit numerous reactions in the autonomic and motor system. Particular success in the quantification of pain sensation was obtained from electroencephalogram (EEG) analysis, especially of the pain-evoked cerebral potentials [8, 9, 10]. These are defined as changes in the EEG induced by painful stimuli; they are the highest representation of sensory perception that can be investigated by electrophysiological methods.

The cerebral potentials evoked by painful intracutaneous stimuli consist of a negativity appearing 150 ms after stimulus onset (N_{150}) and a positivity after about 250 ms (P_{250}). The amplitude difference between both components (evoked potential, EP) is used to quantify the cerebral reaction. These late pain-related components are assumed to reflect cognitive processing mechanisms of the brain, e.g., the evaluation of the painfulness of the stimulus [see 2]. As such, they depend on many factors, e.g., the arousal level, stimulus expectancy, or the attention of the subject to the stimulus event (see [21]); the same holds true for the subjective pain sensation.

Due to a poor signal-to-noise ratio in each single EP, we are forced to average the cerebral reactions over several stimulus repetitions. In our experiments, we apply the stimuli in blocks of eighty stimuli. In order to eliminate the influence of latency variability from trial to trial, we transform each single cerebral potential into the frequency domain and compute the power in the delta band (DP; for details see [22]).

The intensity of the pain experience (E) is measured on a numerical scale: 0, no sensation; 4, pain threshold; 10, highest pain sensation. Both electrophysiological parameters, EP and DP, are highly correlated with subjective pain estimation and react with high sensitivity to analgesic drugs. In addition, spontaneous EEG, auditory EP, and reaction times were evaluated to determine unspecific effects upon the vigilance system. Blood samples were collected to monitor the plasma concentration of the analgesic active agents.

Evaluation of Analgesic Potency

Centrally Acting Analgesics

The utility of cerebral potentials evoked by the intracutaneous stimulus to quantify the analgesic potency is demonstrated in the following example taken from a study in which we compared the efficacy of two centrally acting analgesic drugs competing for a similar market segment [6]. Tramadol (Tramal) is an opioid with low affinity to all opioid receptors. Valoron N is an opioid agonist–antagonist combination. A single therapeutic dosage of this combination contains the opioid agonist tilidine hydrochloride and the opioid antagonist naloxone in a ratio of 100:8. Naloxone is added in order to reduce the abuse potential of nontherapeutic overdosing.

Tramadol and Valoron N were given orally, each at twice the minimum clinical dose of 50 mg. The study was carried out as a double-blind, placebo-controlled, crossover study with 31 healthy male subjects. Each subject received four experimental sessions with intersession intervals of 7 days to avoid carry-over effects. In the first session, the subjects were familiarized with the experimental surroundings. Three medication sessions (placebo and two analgesics) followed in a randomized treatment sequence. Each session lasted for about 5 h. Two blocks of stimuli were given before medication, and three blocks after medication. The first premedication block was given merely for adaptation. The second premedication block was used to determine the premedication baseline values. Before and after each block of stimuli, blood samples were taken. During the interblock intervals, we also determined side effects and the vigilance level.

The effects of the treatments on the cerebral potentials evoked by the intracutaneous stimuli are shown in Fig. 2. The prestimulus values are plotted as dotted lines, the values in the third poststimulus block as continuous lines. After administration of the placebo, no changes could be observed. After administration of tramadol, the peak-to-peak amplitude difference was reduced by up to about 30%.

The effects of Valoron N were more pronounced: in the third postmedication block, the late somatosensory components were reduced by more than 50%. Statistically, the effects of Valoron N were significantly greater than those of tramadol ($p < 0.01$; Bonferroni-corrected t-tests for paired samples).

Nearly the same results were seen in the pain ratings. Again, no effects of the placebo were observed. Under tramadol, pain relief became significant only in the last postmedication block (about 120 min after administration). In contrast, the effects of Valoron N became significant the first postmedication block (40 min after administration). At the end of session, pain relief due to Valoron N was twice that due to tramadol. Again, this difference was statistically significant.

"Peripherally" Acting Analgesics

The intracutaneous pain model could not only be used in the case of strong analgesics; we were also able to detect the effects of weak or so-called peripherally acting analgesics. As an example, the results of two studies with three different nonsteroidal anti-inflammatory drugs (NSAID) are presented [3, 5, 11].

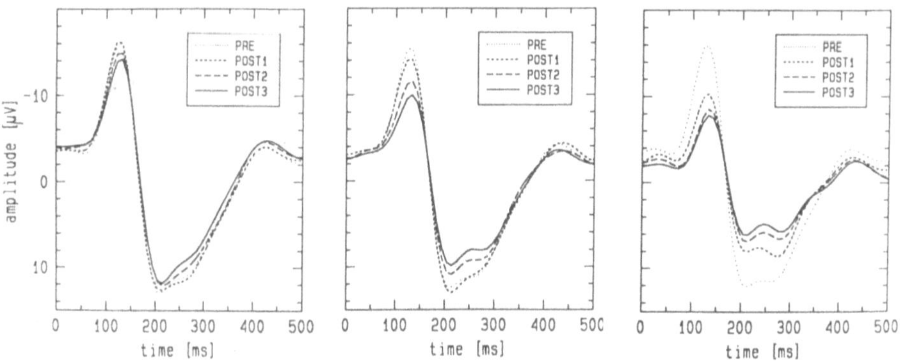

Fig. 2. Pain-related cerebral potentials under centrally acting analgesics. Cerebral potentials evoked by painful intracutaneous stimuli were recorded before (*PRE*) and 40, 80 and 120 min (*POST1, POST2* and *POST3*) after oral administration of placebo (*left*), tramadol (100 mg; *middle*) and Valoron N (100 mg; *right*)

In each study, 32 subjects were included. The effects of acetylsalicylic acid, paracetamol, and phenazone (each drug given orally at a dose of 1000 mg) were investigated in the same standardized experimental design as described above.

The most important effects of acetysalicyclic acid, paracetamol, and phenazone are summarized in Fig. 3. Mean values were obtained just before and 90 min after medication, 90 min being in the middle of the second postmedication stimulus block. As indicated, the premedication response values were stable from session to session. The mean pain ratings (E > 5) document that the stimulus intensities were adequately chosen to induce a clear pain sensation (as explained above, the pain threshold was defined as E = 4). Because the t-test comparisons revealed no significant differences between the placebo effects of both studies, the data of the placebo sessions were pooled in Fig. 3.

Mean pain ratings diminished under acetylsalicylic acid by about 4%. This decrease is numerically small, but significantly larger than the corresponding

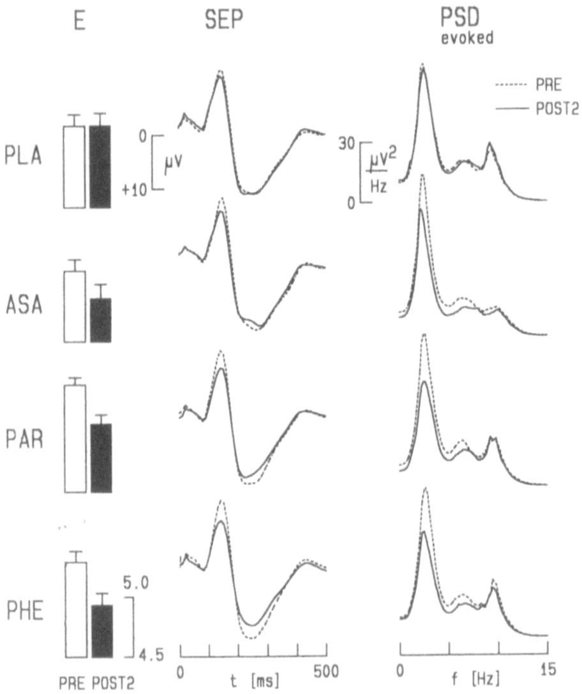

Fig. 3. Specific effects of acetylsalicylic acid (*ASA*), paracetamol (*PAR*) and phenazone (*PHE*). Mean pre- (*PRE*) and postmedication values (*POST2*, 90 min after oral administration) are given for the pain ratings (*E, left*), for the pain-relevant somatosensory evoked potentials (*SEP, middle*) and the power-spectral densities (*PSD$_{evoked}$ right*) of electroencephalogram (EEG) activity in response to the pain-inducing stimuli; for sessions with placebo, ASA (1000 mg) and PHE (1000 mg). The data were collected in two samples of 32 subjects each; the placebo data were pooled from the two studies

placebo effect. Under paracetamol, the mean pain ratings diminished by 6%, and under phenazone by 7%. The latencies of the pain-related late SEP components were not altered by any drug, but the peak-to-peak amplitude decreased by 15%, 20%, and 19%, respectively, after the administration of acetylsalicylic acid, paracetamol, and phenazone. The reduction of amplitudes were due to a similar attenuation of the N_{150} and the P_{250} components. Effects of similar magnitude were seen in the power spectral density function of the evoked EEG. DP declined under acetylsalicylic acid by 20%, under paracetamol by 21%, and under phenazone by 22%. The results suggest that in this paradigm, the agents all possessed similar potency, e.g., approximately that of 1000 mg acetylsalicylic acid.

Whereas all three drugs had a similar analgesic potency, they could be differentiated by their effects on the spontaneous EEG activity. Whereas no changes after the administration of acetylsalicylic acid were seen, paracetamol enhanced the power in the theta range, and phenazone reduced the alpha frequencies significantly compared to placebo. No influences of the drugs could be observed upon the auditory EP and reaction times.

The studies suggest, at least in part, a central mode of analgesic action for acetylsalicylic acid, paracetamol, and phenazone: all drugs reduced nociceptive brain activity in response to pain-inducing stimuli, measured by pain ratings and late cerebral potentials. Since we do not induce inflammation by our stimulation technique and we activate the nociceptive afferent neurons directly by electrical currents, we assume that the effects documented here are the result of a modulation of the conductive nerve membrane or of synaptic impulse transmission. This is in agreement with our additional findings that paracetamol and phenazone significantly modified the spontaneous EEG. As such, the term peripherally acting analgesics (see [19]) should be discarded for these NSAIDs.

Comparative Evaluation of the Potency of Analgesics

An overview of our attempts to determine the efficacy of a variety of different analgesics in the standardized noninflammatory pain model described above is compiled in Fig. 4. The results of 11 studies are presented, involving sixteen analgesic treatments and about 250 subjects [2–6, 8, 9, 11]. All studies had the same experimental design: randomized, placebo-controlled, double-blind, cross-over studies; intersession intervals of 7 days; homogenous samples of healthy male subjects (20–30 years of age). The variables were measured, quantified, and evaluated using a standardized protocol.

On the ordinate scale, percentage drug-induced pain relief is plotted, and on the abscissa scale the drug-induced decrease of late cerebral potential components evoked by the intracutaneous stimuli. Each point in the scatter diagram represents the results of one stimulus block averaged over all subjects in that study.

Near zero is the cluster of the results with placebos. In addition, the well-known ranking of analgesic potency of the tested drugs can be seen in Fig. 4. This figure documents the very high correlation ($r = 0.92$) between pain

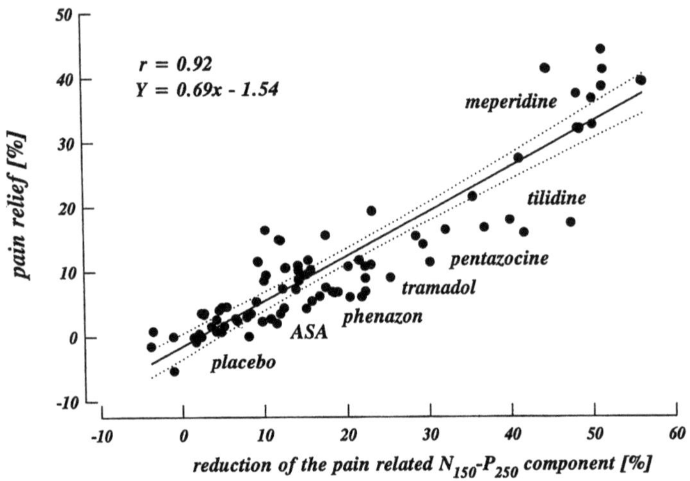

Fig. 4. Synoptic display of the results of 11 experimental studies with the intracutaneous pain model comparing 16 analgesics in 248 subjects. The relation between the pharmacologically induced reduction of the pain-related N_{150}–P_{250} (N_{150}, negativity appearing approximately 150 ms after onset of stimulus; P_{250}, positivity approximately 250 ms after stimulus onset) component and the drug-induced pain relief is shown as percentage of the corresponding premedication value. Each *dot* represents a mean value, averaged over all subjects in that study for one block of stimuli. Also plotted is the regression line of pain relief on amplitude reduction and their 95% confidence interval. In the *inset*, the Pearson product–moment correlation coefficient r and the linear regression equation are given

relief and changes in pain-related cerebral potentials. About 85% of the pharmacologically induced pain relief can be predicted by changes in the cerebral potentials.

Analgesia and Sedation

Instead of classifying the substances by their analgesic effects, we can also classify them by their side effects and their effects on the spontaneous EEG [for review see 17]. In most cases, strong analgesics also induce marked sedation, but there are some analgesics which had the same analgesic potency, but induce quite different changes in vigilance.

In two of the analyzed analgesics, we were able to detect a dissociation between analgesic potency and sedation. Flupirtine (Katadolon), a centrally acting triamino-pyridine derivate, was compared with the mixed opioid agonist–antagonist pentazozine (Fortral). Statistically, the analgesic effects of both drugs were the same: the late pain-related cerebral components and the pain ratings were similarly depressed by both treatments.

In contrast, both drugs affected vigilance differently. This was documented by the drug-induced changes in spontaneous EEG activity (Fig. 5). Mean power

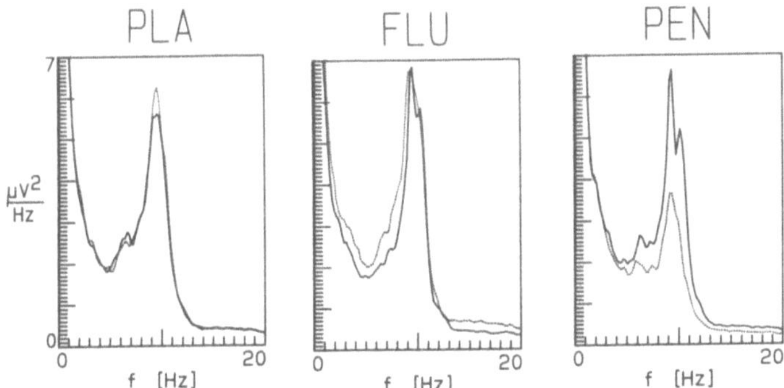

Fig. 5. Power spectral density functions of spontaneous electroencephalogram (EEG) activity before (*continuous lines*) and 30 min after (*dotted lines*) intravenous administration of placebo (*PLA*), 80 mg flupirtine (*FLU*), and 30 mg pentazocine (*PEN*) were presented. $N = 20$ health male subjects. Adapted from [4]

spectral density functions before (continuous lines) and 30 min after medication (dotted lines) are presented. No effects of the placebo are found; under flupirtine, a slowing of the alpha frequency is seen and a slight increase of power in the low-frequency bands. This corresponds to a moderate decrease in vigilance (see [20]). Under pentazozine, we observed a marked decrease in the power in the alpha band, indicating a greater reduction in vigilance. Reaction times and subjective feeling of tiredness were correspondingly changed.

Our pain model was successfully used to objectify subjective pain experience and pain relief due to differently acting analgesics. In the future, we hope to expand the range of application of our model for cases in which a subjective pain report is not available i.e., for the analysis of nociceptive information processing during anesthesia and in comatose patients. The results of first stamps in this direction are reported in the chapter by Kochs (this volume).

Acknowledgements. These studies were supported by the Deutsche Forschungsgemeinschaft (SFB) 115/C25; BR 310/16 1,2,3.

References

1. Arendt-Nielsen L (1990) First pain event related potentials to argon laser stimuli: recording and quantification. J Neurol Neurosurg Psychiatry 53:398–404
2. Bromm B (1989) Laboratory animal and human volunteer in the assessment of analgesic efficacy. In: Chapman CR, Loeser JD (eds) Issues in pain measurement. Raven, New York, pp 117–143
3. Bromm B, Forth W, Richter E, Scharein E (1992) Effects of acetaminophen and antipyrine on non-inflammatory pain and EEG activity. Pain 50:213–221
4. Bromm B, Ganzel R, Herrmann WM, Meier W, Scharein E (1986) Pentazocine and flupirtine effects on spontaneous and cvoked EEG activity. Neuropsychobiology 16:152–156

5. Bromm B, Herrmann WM, Scharein E (1988) Zur analgetischen Wirksamkeit von Paracetamol und Acetysalicylsäure im experimentellen Schmerzmodell. Schmerz Pain Douleur 9:5–11

6. Bromm B, Herrmann WM, Scharein E (1989) Zwei effektive Analgetika im Wirkungsvergleich. Fortschr Med 197:385–389

7. Bromm B, Meier W (1984) The intracutaneous stimulus: a new pain model for algesimetric studies. Methods Find Exp Clin Pharmacol 6:405–410

8. Bromm B, Meier W, Scharein E (1983) Antagonism between tilidine and naloxone on cerebral potentials and pain ratings in man. Eur J Pharmacol 87:431–440

9. Bromm B, Meier W, Scharein E (1986) Imipramine reduces experimental pain. Pain 25:245–257

10. Bromm B, Meier W, Scharein E (1989) Pre-stimulus/post-stimulus relations in EEG spectra and their modulations by an opioid and an antidepressant. Electroencephalogr Clin Neurophysiol 73:188–197

11. Bromm B, Rundshagen I, Scharein E (1991) Central effects of acetylsalicylic acid in healthy men. Arzneimittelforschung Drug Res 41:12–15

12. Bromm B, Scharein E (1990) Analgesics affecting EEG activity in humans. In: Lipton S, Tunks E, Zoppi M (eds) Advances in pain research and therapy, vol 13, Raven, New York, pp 233–256

13. Carmon A, Mor J, Goldberg J (1976) Evoked cerebral responses to noxious thermal stimuli in humans. Exp Brain Res 25:103–107

14. Chapman RC, Jacobson RC (1984) Assessment of analgesic states: can evoked potentials play a role? In: Bromm B (ed) Pain measurements in man: neurophysiological correlates of pain. Elsevier, Amsterdam, pp 189–202

15. Chatrian GE, Canfield RC, Lettich E, Black RG, Knauss TA (1975) Cerebral responses to electrical tooth pulp stimulation in man. Neurology 25:745–757

16. Chudler EH, Dong WK (1983) The assessment of pain by cerebral evoked potentials. Pain 16:221–244

17. Herrmann WM, Schaerer E (1989) Pharmaco-EEG: computer EEG analysis to describe the projection of drug effects on a functional level in humans. In: Lopes da silva FH, Storm van Leeuwen W, Remond A (eds) Handbook of electroencephalography and clinical neurophysiology, rev ser, vol 2, Elsevier, Amsterdam, pp 385–448

18. Kobal G (1985) Pain-related electrical potentials of the human nasal mucosa elicited by chemical stimulation. Pain 22:151–163

19. Lim RKS (1970) Pain. Annu Rev Physiol 32:269–288

20. Matejcek M (1982) Vigilance and the EEG: psychological, physiological and pharmacological aspects. In: Herrmann WM (ed) Electroencephalography in drug research. Fischer, Stuttgart, pp 405–508

21. McCallum WC (1988) Potentials related to expectancy, preparation and motor activity. In: Picton TW (ed) Handbook of electroencephalography and clinical neurophysiology, rev ser, vol 3. Elsevier, Amsterdam, pp 427–534

22. Scharein E, Häger F, Bromm B (1984) Spectral estimators of short EEG segments. In: Bromm B (ed) Pain measurement in man: neurophysiological correlates of pain. Elsevier, Amsterdam, pp 189–202

23. Treede RD, Kief S, Hölzer T, Bromm B (1988) Late somatosensory evoked cerebral potentials in response to cutaneous heat stimuli. Electroencephalogr Clin Neurophysiol 70:429–441

Long-Term Monitoring in Intensive Care Patients: Electroencephalogram, Evoked Responses, and Brain Mapping

E. Facco, M. Munari, F. Baratto, A. U. Behr, and G. P. Giron

Introduction

The great progress computer science has made in the past two decades has led to the birth and development of both averaging and imaging techniques and the introduction in clinical practice firstly of evoked potentials and secondly of brain mapping.

Imaging techniques undoubtedly began with radiology; as far as the brain is concerned, the computed tomography (CT) scan was the first method to provide pictures of cerebral tissue, followed by magnetic resonance imaging (MRI). Nowadays, neuroimaging includes a wide range of investigative techniques that can provide information on different functional aspects of the brain, such as mapping of electroencephalograms (EEG) and evoked potentials, regional cerebral blood flow (rCBF), single photon emission tomography (SPET), positron emission tomography (PET), and magnetoencephalography (MEG). Among these, EEG and evoked potentials are the most widely available and routinely used.

CT scan and MRI have proved to be essential investigative techniques in neurological practice, but they mainly reflect the structural aspect of cerebral damage. In contrast, the other techniques mentioned allow the functional status of the brain to be explored, giving information on the response of the brain to injury, both in focal areas and in distant ones. In other words, they allow functional neuroimaging, which forms a complementary aspect in the assessment of brain damage: the combined use of "structural" and "functional" imaging techniques enables us to gain an insight in vivo and in real time into the pathophysiology of the injury.

Some functional imaging techniques, particularly EEG and rCBF, are more changeable than structural imaging, something which is both the main advantage as well as the disadvantage of these techniques. In fact, changeability decreases the reproducibility of results, but allows the evolution of brain conditions (whether spontaneous or caused by therapy) to be checked, even when the structural aspect of damage, as defined by serial CT scan examinations, appears to be steady. Furthermore, considerable changes in EEG and/or rCBF can occur during the clinical course in some patients with no apparent changes revealed by clinical examination; this suggests that functional neuroimaging may improve patient evaluation, enabling us to detect what is happening to the brain before it results in evaluable changes in the clinical picture.

As conventional EEG and evoked potentials are well-known and widely used techniques, remarks concerning their methodology are unnecessary here (recording methods and wave generators of evoked potentials have been described in details elsewhere, see [23, 26, 27, 28]), while some essential aspects of both EEG and SEP mapping are worth describing before showing their use in comatose patients.

Mapping of Electroencephalograms and Evoked Potentials: Essential Methodological Aspects

The conventional EEG remains essential in clinical practice and is always the first step before signal processing: it allows rhythms to be checked and irregularities to be found, such as spikes, sharp waves, triphasic waves, and burst suppression; visual inspection is also essential in order to recognize and reject artifacts before processing.

Although visual inspection may be the gold standard in routine EEG practice, during which possible irregularities need to be sought and found, it is not a sensitive tool when the quantitative analysis of background activity is required; in fact, visual inspection is more qualitative than quantitative and cannot easily detect and quantify small changes in EEG spectral content. Much more information can be extracted with EEG mapping, obtained by multichannel spectral analysis, interpolation, and spatial topographic display of power spectra. These markedly increase EEG sensitivity, enabling us to detect even small changes in background activity, which are undetectable by visual inspection, and to check the significance of asymmetries or focal abnormalities by statistical analysis; the latter allows a patient's data to be compared with those of controls, significance probability maps to be drawn, and serial recordings to be compared using paired data tests. Another advantage of topographic display is the improvement of information exchange between the expert neurophysiologist and the doctors in charge, who are often not familiar with conventional EEG and evoked potential waves; a map is a much more concise and understandable display of a patient's neurological status.

Different methods of interpolation have been suggested, but the most widely used are the nearest three- or four-point linear interpolations [14, 79]. As far as the reference is concerned, none of the commonly used references (such as linked ears, nose, chin, and noncephalic references) can be considered as "ideal" and always reliable; sometimes it is very useful to check both the original EEG record (obtained with a referential montage) and reconstructed, so-called reference-free montages in order to avoid misleading conclusions [19, 21, 22]. Among these, well-known average reference [67] and Laplatian methods such as source derivation [40] are worth mentioning. We recently saw a patient with unilateral temporal delta focus, showing outstanding distortions of electrical scalp potential yielded by the linked-ear reference; in this case, even the reference-free montages were not able to completely remove the distortion introduced by earlinking, according to the hypothesis of ear-bridging suggested by Katznelson [48], even if they did allow the focus to be correctly localized.

Although EEG mapping makes focus localization easier, its topographic accuracy is rather low in comparison to "anatomical" investigations; this is due to several factors: (a) in EEG recordings performed using the 10–20 International System, the spatial resolution is limited by the relatively low spatial frequency of electrodes; (b) the maps give a bidimensional layout of the topography of EEG rhythms, but cannot provide information on the depth of a focus, while deep generated, slow-wave activity may spread all over the scalp, mostly in frontal areas (such as frontal intermittent rhythmic delta activity, FIRDA); (c) wave polarity is lost in spectral maps and, especially in the case of spike discharge, two distinct focal areas instead of one (corresponding to the positive and negative dipole fields) may be represented; and (d) different phenomena, with different localizing values, indicate a focal brain lesion (such as depression of fast activity, presence of polymorphic and/or rhythmic delta activity, and paroxysmal sharp waves or spikes). Rhytmic delta activity has a lower localizing value than its polymorphic equivalent and tends to be present at some distance from the lesion. The use of different montages (i.e., source derivation) and perhaps coherence analysis may help to solve diagnostic problems.

As far as spatial resolution is concerned, Gevins et al. [34] reported that a 128-channel montage yields a great increase in EEG accuracy in focus localization, but it cannot yet be used in routine clinical practice; since with conventional montages, the interelectrode distance is about 6–7 cm, one can expect some 3 cm leeway in focus localization. Therefore, some lesions on CT scans may be undetected by EEG, especially when they are very small and/or deep. On the contrary, EEG is seldom able to find lesions undetected by CT scan; this may occur in the early stage of brain ischemic lesions or contusions, in which marked changes in electrical activity ensue before tissue density is altered.

Evoked potentials are more accurate than EEG in the localization of brain damage, since they explore specific pathways, but do not investigate the whole brain. The mapping of somatosensory evoked potentials (SEP) seems very promising, as it allows the functional status of both pre- and postrolandic structures to be checked. Recording methods and generators of N20, P22, and N30 are described in detail elsewhere [23, 26, 56]; here is only worth mentioning that the use of the earlobe contralateral to the stimulated side as a reference allows the long-lasting negativity of N18 to be decreased or cancelled out, improving the definition of cortical components.

SEP mapping, of course, is of no use in deeply comatose patients, in whom all cortical components are lost. However, it is very useful to check whether all cortical components are present or an isolated loss of frontal or parietal waves has occurred, thus improving the assessment of comatose patients in comparison to conventional SEP recording; furthermore, SEP mapping allows us to check the topography of waves, which can be altered in response to injury even if central conduction time (CCT) is normal.

The combined analysis of traces and maps is recommended for a careful definition of the presence, latency, and topography of waves and to avoid misleading conclusions based only on perception of map colours; in fact, both abnormal SEP patterns and possible artifacts (particularly in SEP recorded from unrelaxed, uncooperative patients) may change waveshapes.

Short-Latency Evoked Potentials in Coma and Brain Death

Evoked Potentials in Coma

Evoked potentials have proved to be a very useful tool in the evaluation of neurological diseases. One important advantage in the operating room and intensive care unit is the possibility of checking the functional status of the explored pathways even when clinical examination is not reliable. In the past decade, great interest in coma has developed and a wealth of data is now available in the literature; among coma states of different etiologies, severe head injury is the most extensively studied [1, 2, 10, 11, 17, 23, 24, 26, 29, 30, 33, 36–38, 42, 43, 46, 47, 50, 54, 57, 58, 60, 64, 68, 70, 73, 78, 81, 85, 86, 89].

The auditory brain stem responses (ABR) parameters used for early outcome prediction are the interpeak latency between waves I and V (IPL V–I), the amplitude ratio (AR) of wave V/I, and the absence of waves, but sometimes even absolute amplitudes and latencies have been used. Unfortunately, ABR gradings reported in the literature are nonhomogeneous (Table 1) and the different criteria make it difficult or even impossible to compare series. Further differences exist between methods, such as intensity and frequency of the stimulus, presence/absence of contralateral masking, and the side taken into account (the "better" or "worse" one) for prognostic prediction. Exactly the same problem exists for SEP (Table 2), the main methodological controversy of which has been the reference: although the frontal reference has been extensively used for many years, it should not be used, since it cancels out the far-field components and distorts waves' shape; the noncephalic reference is far superior in conventional SEP recording, while, as already mentioned, the contralateral earlobe is better for SEP mapping [23, 56].

In our experience, the distribution curve for IPL V–I in patients that survive appears to be a gaussian one: the chances of coming out of posttraumatic coma drastically decrease when the IPL V-I exceeds 4.5 ms, a value corresponding to the mean plus two standard deviations of surviving patients [15]. Likewise, outcome prospects worsens when AR V/I falls below 0.5, both in head injury and subarachnoidal hemorrhage, while the combined use of both parameters appears to improve early prognosis in comparison to the use of only one of them

Table 1. Main gradings of auditory brain stem responses in severe head injury

Authors	Year	Reference	Grading
Greenberg et al.	1977	36	1 = Normal 2 = VI Absent; O IPL V-I = 4.6 ± 0.1 ms 3 = Only I and V present, with latencies of 1.5 and 6.0 ± 0.1 ms, respectively 4 = Only wave I is present
Seales et al.	1979	78	1 = Normal 2 = Abnormal (Latency > M + 2DS of controls, IPL V–I > 4.40 ms or AR V/I < I or low-voltage V wave
Karnaze et al.	1982	46	1 = Normal 2 = Mildly abnormal (4.5 < IPL V–I < 4.9 ms and/or 0.4 < AR V/I < 0.5) 3 = Moderately abnormal (4.9 < IPL V–I < 5.3 ms and/or 0.25 < AR V/I < 0.4) 4 = Markedly abnormal (IPL V–I > 5.3 ms and/or AR V/I < 0.25) 5 = Severely abnormal (absent V, III–V)
Mjøen et al.	1983	58	1 = Normal 2 = Increased IPL V–I 3 = Absent V 4 = Only I present 5 = Absent response
Anderson et al.	1984	[2]	1 = Normal 2 = Increased IPL V–I 3 = Decreased amplitude of V or III–V, AR V/I < 0.8 4 = Only I present
Facco et al.	1985	15	1 = Normal 2 = IPL V–I > 4.50 ms; absent V
Ottaviani et al.	1986	68	= Mjøen 1983
Cant et al.	1986	10	1 = Normal 2 = Abnormal, with present V (increased IPL V–I or absentt I–III) 3 = Absent V
Fischer et al.	1988	30	1 = Normal 2 = Increased IPL V–I 3 = Inversion of AR V/I 4 = Desynchronization peaks of IV and V 5 = Absent IV and V 6 = Only I present 7 = Absent ABR

AR, amplitude ratio; ABR, auditory brain stem response; M + 2DS, mean + two standard deviations; IPL V–I, interpeak latency between waves I and V.

E. Facco et al.

Table 2. Main somatosensory evoked potential gradings in severe head injury

Author	Year	Reference	Grading
Greenberg et al.	1977	36	1 = Normal, but absence of components following 100 ms 2 = Absence of waves following 40 ms 3 = „ „ „ „ 30 ms 4 = „ „ „ „ P15
Greenberg et al.	1981	38	= Greenberg et al. 1977
Lindsay et al.	1981	50	= Greenberg et al. 1977
Rumpl et al.	1983	75	GO = Normal CCT MD = CCT = 7.0q1.1 (MqDS) mono- or bilaterally SD = CCT = 7.5q0.2 „ „ „ D = CCT = 8.0q1.5 „ „ „
Anderson et al.	1984	2	1 = Normal 2 = Delayed CCT; present N20 and absent P24 3 = Monolateral absence of N20 4 = Bilateral absence of N20
Cant et al.	1986	10	1 = Normal 2 = Delayed CCT 3 = Mono- or bilateral absence of N20
Fischer et al.	1988	30	1 = Normal 2 = Bilateral absence of N20 3 = Monolateral absence of N20 4 = Only Erb's point potential 5 = Absent SEP
Nau et al.[a]	1988	65	1 = Normal 2 = Slightly abnormal, delayed latencies 3 = Severely abnormal, delayed latencies 4 = Absent response
Facco et al.	1990	23	1 = Normal 2 = N13–N20 > 8 ms 3 = Absent N20
Hutchinson et al.	1991	43	1 = Normal CCT 2 = One or both CCT abnormal 3 = One or both N20 absent

GO, good outcome; MD, moderate disability; SD, severe disability; D, death; CCT, central conduction time.

[a] Series including nontraumatic coma.

[17,18,24]. Furthermore, both ABR grading and effectiveness appear to be similar in children and in adults [62].

SEP are powerful prognostic indicators in both severe head injury and subarachnoidal hemorrhage; the risk of poor outcome dramatically increases when bilateral increase of N13–N20 above 8 ms or disappearance of N20 occurs during the acute phase of coma [23], while the combined use of ABR and SEP may further improve the accuracy of predictions, decreasing the number of false negatives [18, 24]. In Tables 3–6, the relationship between evoked potentials and outcome in a series of 76 head-injured patients has been analyzed: ABR appear to give better prognostic predictions when the worse side is taken into account, and both ILP V–I and AR V/I are significantly related to the outcome, while SEP show less false negative predictions than ABR. Combining the results of both modalities (Table 6), even better prognostic accuracy can be achieved: in fact, the relative risk (RR) of poor outcome in patients with abnormal evoked potentials is around 5 for ABR, 16 for SEP, and upto 29 when a combined evaluation of both modalities is used.

A close correlation between brain stem reflexes (apart from corneal ones) and ABR has been reported [53], but no significant relation between ABR, SEP, and the Galsgow Coma Scale exists, the latter being a much less accurate

Table 3. Auditory brain stem response in 76 cases of head injury. The IPL V–I is significantly related to the outcome; the best prognostic prediction is obtained when the worse side is taken into account

Best side (ms)	GR + MD		SD		D + PVS	
	No. cases	%	No. cases	%	No. cases	%
IPL V–I < 4.50	23	92	3	60	26	56.5
IPL V–I > 4.50	2	8	2	40	4	8.7
Absent V wave	–	–	–	–	16	34.8

Worst side (ms)	GR + MD		SD		D + PVS	
	No. cases	%	No. cases	%	No. cases	%
IPL V–I < 4.50	21	84	3	60	18	39.1
IPL V–I > 4.50	4	16	2	40	7	15.2
Absent V wave	–	–	–	–	21	45.7

Best side: Relative risk of poor outcome RR; PVS + D), S; χ^2, 17.894, $P < 0.01$
Worst side: RR, 6.2; χ^2, 21.264; $P < 0.001$
GR, good recovery; MD, moderate disability; SD, severe disability; D, death; PVS, persistent vegetative state; IPL V I, interpeak latency between waves V and I.

Table 4. Auditory brain stem response in 76 severely head-injured patients; the amplitude ratio (AR) V/I is significantly related to the outcome

AR V/I (worst side)	GR + MD		SD		D + PVS	
	No. cases	%	No. cases	%	No. cases	%
AR V/I > 0.5	17	68	5	100	19	41.3
AR V/I < 0.5	8	32	–	–	6	13
Absent V wave	–	–	–	–	21	45.7

Relative risk of poor outcome (PVS + D), 3.9; χ^2, 22.942; $P < 0.001$. GR, good recovery; MD, moderate disability; SD, severe disability; D, death; PVS, persistent vegetative state.

Table 5. Somatosensory evoked potentials from median nerve in 76 patients with severe head injury; there is a close correlation betwen the N13–N20 and outcome

Best side	GR + MD		SD		D + PVS	
	No. cases	%	No. cases	%	No. cases	%
N13–N20 < 8 ms	23	92	3	60	13	28.3
N13–N20 > 8 ms	2	8	2	40	6	13
Absent N20	–	–	–	–	27	58.7

Relative risk of poor outcome (PVS + D), 16.5; χ^2, 33.755; $P < 0.001$. GR, good recovery; MD, moderate disability; SD, severe disability; D, death; PVS, persistent vegetative state.

Table 6. Combined use of auditory brain stem responses of (ABR) and Somatosensory evoked potentials (SEP) in severe head injury

AR V/I ABR + SEP	GR + MD		SD		D + PVS	
	No. cases	%	No. cases	%	No. cases	%
Both normal	20	80	2	40	4	8.7
Only ABR abnormal	3	12	1	20	9	19.6
Only SEP abnormal	1	4	1	20	6	13
Both abnormal	1	4	1	20	27	58.7

Relative risk of poor outcome (PVS + D) when at least one modality is abnormal 28.87, χ^2, 39.569; $P < 0.001$. GR, good recovery; MD, mild disability; SD, severe disability; PVS, persistent vegetative state D, death.

predictor of outcome [23, 29, 53]. Since evoked potentials appear to be superior to clinical evaluation in early outcome prediction, they may be very helpful in some cases to check the real neurological condition of the patient, avoiding false conclusions and the risk of a negative therapeutical approach. For example, Fig. 1 shows a large brain stem contusion in a 10-year-old child who had a Glasgow score of 5 (E1V1M3), bilateral midriasis and absent light reflex. The evoked potentials (Fig. 2) showed completely normal ABR and MEP, while in the SEP a low voltage cortical N20 was still present with normal CCT. Despite the presence of bilaterally fixed pupils for several days, the patient recovered, with diplopia being the only neurological deficit. In other words, the patient had a brain stem contusion involving the IIIrd cranial nerve bilaterally, without affecting lemniscal and pyramidal pathways; as a result, a clinical and CT picture of severe brain stem damage suggesting a poor prognosis was present, while the evoked potentials were the only means of correctly predicting a good outcome.

It is worth remembering that, although an increased CCT in ABR and/or SEP or even bilateral absence of N20 strongly suggests severe brain dysfunction, a great deal of caution should be applied in predicting a poor outcome on the basis of a single recording performed in the early stage; some patients may undergo progressive improvement with restoration of both the N20 and CCT and recover, if properly treated [23], especially when the abnormalities of ABR and/or SEP depend upon secondary brain damage, such as brain edema and intracranial hypertension.

As far as MEP are concerned, only a few data are so far available in the literature. MEP from electrical stimulation do not appear to be related to the

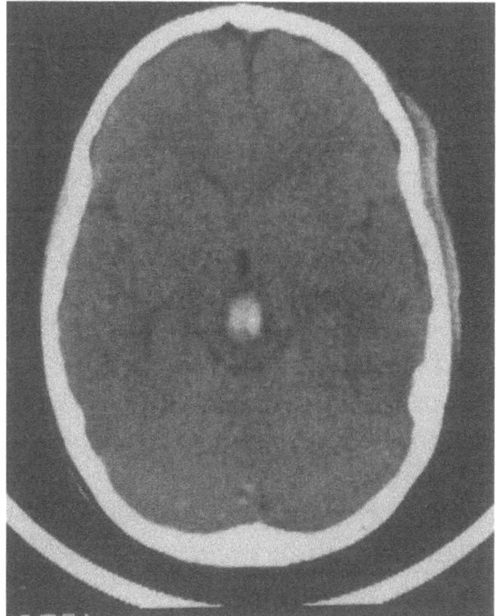

Fig. 1. Computed tomography scan in a severely head-injured child, showing a brain stem hemorrhagic contusion

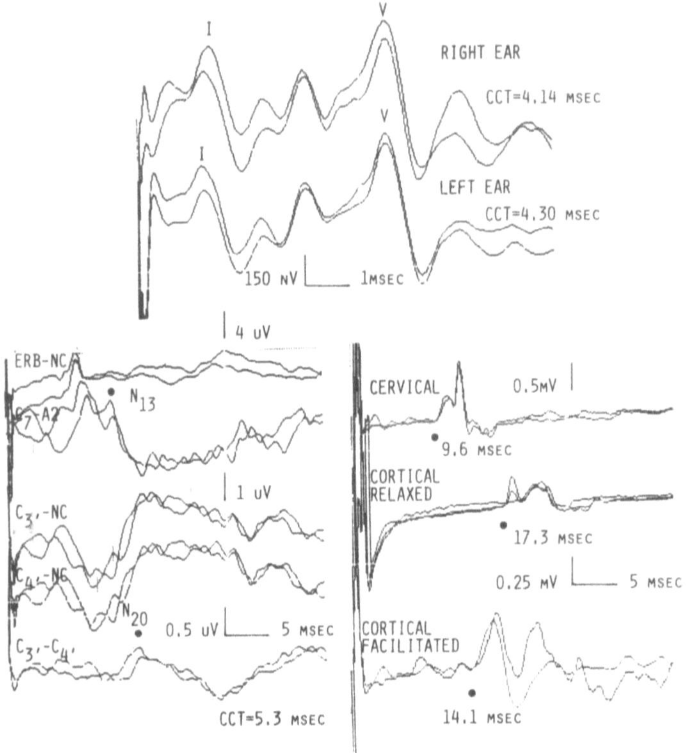

Fig. 2. Auditory brain stem response (ABR; *top*) somatosensory evoked response (SEP; *bottom left*), and motor evoked response (MEP; *bottom right*) in a severely head-injured child (same patient as in Fig. 1). Both ABR and MEP are absolutely normal, while the SEP shows a low voltage, but is still present; N20 with normal central conduction time (*CCT*). The patient recovered

outcome [65, 87, 88], while the magnetic stimulation is much more promising [5, 29]. In fact, in our experience MEP from transcranial magnetic stimulation were closely related to the outcome and appear to be more sensitive detectors of motor pathway dysfunction than clinical evaluation; the combined use of SEP and MEP allowed us to check sensorimotor function and it improved the outcome prediction in comparison to the use of only one modality. The main limitations were a significant rate of both false negatives (likewise ABR and SEP) and false positives; however, the latter probably decrease with routine use of facilitation, since it is sometimes possible to elicit a normal MEP in patients with absent relaxed response [29].

As far as postanoxic coma is concerned, the brain stem is notably more resistant to anoxia than the cortex and, therefore, can be preserved even when irreversible cortical damage has occurred. As a consequence, good prognostic accuracy of SEP can be expected, but not of ABR; in general, a normal SEP is a good predictor of good outcome, while an absent N20 with normal ABR is an indicator of persistent vegetative state. SEP are very early predictors able to give correct information even a few hours following cardiac arrest, while EEG is not reliable during the first 2–3 postresuscitation days. Of course, although the first SEP recorded in the early stage strongly suggest a poor outcome, at least a second test after 1 week is required to confirm the prognosis in order to avoid falsely pessimistic predictions, particularly in children.

Somatosensory Evoked Potential Mapping in Coma

Although substantial agreement on the reliability of SEP in predicting the outcome of coma is present in the literature, despite the mentioned differences in methods and gradings (see [23,27,28] for reviews), remarkable differences and discrepancies in results and conclusions can be found in some series, mainly regarding the rate of false negatives (namely, patients with normal SEP and poor outcome).

Our experience in SEP mapping in coma [26] suggests that multichannel recordings can improve prognostic value in comparison to the use of parietal derivations only. In fact, SEP mapping can recognize patients with severe frontal brain damage sparing the postrolandic structures, who may have normal N20 and N13–N20 interval and selective loss of frontal N30; thus, SEP mapping allows us to decrease the rate of false negatives in comparison to conventional SEP recordings.

The inverse pattern, namely, the absence of parietal N20–P27 with preserved P22–N30, may occur as well, although so far we have never found such a pattern in our patients; it has already been described in focal cortical and subcortical perirolandic lesions [55,80].

It is worth recalling that the absence of both N20 and N30 in the acute phase occasionally might still be compatible with a good outcome; therefore, predictions made on the basis of single SEP recordings may be hazardous, particularly when the absence of waves depends upon secondary, still reversible brain damage (such as brain edema and intracranial hypertension). In contrast, the presence of N20 during the chronic phase does not appear to be a reliable sign of recovery, since it can reappear in the late clinical course of patients remaining vegetative.

The spatial mapping of SEP also appears to be promising in the assessment of abnormal topography of waves; in some patients the midline shift on CT scans appears to yield a parallel shift of N20 topography, which probably reflects brain tissue dislocation due to mass lesions [26]. In few patients with severe brain lesions, all waves (N20, P22, N30) are present with normal CCT, but abnormal topography and poor outcome. Thus, in some instances the abnormal shape of waves might be the only sign of severity in patients with waves present and normal latencies; however, the sensitivity and specificity of this pattern is still to be determined.

Auditory Brain Stem Responses and Somatosensory Evoked Potentials in the Confirmation of Brain Death

Short-latency evoked potentials are an essential tool for the confirmation of brain death, since the most critical aspect of its diagnosis is the assessment of the death of the brain stem. In fact, they are the only means of providing direct and objective evidence of the arrest of conduction in the brain stem and, being roughly unaffected by anesthetics and toxic and metabolic factors, are able to recognize false positives (namely, patients with reversible coma that appear brain dead), even when both EEG and clinical evaluation are no longer able to provide information. Furthermore, they allow us to explore the brain stem even in patients with peripheral lesion of the eyes, in whom brain stem reflexes cannot be elicited (such as in craniofacial trauma), and to check the lemniscal pathways, which cannot be evaluated by clinical examination. There is wealth of data that enables us to evaluate the role of ABR and SEP in the diagnosis of brain death [3, 6, 13, 16, 25, 31, 32, 35, 52, 54, 82].

The absence of wave I in most brain-dead patients led many authors to assign a limited value of ABR; however, this limitation may be overcome by serial monitoring during preterminal states, thus showing the progressive disappearance of waves [32].

SEP are much more reliable than ABR when recorded using a noncephalic reference (provided that there are no cervical spinal cord lesions); the cervical N9–N13 is present in most, if not all patients and is associated with the absence of all components following the far-field P11 or P13, thus showing the arrest of conduction at the level of the foramen magnum (for further details on evoked potentials in coma and brain death, see [16, 23, 25, 27, 28]).

Spectral Analysis and Mapping of Electroencephalograms in Coma

Although many studies on EEG in head trauma, stroke, and anoxia are available in the literature [4, 7, 8, 9, 31, 43, 49, 63, 72, 74, 76, 77, 83], only a small proportion of these deal with EEG mapping, while, among the latter, only a few single studies include patients with impaired consciousness in their series.

Although a good correlation between EEG and outcome in post-traumatic and postanoxic coma was reported by some authors [7, 8, 72, 77, 83], in other studies it appeared to be less reliable [4, 41, 43, 74], particularly if predictions were made on the basis of a single EEG recorded in the acute phase of the clinical course [59]. When EEG and SEP were compared with each other, the latter proved to be superior [43, 74].

In general, two EEG parameters appear to be related to a good outcome: (a) reactivity to painful stimuli and (b) the presence of fast activity in the alpha or beta band. The EEG in severe head injury has been graded into the following five patterns showing a good correlation with outcome (Figs. 3, 4) [7, 8]: (1)

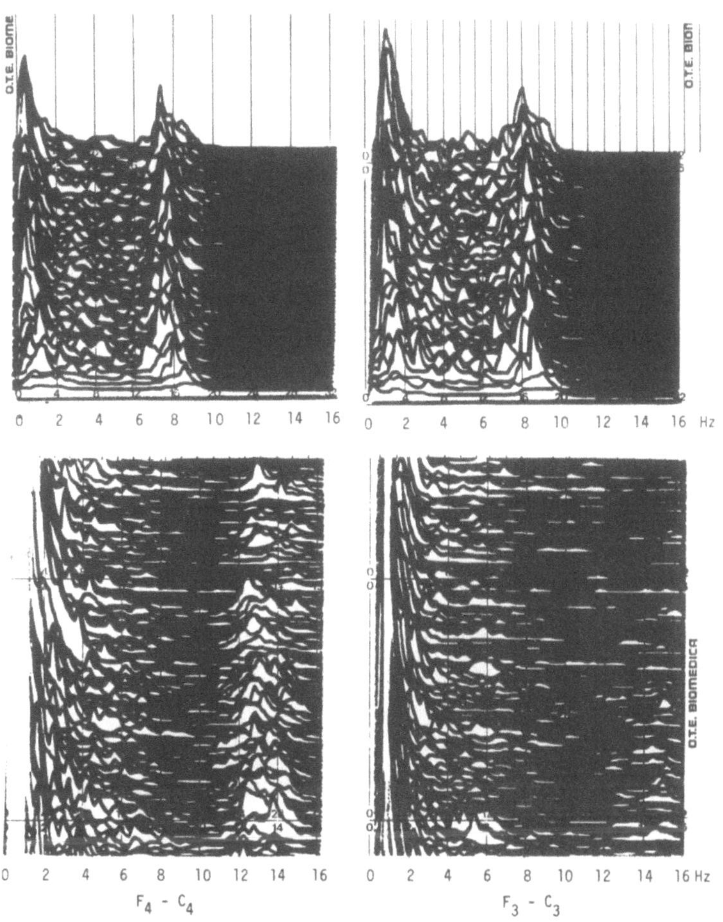

Fig. 3. Spectral analysis of the electroencephalogram (EEG) in severe head injury: a diphasic EEG (*top*) and a sleep-like (*bottom*) pattern are shown, which differ in the peak frequency of fast activity

diphasic, (2) sleep-like, (3) borderline (or alpha coma), (4) slow-wave and (5) silent (or flat). Grades 4 and 5 are a strong index of poor outcome, while patients with grade 3 have about 30%–40% probability of coming out of coma [39]; in contrast, grades 1 and 2 are indicative of good outcome. Sleep-like and diphasic patterns imply the presence of two distinct peak frequencies in the delta and beta or alpha band, respectively, which can be easily detected by spectral analysis (Fig. 3). The use of EEG mapping may improve the evaluation of this EEG pattern, allowing its topographic analysis.

In our experience [21, 61], the peak in the alpha band of patients with diphasic EEG is often localized in the anterior regions. When patients come out of coma, the alpha rhythm reappears in parieto-occipital regions (Fig. 5). This

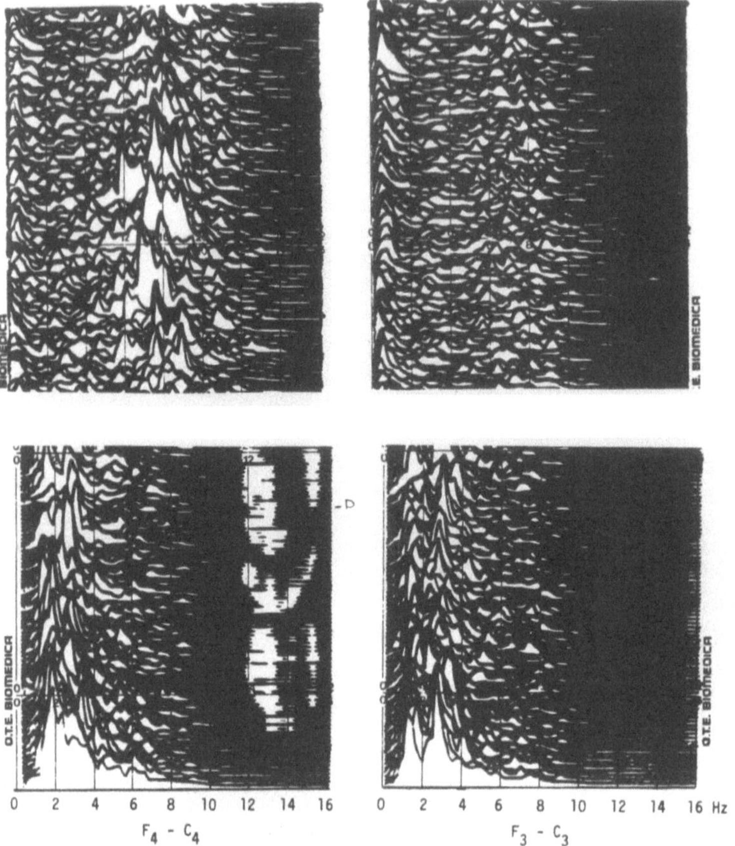

Fig. 4. Spectral analysis of the electroencephalogram in severe head injury: border line (*top*) and slow-wave patterns (*bottom*)

suggests that the topography of alpha-band activity reflects the level of consciousness, since the absence of alpha rhythm does not depend upon the presence of focal lesions is parieto-occipital regions. Even in patients with alpha-pattern coma [39], the nonreactive alpha-band activity shows higher voltage in anterior and central areas, unlike the true alpha rhythm, as defined by Chatrian et al. [12].

According to Plum and Posner [71] consciousness is lost entirely when the diencephalon is involved or a diffuse damage of gray or white matter has occurred. Although the generators of alpha rhythm are not yet well known, there is strong evidence that circuits involving both the thalamus and cortex are responsible for the generation and modulation of both spindles and alpha activity [51]. Furthermore, the alpha rhythm appears to depend upon clusters of generators [51], while coherence analysis suggests the existence of at least two distinct components of alpha rhythm, spread and localized, respectively

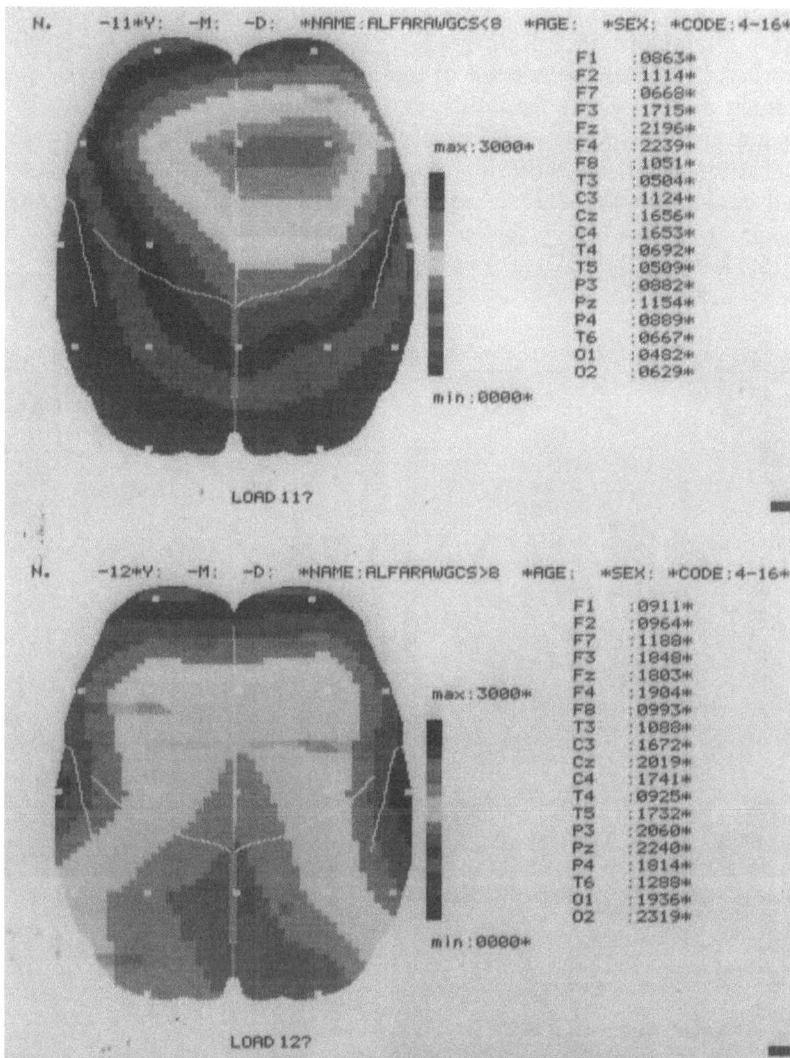

Fig. 5. Serial electroencephalogram mapping of alpha-band activity in a group of ten patients comatose following severe head injury; the alpha-band activity was localized in frontal areas during the clinical course of coma (*top*), whereas the occipital alpha rhythm reappeared when the patients had come out of coma (*bottom*)

[44, 45, 69]; as a consequence, in each derivation both endogenous and exogenous components of alpha rhythm were recognized. In anterior regions, the former appears to be lower than the latter, while the reverse occurs in posterior ones; as far as the peak frequency is concerned, it may be slightly lower in the anterior regions than in posterior ones.

These data agree with those observed in coma patients by us; in fact, the depression of alpha rhythm in our patients seems to be related to the level of consciousness, while the persistence of some anterior alpha-band activity confirms that it may originate from different generators. Moreover, the anterior alpha-band activity showed a lower peak frequency than the alpha rhythm recorded after the patients came out of coma. In other words, these data suggest that both consciousness and alpha rhythm may be depressed at the same time by the functional, reversible involvement of the diencephalon occuring in coma and

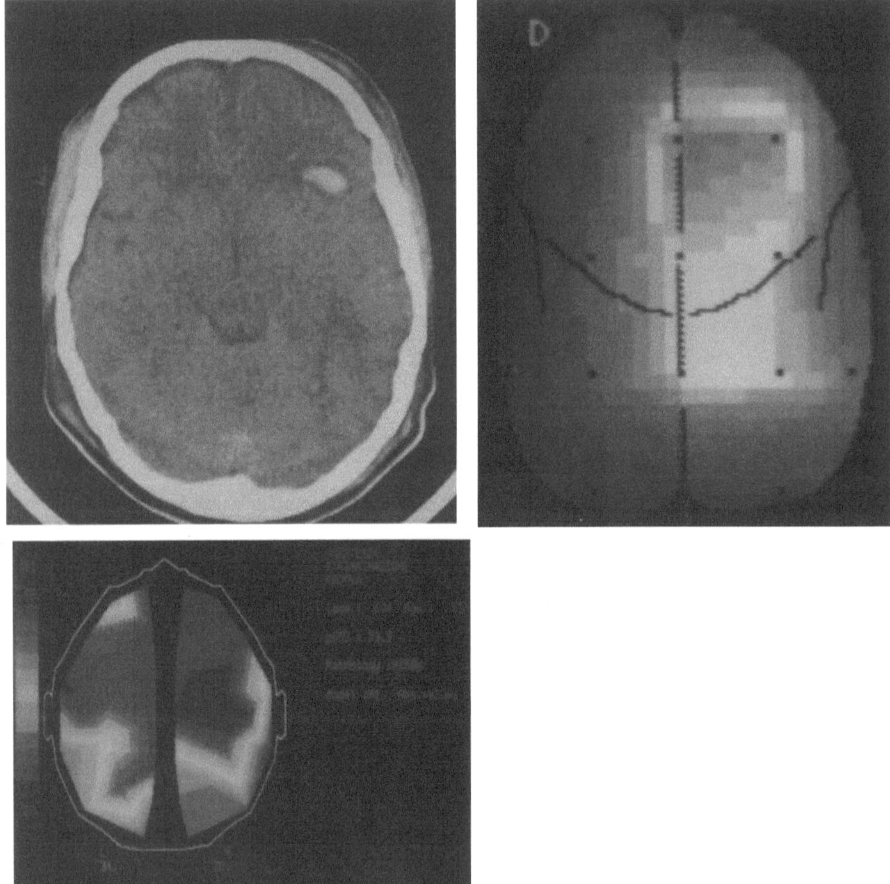

Fig. 6. Early phase of clinical course of a patient with postraumatic coma. The computer tomography scan (*top left*) shows a right frontotemporal contusion, associated to a delta focus (*top right*) on the electroencephalogram map and a marked decrease of regional cerebral blood flow (rCBF) (*bottom*) in the same region. (The right side in all parts of the figure corresponds to the patient's right side)

that they are restored when the patient recovers. However, this is only a hypothesis and calls for further studies; the reproducibility of this EEG pattern is yet to be confirmed, and it is not clear why anterior alpha activity persists in coma, unlike occipital alpha rhythm, or what its pathophysiological significance is.

As far as focal lesions are concerned, EEG mapping is more sensitive in their detection than the conventional EEG, while the topography of scalp-recorded abnormalities shows a close correlation with both CT scan and clinical data in patients with cerebrovascular disorders [76]: in this regard, delta activity is more reliable than theta, although in some cases the former may be located in different position (e.g., FIRDA). More recently, Nagata [63] reported a significant correlation between EEG power spectra and PET in patients with cerebral infarction; delta and theta activity were negatively related to CBF, while alpha activity was positively related to CBF.

Our preliminary experience in patients comatose as a result of head injury or stroke confirms the relationship between EEG, CT scan, and rCBF obtained by the method [20] (Fig. 6) when a focal damage is present. The focal areas defined by each technique do not correspond exactly to each other, but the functional ones are usually larger for two reasons: (1) the larger extent of dysfunctioning area in comparison to the extent of structural damage and (2) the aliasing introduced by interpolation (dependent on interelectrode distance). However, in case of small and/or deep contusions, the EEG sometimes appears to be more sensitive than rCBF, due to the so-called look-through phenomenon occurring in rCBF measurement.

Conclusions

A vast amount of data available in the literature on both evoked potentials and EEG clearly shows their usefulness in the assessment of comatose patients, improving the diagnosis and early outcome prediction. Moreover, the EEG is routinely used in many countries (and in some of them is mandatory by law) in the diagnosis of brain death, while evoked potentials are a unique means of giving objective confirmation of the death of the brain stem.

Brain mapping in coma is very promising, since it improves our knowledge of brain functional status in comparison to the use of conventional investigations. It will not replace conventional EEG and evoked potentials, but rather represents a new way of interpretation, which is to be added to the classical ones, increasing their diagnostic and prognostic power. Conventional EEG remains essential in the evaluation of transients and for analysis in the time domain, while conventional evoked potentials are not replaceable in the assessment of brain stem function (by analysis of far-field potentials and (CT).

When maps of EEG, evoked potentials, and, possibly, rCBF are simultaneously recorded, serially monitored, and combined with CT scan and clinical data, a noninvasive insight into the pathophysiology of brain damage in vivo, in real time, may be achieved. In fact, both the topography of structural damage and the response of the brain to injury in focal, perifocal, and distant areas may be checked, analyzing the relationship between perfusion and neural function. Moreover, as functional investigations are more sensitive to drugs than the clinical and CT scan data, they provide a new opportunity for evaluating the effects of therapy, thus improving both patient management and therapeutic trials. However, the use of brain mapping is still at an early stage and this field will be extended more and more in the future, as brain mapping becomes more ubiquitous and less separated from conventional recordings.

References

1. Adler G, Schwerdtfeger K, Lang E, Kivelitz R, Nacimiento AC, Loew F (1985) The use of somatosensory, brainstem auditory and visual evoked potentials for prognostic and localization purposes in the assessment of head injury. In: Morocutti C, Rizzo PA (eds) Evoked potentials. Neurophysiological and clinical aspects. Elsevier, Amsterdam, pp 361–367
2. Anderson DC, Bundlie S, Rockswold GL (1984) Multimodality evoked potentials in closed head trauma. Arch Neurol 41:369–374
3. Anziska BJ, Cracco RQ (1980) Short latency somatosensory evoked potentials in brain dead patients. Arch Neurol 37:222–225
4. Asheval S, Schneider S (1979) Failure of electroencephalography to diagnose brain death in comatose children. Ann Neurol 6:512–517
5. Baratto F, Munari M, Dona' B, Behr AU, Pasini L, Facco E (1990) I potenziali evocati motori (MEP) nel trauma cranico. Min Anest 56:1307–1310
6. Belsh JM, Chokroverty S (1987) Short-latency somatosensory evoked potentials in brain-dead patients. Electroencephalogr Clin Neurophysiol 68:75–78
7. Bricolo A (1976) Electroencephalography in neurotraumatology. Clin Electroencephalogr 7:184–197
8. Bricolo A, Turazzi S, Faccioli F (1979) Combined clinical and EEG examinations for assessment of severity of acute head injuries. Acta Neurochir (Wein) 24:40–42
9. Cant BR, Shaw NA (1984) Monitoring by compressed spectral array in prolonged coma. Neurology 34:35–39
10. Cant BR, Hume AL, Judson JA, Shaw NA (1986) The assessment of severe head injury by short-latency somatosensory and brain-stem auditory evoked potentials. Electroencephalogr Clin Neurophysiol 65:188–195
11. Cant BR (1987) Evoked potential monitoring of post-traumatic coma and its relation to outcome. Electroencephalogr Clin Neurophysiol [Suppl] 39:250–254
12. Chatrian GE, Bergamini L, Dondey M, Klass DW, Lennox-Buchthal M, Peterson I (1974) A glossary of terms most commonly used by clinical electroencephalographers. Electroencephalogr Cl in Neurophysiol 37:538–548
13. Dear PRF, Godfrey DJ (1985) Neonatal auditory brainstem responses reliably diagnose brainstem death. Arch Dis Child 60:17–19
14. Duffy FH, Burchfiel JL, Lombroso CT (1979) Brain electrical activity mapping (BEAM): a method for extending the clinical utility of EEG and evoked potential data. Ann Neurol 5:309–312

15. Facco E, Martini A, Zuccarello M, Agnoletto M, Giron GP (1985) Is the auditory brain-stem response (ABR) effective in the assessment of post-traumatic coma? Electroencephalogr Clin Neurophysiol 62:332–337
16. Facco E, Caputo P, Casartelli Liviero M et al. (1988) Auditory and somatosensory evoked potentials in brain-dead patients. Riv Neurol 58(4):140–145
17. Facco E, Munari M, Casartelli Liviero M et al. (1988) Serial recordings of auditory brainstem responses in severe head injury: relationship between test timing and prognostic power. Int Care Med 14:422–428
18. Facco E, Toffoletto F, Caputo P et al. (1989) Multimodality evoked potentials in subarachnoidal hemorrhage. Neurology 39 [Suppl 1]:293 (abstract)
19. Facco E, Munari M, Colombis G, Baggio C, Bolcioni G (1990) EEG mapping versus high resolution spectra of power, coherence and phase: reference artifacts in patients with focal brain damage. Brain Topogr 2:305–306
20. Facco E, Munari M, Baratto, F, Don B, Pasini L, Giron GP (1990) Relationship between regional cerebral blood flow and EEG mapping in severe head trauma: preliminary report. Int Care Med 16 [Suppl 1]: S115 (abstract)
21. Facco E, Munari M, Baggio C, Baratto F, Casartelli Liviero M, Giron GP (1990) EEG mapping in comatose patients: relationship between topography of alpha rhythm and level of consciousness. Int Care Med 16 [Suppl 1]: S44 (abstract)
22. Facco E, Don B, Colombis G, Baggio C, Bolcioni G, Giron GP (1990) Pitfalls in the EEG evaluation of stupor and coma: usefulness of coherence and phase analysis. Int Care Med 16 [Suppl 1]: S13 (abstract)
23. Facco E, Munari M, Baratto F, Don B, Giron GP (1990) Somatosensory evoked potentials in severe head trauma. In: Rossini PM, Mauguiere F (eds) New trends and advanced techniques in clinical neurophysiology (EEG Suppl 41). Elsevier Science, Amsterdam, pp 330–341
24. Facco E, Munari M, Toffoletto F et al. (1990) Multimodality evoked potentials in severe head injury: a neurophysiological scale for early prognosis. Electroencephalogr Clin Neurophysiol scale for early prognosis. Electroencephalogr Clin Neurophysiol 75:S42 (abstract)
25. Facco E, Casartelli Liviero M, Munari M, Toffoletto F, Baratto F, Giron GP (1990) Short latency evoked potentials: new criteria for brain death? J Neurol Neurosurg Psychiat 53:351–353
26. Facco E, Munari M, Dona' B et al. (1991) Spatial mapping of SEP in comatose patients: improved outcome prediction by combined parietal N20 and frontal N30 analysis. Brain Topogr 3:447–455
27. Facco E, Munari M, Baratto F, Behr AU, Giron GP (1993) Multimodality evoked potentials (auditory, somatosensory and motor) in coma. Clin Neurophysiol (in press)
28. Facco E, Martini A, Munari M, Baratto F, Giron GP (1993) I potenziali evocati in terapia intensiva. In: Grandori F, Martini A (eds) Potenziali evocati uditivi: basi teoriche ed applicazioni pratiche. Piccin, Padova (in press)
29. Facco E, Baratto F, Munari M et al. (1991) Sensorimotor central conduction time in comatose patients. Electroencephalogr Clin Neurophysiol 80:469–476
30. Fischer C, Ibanez V, Jourdan C, Grau A, Mauguiere F, Artru F (1988) Potentiels evoques auditifs precoces (PEAP), auditifs de latence moyenne (PALM) et somesthesiques (PES) dans le prognostic vital et fonctionnel des traumatismes craniens graves en reanimation. Agressologie 29:359–363
31. Ganes T, Lundar T (1988) EEG and evoked potentials in comatose patients with severe brain damage. Electroencephalogr Clin Neurophysiol 69:6–13

32. Garcia-Larrea L, Bertrand O, Artru F, Pernier J, Mauguiere F (1987) Brainstem monitoring in coma II: dynamic interpretation of preterminal BAEP changes observed until brain death in deeply comatose patients. Electroencephalogr Clin Neurophysiol 68:446–457

33. Garcia-Larrea L, Artru F, Bertrand O, Jourdan C, Deschamps J, Mauguiere F (1988) Monitorage des potentiels évoqués auditifs du tronc cerebral lors des alterations aigues de la pression intra-cranienne. Agressologie 29:329–332

34. Gevins A, Le J, Brickett P, Reutter B, Desmond J (1991) Seeing through the skull: advanced EEGs use MRIs to accurately measure cortical activity from the scalp. Brain Topogr 4:125–131

35. Goldie WD, Chiappa KH, Young RR, Brooks EB (1981) Brainstem auditory and short latency somatosensory evoked responses in brain death. Neurology 31:248–256

36. Greenberg RP, Mayer DJ, Becker DP, Miller JD (1977) Evaluation of brain function in severe human head trauma with multimodality evoked potentials, part 1. J Neurosurg 47:150–162

37. Greenberg RP, Becker DP, Miller JD, Mayer DJ (1977) Evaluation of brain function in severe human head trauma with multimodality evoked potentials, part 2. J Neurosurg 47:163–177

38. Greenberg RP, Newlon PG, Hyatt MS, Narayan RK, Becker DP (1981) Prognostic implications of early multimodality evoked potentials in severely head-injured patients. A prospective study. J Neurosurg 55:227–236

39. Grindal AB, Suter C, Martinez AJ (1977) Alpha pattern coma: 24 cases with 9 survivors. Ann Neurol 1:371–377

40. Hjorth B (1975) An on-line transformation of scalp potentials into ortogonal source derivations. Electroenceph Clin Neurophysiol 39:526–530

41. Hughes JR (1978) Limitations of the EEG in coma and brain death. Ann N Acad Sci 315:121–135

42. Hume AL, Cant BR (1981) Central somatosensory conduction after head injury. Ann Neurol 10:411–419

43. Hutchinson DO, Frith RW, Shaw NA, Judson JA, Cant BR (1991) A comparison between electroencephalography and somatosensory evoked potentials for outcome prediction following head injury. Electroencephalogr Clin Neurophysiol 78:228–233

44. Inouye T, Shinosaki K, Yagasaki A (1983) The direction of spread of alpha activity over the scalp. Electroencephalogr Clin Neurophysiol 55:290–300

45. Inouye T, Shinosaki K, Yagasaki A, Shimizu A (1986) Spatial distribution of generators of alpha activity. Electroencephalogr Clin Neurophysiol 63:353–360

46. Karnaze DS, Marshall LF, McCarthy CS, Klauber MR, Bickford RG (1982) Localizing and prognostic value of auditory evoked responses in coma after closed head injury. Neurology 32:299–302

47. Karnaze DS, Weiner JM, Marshall LF (1985) Auditory evoked potentials in coma after closed head injury: a clinical-neurophysiologic coma scale for predicting outcome. Neurology 35:1122–1126

48. Katznelson RD (1981) EEG recording, electrode placement, and aspects of generators localization. In: Nunez PL (ed) Electric fields of the brain. Oxford University Press, Oxford, pp 176–213

49. Labar DR, Fisch BJ, Pedley TA, Fink ME, Solomon RA (1991) Quantitative EEG monitoring for patients with subarachnoid hemorrhage. Electroencephalogr Clin Neurophysiol 78:325–332

50. Lindsay KW, Carlin J, Kennedy I, Fry I, McInnes A, Teasdale GM (1981) Evoked potentials in severe head injury : analisis and relation to outcome. J Neurol Neurosurg Psychiatry 44:796–802
51. Lopes da Silva F (1991) Neural mechanisms underlying brain waves: from neural membranes to networks. Electroencephalogr Clin Neurophysiol 79:81–93
52. Machado C, Valdes P, Garcia-Tigera J et al. (1991) Brain-stem auditory evoked potentials and brain death. Electroencephalogr Clin Neurophysiol 80:392–398
53. Martini A, Zuccarello M, Agnoletto M, Facco E, Molinari G (1984) Auditory brainstem responses in the clinical evaluation of post-traumatic coma. Audiol Ital 1:273–280
54. Mauguiere F, Grand C, Fischer C, Courjon J (1982) Aspects des potentiels évoqués auditifs et somesthesiques précoces dans les comas neurologiques et la mort cerebrale. Rev EEG Neurophysiol 12:280–286
55. Mauguiere F, Desmedt JE, Courjon J (1983) Astereognosis and dissociated loss of frontal and parietal components in somatosensory evoked potentials in hemispheric lesions. Brain 106:271–311
56. Mauguiere F, Ibanez V, Deiber MP, Garcia-Larrea L (1987) Noncephalic reference recording and spatial mapping of short-latency SEPs to upper limb stimulation: normal responses and abnormal patterns in patients with nondemyelinating lesions of the CNS. In: Barber C, Blum T (eds) Evoked potentials III. Butterworths, Stoneham, pp 40–55
57. Mauguiere F, Garcia-Larrea L, Fischer C (1988) Le monitorage des potentiels évoqués au cours des comas traumatiques. Agressologie 29:351–357
58. Mjoen S, Nordby HK, Torvik A (1983) Auditory evoked brainstem responses (ABR) in coma due to severe head trauma. Acta Otolaryngol (Stockh) 95:131–138
59. Moller M, Holm B, Sindrup E, Nielson BL (1978) Electroencephalographic prediction of anoxic brain damage after resuscitation from cardiac arrest in patients with acute myocardial infarction. Acta Med Scand 203:33–37
60. Munari M, Facco E, Caputo P et al. (1988) Auditory brainstem responses in severe head injury: prognostic value of amplitude ratio of waves V/I. Int Care Med 14 [Suppl 1]: 298
61. Munari M, Baratto F, Casartelli-Liviero M, Dona' B, Facco E (1990) Correlazione tra EEG e stato di coscienza nel trauma cranico grave. Minerva Anestesiol 56(10): 1299–1302
62. Munari M, Toffoletto F, Baratto F, Meroni M, Pasini L, Facco E (1990) Early prognosis in severely head-injured children by auditory brain stem responses (ABR). Int Care Med 16 [Suppl 1]: S73 (abstract)
63. Nagata K (1989) Topographic EEG mapping in cerebrovascular disease. Brain Topogr (1989) 2:119–128
64. Narayan RK, Greenberg RP, Miller DJ et al. (1981) Improved confidence of outcome prediction in severe head injury: a comparative analysis of the clinical examination, multimodality evoked potentials, CT scanning and intracranial pressure. J Neurosurg 54:751–762
65. Nau HE, Wiedemayer H, Dalbah A, Engel W, Mais J (1988) Zum Wert evozierter Potentiale auf der neurochirurgischen Intensivstation. Neurochirurgia (Stuttg) 31 [Suppl 1]: 170–174
66. Newlon PG, Greenberg RP (1984) Evoked potentials in severe head injury. J Trauma 24:61–66

67. Offner FF (1950) The EEG as potential mapping: the vlaue of the average monopolar reference. Electroencephalogr Clin Neurophysiol 2:215–216
68. Ottaviani F, Almadori AB, Calderazzo AB, Frenguelli A, Paludetti G (1986) Auditory brain-stem (ABRs) and middle latency auditory responses (MLRs) in the prognosis of severely head-injured patients. Electroencephalogr Clin Neurophysiol 65:196–202
69. Ozaki H, Suzuki H (1987) Transverse relationships of the alpha rhythm on the scalp. Electroencephalogr Clin Neurophysiol 66:191–195
70. Papanicolaou AC, Loring DW, Eisenberg HM, Raz N, Contreras FL (1986) Auditory brain stem evoked responses in comatose head-injured patients. Neurosurgery 18:173–175
71. Plum F, Posner JB (1983) Diagnosis of Stupor and coma. Davis, Philadelphia (1983)
72. Prior PT (1985) EEG monitoring and evoked potentials in brain ischemia. Br J Anaesth 57:63–81
73. Rappaport M, Hopkins HK, Hall K, Belleza T (1981) Evoked potentials and head injury. 2. Clinical applications. Clin Electroencephalogr 12:167–176
74. Rothstein TL, Thomas EM, Sumi SM (1991) Predicting outcome in hypoxic-ischemic coma. A prospective clinical and electrophysiologic study. Electroencephalogr Clin Neurophysiol 79:101–107
75. Rumpl E, Prugger M, Gerstenbrand F, Hackl JM, Pallua A (1983) Central somatosensory conduction time and short latency somatosensory evoked potentials in post-traumatic coma. Electroencephalogr Clin Neurophysiol 56:583–596
76. Samson-Dollfus D, Delmer C, Vaschalde Y, Dreano E, Fodil D (1989) Topography of background EEG rhythms in normal subjects and in patients with cerebrovascular disorders. In: Maurer K (ed) Topographic brain mapping of EEG and evoked potentials. Springer, Berlin Heidelberg New York, pp 185–191
77. Sanamann ML (1983) The use of EEG in the prognosis of coma. Clin Electroencephalogr 14:47–52
78. Seales DM, Rossiter VS, Weinstein ME (1979) Brainstem auditory evoked responses in patients comatose as a result of blunt head trauma. J Trauma 19:347–353
79. Shepard D (1968) A two dimensional interpolation function for irregular-spaced data. Proc ACM Nat Conf 517–524
80. Slimp JC, Tamas LB, Stolov WC, Wyler AR (1986) Somatosensory evoked potentials after removal of somatosensory cortex in man. Electroencephalogr Clin Neurophysiol (1988) 65:111–117
81. Sonnet ML, Buffet G, Godard J et al. (1988) Les potentiels évoqués somesthesiques (PESp) et auditifs précoces (PEAp) chez les traumatisés craniens graves. Agressologie 29:371–373
82. Starr A (1976) Auditory brainstem responses in brain death. Brain 99:543–554
83. Synek VM (1990) Value of revised EEG coma scale for prognosis after cerebral anoxia and diffuse head injury. Clin Electroencephalogr 21:25–30
84. Trojaborg W, Jorgensen EO (1973) Evoked cortical potentials in patients with "isoelectric" EEGs. Electroencephalogr Clin Neurophysiol 35:301–309
85. Tsubokawa T, Nishimoto H, Yamamoto T, Kitamura M, Katayama Y, Moriyasu N (1980) Assessment of brainstem damage by the auditory brainstem response in acute severe head injury. J Neurol Neurosurg Psychiatry 43:1005–1011
86. Uziel A, Benezech J, Lorenzo S, Monstrey Y, Duboin MP, Roquefeuil B (1982) Clinical applications of brainstem auditory evoked potentials in comatose patients. Adv Neurol 32:195–202

87. Zentner J, Ebner A (1988) Somatosensibel und motorisch evozierte Potentiale bei der prognostischen Beurteilung traumatisch und nichttraumatisch komatoser Patienten. EEG EMG 18:267–271
88. Zentner J, Ebner A (1988) Prognostic value of somatosensory- and motor-evoked potentials in patients with a non-traumatic coma. Eur Arch Psychiatry Neurol Sci 237:184–187
89. Zuccarello M, Fiore DL, Pardatscher K et al. (1983) Importance of auditory brainstem responses in the CT scan diagnosis of traumatic brainstem lesions. AJNR 4:481–483

Part V

Present and Future Trends in Cerebral Monitoring

Present and Future Trends in Multimodal Cerebral Monitoring in Anesthesia and Intensive Care

G. Litscher, G. Schwarz, W. Marte, G. Pfurtscheller, and W. F. List

Introduction

The activation of neuronal structures and systems in the brain is accompanied by changes of the electrical potentials recorded from the scalp. Evoked potentials (EP) can be used to study brain functions in the intensive care unit and the operating room, in addition to spontaneous cerebral electrical activity (electroencephalography, EEG) [1–5].

Different sensory systems can be investigated by recording auditory (AEP), visual (VEP), and somatosensory (SEP) EP. In the auditory and somatosensory system, either the pathway between peripheral receptors and the primary area, or integrative cortical processes can be investigated. In the first case, early EP components generated from primary cortical areas within the brain stem have to be analyzed, and in the second case, mid- and long-latency components. Early EP components are the brain stem AEP (BAEP) recorded as far-field potentials from the scalp and the cervical (N13) and primary cortical (N20) components (SSEP) after electrical stimulation, for example, of the median nerve. All these early or short-latency (< 25 ms) components are relatively independent of pharmacological influences [2, 3] and especially suitable for proving the function of the brain stem and the corresponding pathway, respectively.

Mid- and long-latency EP components depend on the degree of cortical integrity and are influenced by drugs. Long-latency SEP and VEP are extremely variable in coma, but especially suitable for assessing and monitoring cortical functions during emergence from coma [12]. When the auditory or somatosensory pathway in the brain stem is affected or the primary area is disturbed, no long-latency components can be recorded.

In the present paper, two clinical examples are described and the advantages of the simultaneous monitoring of early EP components (SSEP and BAEP) together with the EEG is demonstrated.

Methods

Two channels of EEG, BAEP, and SSEP were monitored in parallel, together with different cardiocirculatory and respiratory parameters, using a multivariable monitoring system developed in Graz [5–9, 13, 14, 17]. Two symmetrical EEG channels from bipolar electrodes were acquired using standard gold disk electrodes (Grass E6GH). Gentle scratching of the skin surface was used to reduce the interelectrode impedances to less than 2 kΩ. Two-second periods of EEG activity were analyzed using a fast Fourier transform algorithm and plotted as a compressed spectral array.

For acoustic stimulation, clicks with a duration of 0.2 ms were generated and presented to the patient monaurally through small earphones at a rate of ten per second. The pulse polarity was reversed alternately to decrease the amplitude of stimulus artifacts during the averaging process. The non-stimulated ear was masked with white noise. Click intensity was adjusted to 85 dB sound pressure level. Analysis time was 10 ms and the sampling rate was 5.2 kHz. Each trial consisted of 1800 on-line stimulus presentations in patients in the intensive care unit and of 600 stimuli in intraoperative monitoring. SSEP, following median nerve stimulation at the wrist, were recorded over the second cervical vertebrate and from the scalp over the contralateral somatosensory area with a sampling rate of 2.6 kHz. A frontal reference and a bandpass of 10–1500 Hz were used. Stimuli with a duration of 0.1 ms were delivered to the median nerve at a rate of five per second. The stimulus intensity was adjusted to produce a vigorous twich of the thumb. For on-line intraoperative monitoring, 300 repetitions were averaged, and for monitoring comatose patients in the intensive care unit, 900 stimulus responses.

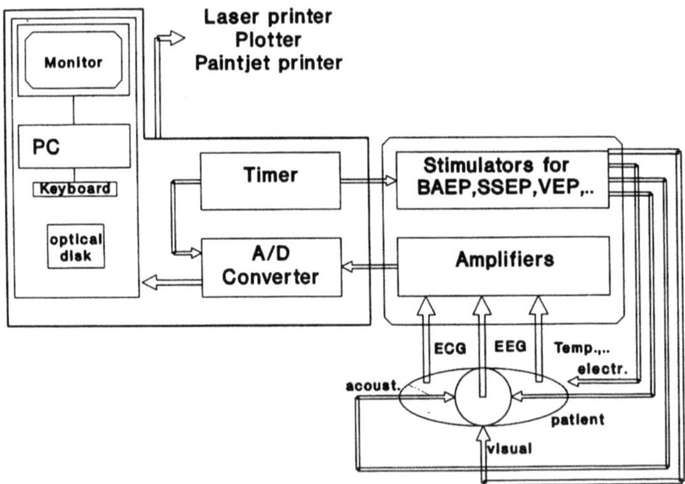

Fig. 1. The multivariable cerebral monitoring system. *PC*, personal computer; *BAEP*, brain stem auditory evoked potential; *SSEP*, cervical (N13) and primary cortical (N20) somatosensory evoked potential components; *VEP*, visual evoked potential; *ECG*, electrocardiogram; *EEG*, electroencephalogram

An updated version of combined EEG–EP data processing was used for the investigation. The system is based on a personal computer (80386 or 80486) with an amplifier unit located in the EEG headbox, a stimulation unit, an optical disk drive, a laser, and a paint jet printer (see Fig. 1). Further methodological details were presented in earlier investigations [4–9, 13, 14, 16, 17].

Results

In order to demonstrate the clinical value of the multimodal EEG–EP monitoring, two examples, the first from the intensive care unit and the second from the operating theatre, are given below.

Patient 1

A 23-year-old pregnant female (27th week of pregnancy) was involved in an automobile accident resulting in a severe head injury. After hospitalization at the anesthesiological intensive care unit, the patient was in a coma with a Glasgow Coma Score (GCS) rating of 6.

The first computed tomography (CT) scan (23 September 1992, 2 P.M.) showed small multiple contusion hemorrhages in the frontal areas, a traumatic subarachnoideal hemorrhage, compression of ventricular system (Fig. 2), and obliterations of perimesencephal cistern. The next morning the neurological situation showed a clinical improvement (GCS, 7).

The first multivariable electrophysiological monitoring was started that morning at 10:30 A.M. At the beginning of the neuromonitoring, the pupils were equal in size without reaction to light. EEG, BAEP, and SSEP were recorded simultaneously and continuously. Part of the results are shown in Fig. 3.

EEG power spectra showed clear suppression in all frequency bands. In the spectra of the left hemisphere, electrocardiographic artifacts were detected. BAEP consisted of all components (wave I to wave V). However, the interpeak latency between wave I and V (IPL I–V, 4.60 ms) was bilaterally significantly (> 2.5 SD) prolonged. The latencies and amplitudes of the cervical SSEP (N13) recorded over C_2 after electrical median nerve stimulation were within the normal range. However, the cortical N20 component was bilaterally absent (see Fig. 3, marked with arrows).

Due to these findings (loss of cortical responses of SSEP), a control of pupil size was performed again immediately. At that moment, the pupils were bilaterally dilated without reaction to light. The patient's GCS was 3.

A CT scan was performed without delay and shows a subdural hematoma (right) and a midline shift (Fig. 2) compared with the scan performed the day before.

After performing a craniotomy, a constriction of the pupils to their normal size with a small reaction to light unilaterally was found postoperatively. The

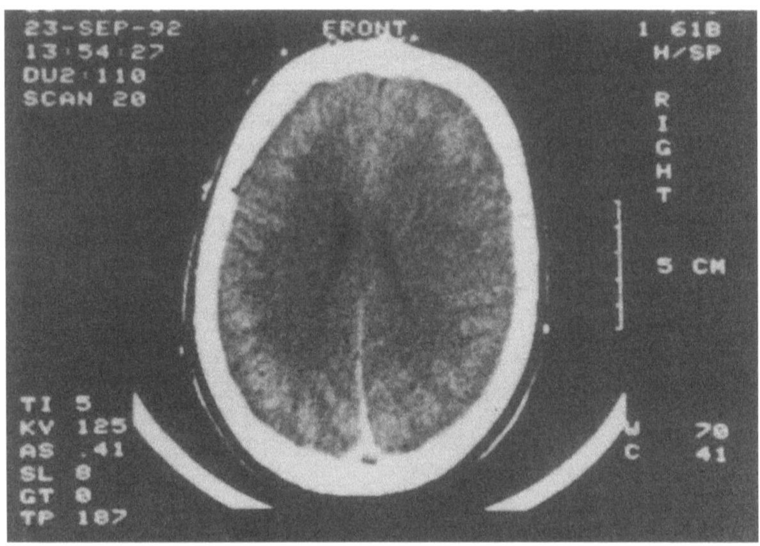

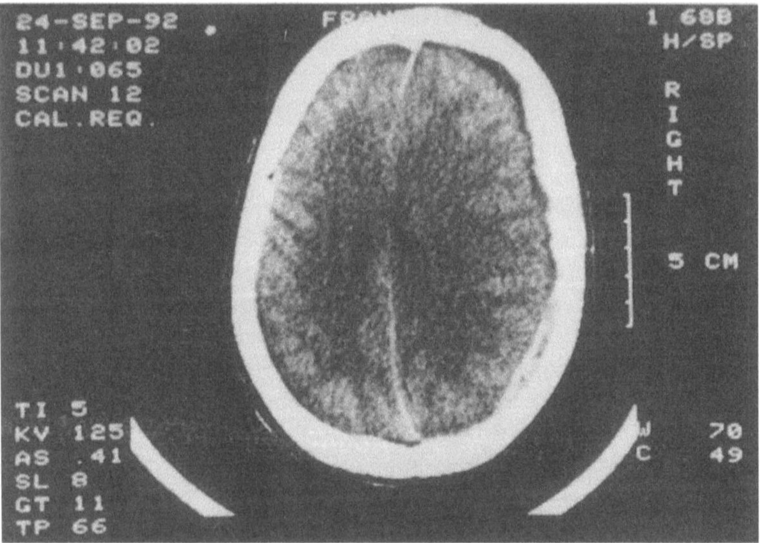

Fig. 2. Initial cranial computed tomography (CT) scan (23 September 1992, 2 P.M.; *top*) and control CT scan (24 September 1992, 11:40 A.M.; *bottom*)

postoperative findings (24 September 1992, 9:30 P.M.) of the EEG and multimodal evoked potentials are presented graphically in Figs. 4–6.

The total EEG power increases, I–V IPL of the BAEP decreases by 0.5 ms to the normal range, and clear N20 components on both sides could be detected. The central conduction time (CCT) was 5.0 ms on the left and 6.5 ms on the right

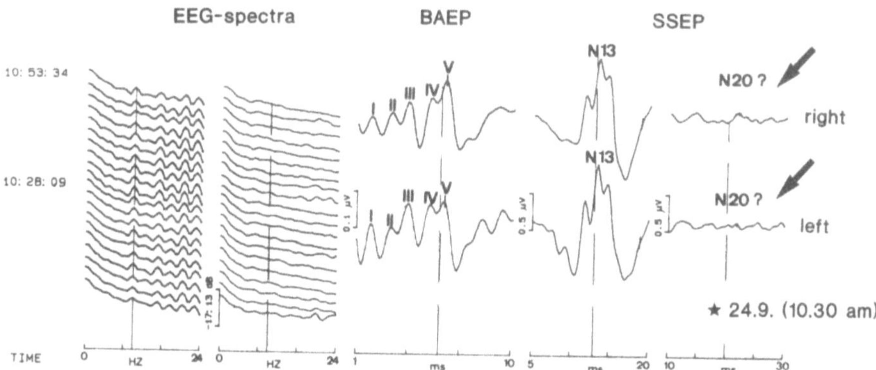

Fig. 3. Simultaneous electroencephalogram (*EEG*) and multimodality evoked potential (EP) recording in a 23-year-old comatose patient with severe head injury (24 September 1992, 10:30 A.M.). From left to right; EEG power spectra of the left and right hemisphere; brain stem auditory EP (BAEP, left, *bottom*; right, *top*); cervical (N13) and primary cortical (N20) components (SSEP; left, *bottom*, and right, *top*). Note that the cortical components of the SSEP are bilaterally absent (marked with *arrows*)

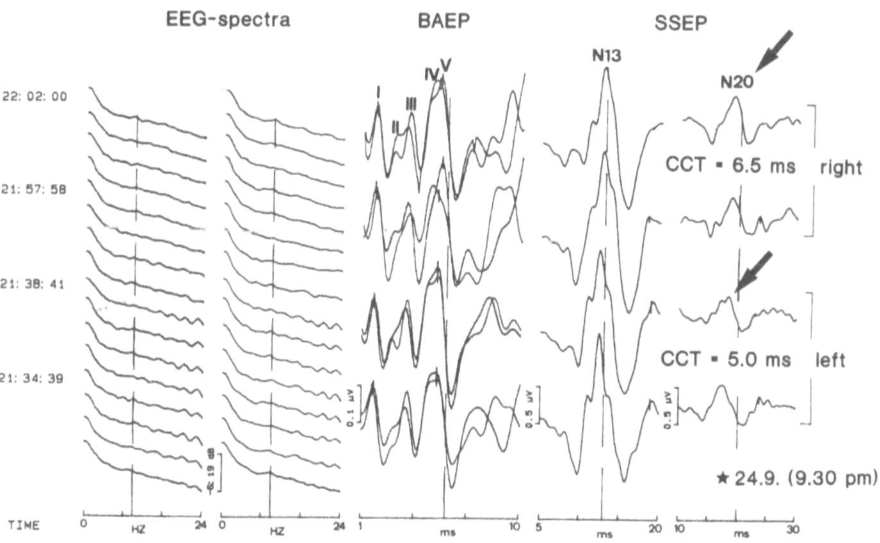

Fig. 4. Simultaneous electroencephalogram (EEG)–brain stem auditory evoked potential (BAEP)–cervical (N13) and primary cortical (N20) somatosensory evoked potential components (SSEP)-investigation in the same comatose patient as in Fig. 3 (24 September 1992, 9:30 P.M.). Note the bilateral presence of the cortical SSEP after surgical intervention. *CCT*, central conduction time

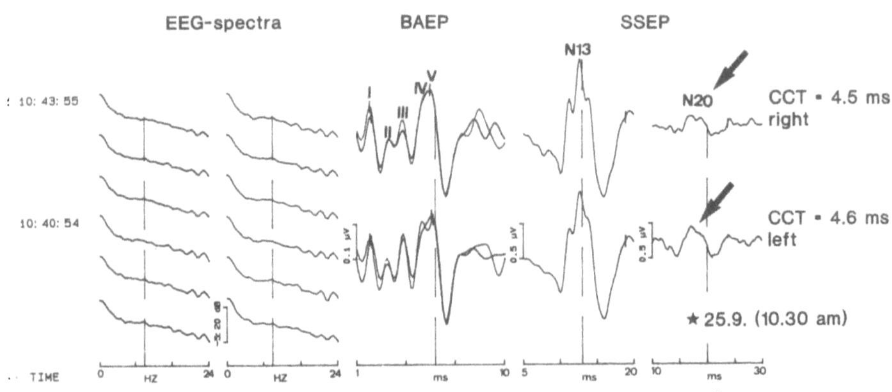

Fig. 5. Control measurement in the same 23-year-old comatose female as in Fig. 3 (25 September 1992, 10:30 A.M.). *EEG*, electroencephalogram; *BAEP*, brain stem auditory evoked potential; *SSEP*, cervical (N13) and primary cortical (N20) somatosensory evoked potential component; *CCT*, central conduction time

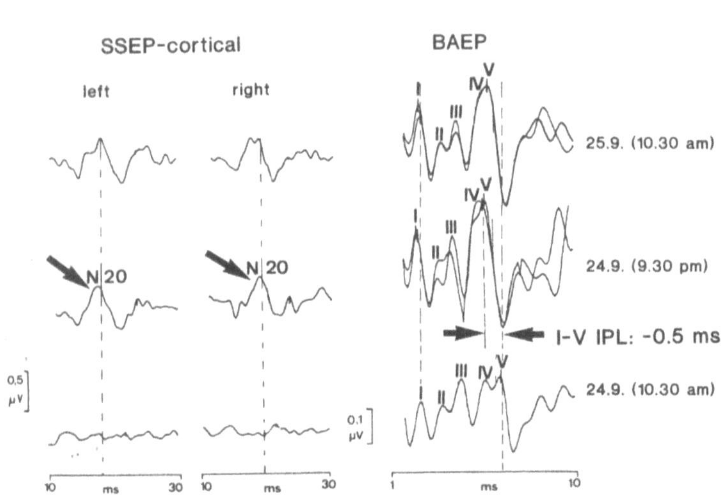

Fig. 6. Primary cortical (N20) Somatosensory evoked potential components (SSEP) and brain stem auditory evoked potential (*BAEP*) data from three different measurements in the same patient as in Fig. 3. The reappearance of the cortical SSEP components and the decrease of the interpeak latency between waves I and V (*I–V IPL*) of the BAEP are the most prominent findings

side. Similar results were found in a further control investigation on the next day at 10:30 A.M. (Figs. 5, 6) CCT decreased again. Nevertheless, after the temporary clinical improvement and performance of a successful caesarean section, the patient died on the ninth day after admission to the intensive care unit.

Patient 2

A 60-year-old man suffering from an ischemic injury presented with a massive occlusion of both carotid arteries in ultrasonographic and neuroradiological investigations. In addition, a vertebral stenosis on the left side was detected.

In order to diminish the combined anesthesiological and surgical risk, the neurosurgical intervention (extra-intra-arterial bypass) was performed under hyperbaric oxygenation (HBO). Figure 7 shows a protocol of the multivariable monitoring under HBO and the steady state conditions of enflurane anesthesia.

During surgical treatment, an increase in latency of the cortical SSEP component N22 was observed. The cervical N15 peak latency remained unchanged, as expected. The N22 amplitude did not decrease. This was thought to be related to the effects of HBO. The cortical SSEP reacquired the original characteristics in latency in the period after decompression. No significant changes in BAEP latencies were detected during the monitoring session. EEG power spectra demonstrate alterations as early as 5 min before compression (10 m seawater). The dominant frequency in the alpha band decreased from 10 to 8 Hz. These changes are due to the administration of 0.2 mg fentanyl marked with A on Fig. 7. After surgical intervention, clinical examination showed no neurological deficit.

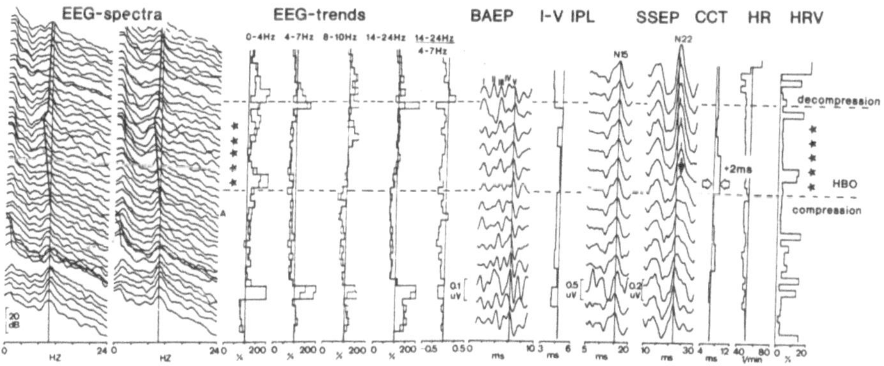

Fig. 7. Intraoperative monitoring of a 60-year-old patient during surgical intervention performed under hyperbaric oxygenation. The protocol shows data recorded over a 1-h period. From *left* to *right*, Compressed electroencephalogram (*EEG*) spectral array from the left and right hemisphere; EEG power of different bands (0–4 Hz, 4–7 Hz, 8–10 Hz, and 14–24 Hz); EEG ratio 14–24 to 4–7 Hz; brain stem auditory evoked potential (*BAEP*); interpeak latency between waves I and V (*I–V IPL*); cervical (N15) and cortical primary (N22) somatosensory evoked potential components (SSEP); central conduction time (*CCT*); heart rate (*HR*), and heart rate variability (*HRV*). The hyperbaric situation is marked by *HBO*.

Discussion

The evaluation of comatose patients by means of simultaneous multimodality EP recordings together with EEG can make the diagnosis and prognostic assessment more precise. The first case obviously showed that continuous multimodal EP monitoring is able to reflect massive acute pathophysiological events, which should subsequently have effective therapeutic consequences (in this case, a craniotomy). Beyond that, the case showed that each clinical investigation performed in very short intervals (monitoring of pupils) has a step by step character. On the other hand, electrophysiological monitoring can actually be performed continuously.

The example in Fig. 7 shows an intraoperative recording under extreme conditions. EEG and EP have not been used extensively in the exploration of the hyperbaric environment [6]. In the past, numerous technical problems have been associated with data acquisition, especially in the operating theatre [2, 3, 11, 15]. Our second case shows that it is possible to record different modalities of EP in parallel, even under difficult clinical conditions.

Serial recording of multimodality EP in the intensive care unit was a time-consuming method. For example, 10 years ago about 4 h were necessary for one recording session including three to four modalities of EP [1, 10]. Using new simultaneous recording techniques, e.g., EEG, BAEP, and SSEP monitoring [4, 6, 13, 16] or median nerve SSEP and tibial nerve SSEP monitoring, or EEG, AEP and VEP monitoring [9], it is now possible to perform multimodality recordings in intensive care units within 1 h. However, there are still limitations in clinical interpretation and artifact detection. Although our example of the first patient has shown that changes in EP have direct clinical consequences, there are still great problems in the choice of automatic alarm criteria.

Our experiences show that it is evident that any change in EP must be interpreted both by a physician and a biomedical engineer in order to encompass multiple points of view. Using new modern devices and global assessments of interpretation, multivariable cerebral monitoring is useful for both clinical decision making and scientific research, as our examples show.

Acknowledgements. These investigations were supported by the Jubiläumsfonds der Österreichischen Nationalbank (project 4205) and the Austrian Science Foundation (project S49-03). The authors are grateful to Dr. H. Seitlinger for part of the electrophysiological data recording.

References

1. Dauch WA (1991) Prediction of secondary deterioration in comatose neurosurgical patients by serial recording of multimodality evoked potentials. Acta Neurochir (Wien) 111:84–91

2. Freye E (1990) Cerebral monitoring in the operating room and the intensive care unit. Kluwer, Dordrecht

3. Grundy BL, Villani RM (1988) Evoked potentials. Intraoperative and ICU monitoring. Springer, Vienna New York

4. Hilz MJ, Litscher G, Weis M, Claus D, Druschky KF, Pfurtscheller G, Neundörfer B (1991) Continuous multivariable monitoring in neurological intensive care patients – preliminary reports on four cases. Intensive Care Med 17:87–93

5. Litscher G, Schwarz G, Schalk HV, Pfurtscheller G (1988) Polygraphisches Neuromonitoring nach Medikamentenintoxikation. Intensivbehandlung 13(4):154–158

6. Litscher G, Friehs G, Maresch H, Pfurtscheller G (1990) Electroencephalographic and evoked potential monitoring in the hyperbaric environment. J Clin Monit 6:10–17

7. Maresch H, Pfurtscheller G, Schuy S (1983) Brain function monitoring: a new method for simultaneous recording and processing of EEG power spectra and brainstem potentials (BAEP). Biomed Tech (Berlin) 28:117–122

8. Maresch H, Gonzalez A, Pfurtscheller G (1985) Intraoperative patient monitoring including EEG and evoked potentials. Proceedings of XIV ICMBE and VII ICMP, Espoo, pp 776–777

9. Maresch H, Litscher G, Pfurtscheller G (1992) Methods and applications of continuous and simultaneous EEG and evoked potential monitoring. In: Lun KC, Degoulet P, Piemme TE, Rienhoff O (eds) MEDINFO 92, 7th world congress on medical informatics, 6–10 Sep 1992, Geneva. Elsevier, Amsterdam, pp 844–850

10. Mecum CR, Greenberg RP, Becker DB (1983) Evoked potentials in severe head injury patients. Am J EEG Technol 22:125–134

11. Nuwer M (1986) Evoked potential monitoring in the operating room. Raven, New York

12. Pfurtscheller G, Schwarz G, Gravenstein N (1985) Clinical relevance of long-latency SEPs and VEPs during coma and emergence from coma. Electroencephalogr Clin Neurophysiol 62:88–98

13. Pfurtscheller G, Schwarz G, Schroettner O, Litscher G, Maresch H, Auer L, List WF (1987) Continuous and simultaneous monitoring of EEG spectra and brainstem auditory and somatosensory evoked potentials in the intensive care unit and the operating room. J Clin Neurophysiol 4(4):389–396

14. Pfurtscheller G (1993) Special uses of EEG computer analysis in clinical environments. In: Niedermeyer E, Lopes da Silva F (eds) Electroencephalography, 3rd edn. Williams and Wilkins Publ., New York

15. Schramm J, Moller AR (1991) Intraoperative neurophysiologic monitoring in neurosurgery. Springer, Berlin Heidelberg New York

16. Schwarz G, Litscher G, Pfurtscheller G, Schalk HV, Rumpl E, Fuchs G (1992) Brain death: timing of apnea testing in primary brain stem lesion. Intensive Care Med 18:315–316

17. Steller E, Litscher G, Maresch H, Pfurtscheller G (1990) Multivariables Langzeit-Monitoring von zerebralen und kardiovaskulären Größen mit Hilfe eines Personal Computers. Biomed Tech (Berlin) 35:90–97

Jugular Bulb Venous Oxygen Saturation and Transcranial Doppler Ultrasonography in Neurosurgical Patients

N. M. Dearden

Introduction

Ischaemic brain damage from systemic hypotension, intracranial hypertension, hypoxaemia and cerebral vasospasm is commonplace in critically ill neurosurgical patients and contributes to morbidity in patients who survive [1]. Reduction of cerebral perfusion pressure (CPP) from combinations of systemic hypotension and elevation of intracranial pressure (ICP) lowers cerebral blood flow (CBF), while regional ischaemia may result from vasospasm or secondary to intracranial hypertension that causes compression of the posterior cerebral artery against the tentorium cerebelli, leading to medial occipital lobe ischaemia (Fig. 1).

In view of the incidence of cerebral ischaemia in critically ill neurosurgical patients, monitoring and correction of cerebral oxygen delivery during their initial resuscitation and subsequent intensive care offers considerable potential benefit. Such patients may have cerebral vascular distortion from intracranial shifts, deranged autoregulation and cerebral vasospasm in association with intracranial hypertension and may therefore require elevation of their CPP considerably above the arbitrary, but commonly quoted, level of 60 mm Hg to avert cerebral ischaemia. Recently, it has been possible to monitor cerebral venous oxygen saturation continuously in an effort to identify and correct global, and to a lesser degree, regional cerebral ischaemia, while the emergence of transcranial Doppler ultrasonography has allowed continuous non-invasive insonation of major intracranial vessels, thereby facilitating diagnosis of regional hypoperfusion and vasospasm. This article reviews the potential benefits and limitations of these two monitoring techniques in critically ill neurosurgical patients.

Monitoring of Jugular Bulb Venous Oxygen Saturation

Patients at risk of cerebral ischaemia during intensive care after intracerebral haemorrhage and brain trauma have been studied in the past using intermittent measurements of CBF, arterial–jugular venous oxygen content difference ($AJDO_2$) and arterial–jugular venous lactate content difference (AJDL)[2]. Measurement of the $AJDO_2$ from arterial and jugular venous blood oxygen

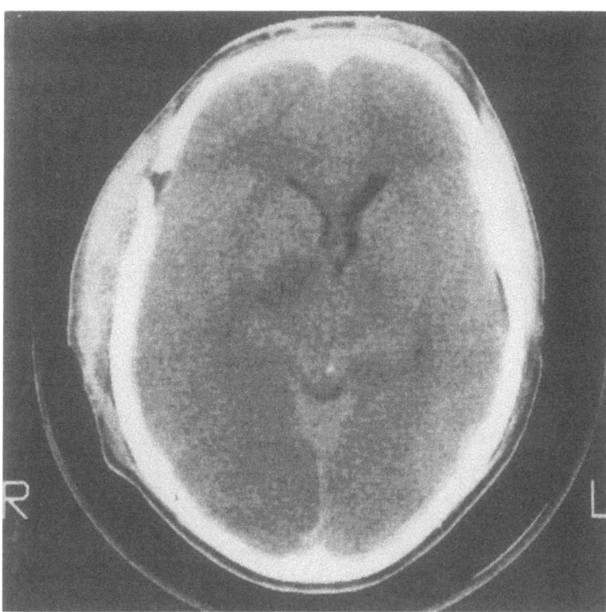

Fig. 1. Brain computed tomography scan following evacuation of a right-sided extradural haematoma. The right cerebral hemisphere (R) is swollen and the medial occipital lobe territories supplied by both posterior cerebral arteries and the posterior right internal capsule appear ischaemic. L, left ·

saturation allows description of the relationship between global cerebral metabolic rate for oxygen ($CMRO_2$) and CBF according to the Fick principle ($AJDO_2$ in ml oxygen/100 ml blood = $CMRO_2$ in ml O_2/100 g per min) $\times$ 100/CBF in ml/100 g per min). Since $AJDO_2$ = haemoglobin concentration $\times$ 1.34 $\times$ arterio venous oxygen saturation difference/100 + ((PaO_2 mm Hg − PjO_2 mm Hg) $\times$ 0.3/100), if SaO_2, PaO_2, haemoglobin level and the position of the haemoglobin dissociation curve remain constant, the ratio of global CBF to $CMRO_2$ is proportional to the jugular bulb venous oxygen saturation (SjO_2). CBF and $CMRO_2$ are coupled in health, $AJDO_2$ remains between 4 and 9 ml% and SjO_2 is within the normal range of 54%-75%. After brain trauma or subarachnoid haemorrhage (SAH) $CMRO_2$ falls with the degree of unconsciousness. CBF is usually also reduced, but does not always remain coupled to $CMRO_2$. In the absence of anaemia or a sudden rise in PaO_2, increases in SjO_2 over 75% ($AJDO_2 < 4$ ml%, OER $\leq$ 22%) suggest relative (CBF < 40 ml/100 g per min) or absolute (CBF > 40 ml/100 g per min) "luxury perfusion" (although areas of regional ischaemia or infarction may still be present). SjO_2 below 50% ($AJDO_2 > 7.5$ ml%) indicates relative hypoperfusion. At SjO_2 less than 40% ($AJDO_2 \geq 9.0$ ml%), global cerebral ischaemia is likely, with increased producion of lactic acid and a lactate oxygen index (LOI) above 0.08 (LOI = − $AJDL$/$AJDO_2$; Fig.2) [2, 3]. Although regional cerebral ischaemia is

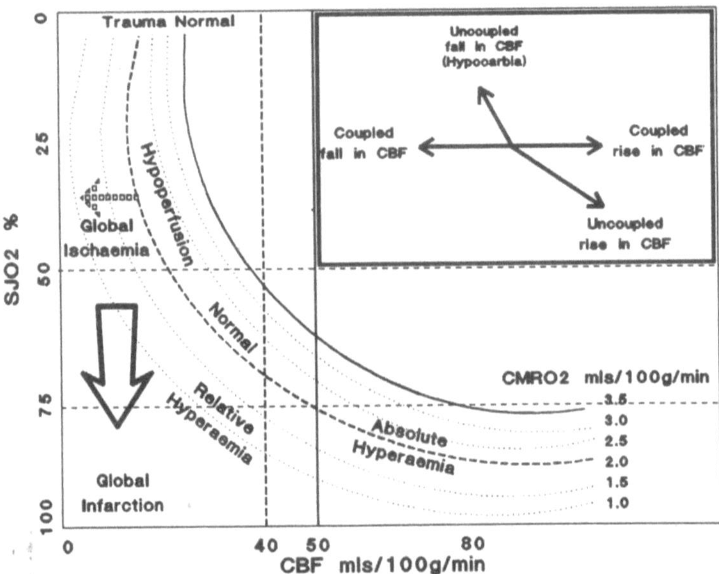

Fig. 2. Relationship between cerebral blood flow (*CBF*), cerebral metabolic rate for oxygen (*CMRO$_2$*) and jugular bulb venous oxygen saturation (*SjO$_2$*) after brain trauma. CMRO2 is reduced, while CBF may be elevated, normal or reduced as reflected by high, normal or low SjO$_2$, respectively. Below SjO$_2$ of 40%, global cerebral ischaemia is associated with an increase in cerebral lactic acid production. Global infarction leads to cessation of cerebral oxygen metabolism and high SjO$_2$

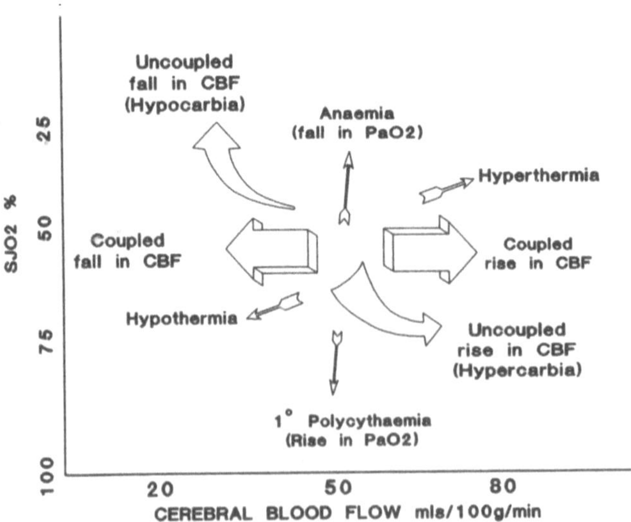

Fig. 3. Factors which modify the relationship between cerebral blood flow (*CBF*) and jugular bulb venous oxygen saturation (*SjO$_2$*)

associated with increased lactate production, SjO_2 may not be reduced. Many other factors influence the relationship between CBF and SjO_2, and the clinician must consider them when interpreting changes in SjO_2 (Fig. 3).

Anatomical Considerations for Jugular Bulb Venous Oxygen Saturation Monitoring

The jugular bulb is a dilatation of the rostral internal jugular vein just below the jugular foramen. Although 85% of blood usually drains from both cerebral hemispheres via intracranial venous sinuses through the sigmoid sinuses and onward into the right internal jugular vein, predominant drainage to the left also occurs [4]. In the presence of focal intracranial pathology, patterns of drainage change and differences in the SjO_2 are occasionally evident between right and left [5, 6]. Many investigators choose to measure SjO_2 from the internal jugular vein on the side of focal pathology [3], but there is little evidence to support this approach. An alternative method in neurosurgical patients with ICP monitoring is to sequentially manually compress the internal jugular veins and to select the side of greater rise in ICP as representative of predominant venous drainage. In the event of equal rises in ICP, which may indicate free communication across the transverse sinus, the right side is preferred as it is easier to cannulate. In patients with suspected intracranial aneurysms, the side of predominant venous drainage from the pathological area may be determined angiographically (Fig. 4).

The internal jugular vein descends lateral to the internal and common carotid arteries in the carotid sheath. It ends behind the medial part of the clavicle, forming the brachiocephalic vein by joining the subclavian vein. Many veins join the internal jugular vein throughout its passage down the neck, and accurate measurement of cerebral venous oxygen saturation necessitates radiological confirmation of the position of the catheter tip high in the jugular bulb. Contraindications to the technique include bleeding diathesis, local infection, local neck trauma and any impairment to cerebral venous drainage. Provided these contraindications are observed and the system is continuously flushed with 3 ml/h of 2 units per ml heparinised saline, impaired cerebral venous drainage is negligible and the risk of significant venous thrombosis is below 5%. If, during monitoring, blood cannot be freely aspirated, the catheter should be removed.

Theory of Continuous In Vivo Jugular Bulb Venous Oxygen Saturation Monitoring

Continuous venous oximetry offers a greater chance of detecting and correcting global cerebral ischaemic insults, since intermittent measurements only monitor the patient at a particular moment in time and transient ischaemic events will be

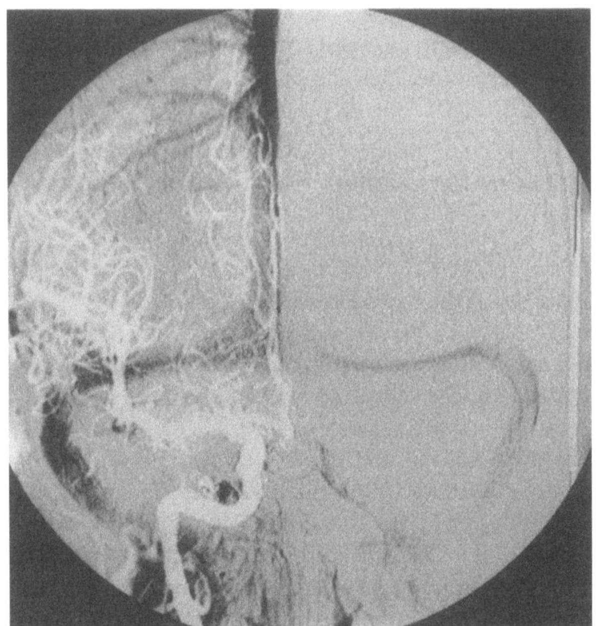

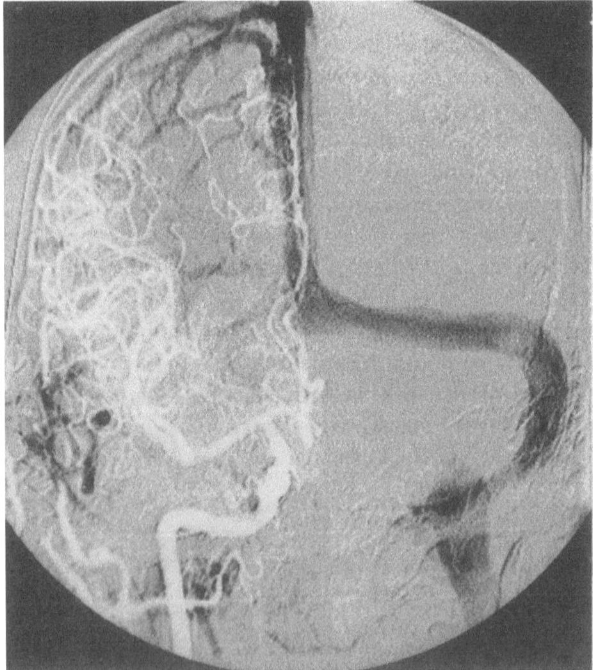

Fig. 4a,b. Right carotid angiograms from two patients showing arterial (*white*) and venous (*black*) phases. Venous drainage is predominantly to the right (**a**) and left (**b**) internal jugular veins, respectively

missed [26]. Although in vivo oximetry has been possible for over 30 years, only recently have accurate, practical and reliable systems emerged for clinical use. In vivo oximetry instruments use light of selected wavelengths, chosen so that the absorption characteristics of oxyhaemoglobin and haemoglobin are different. Light is passed down fibre-optic cables and the intensity of light reflected off erythrocytes back along the fibres is measured by a photodetector. The amount of light detected varies with each wavelength depending on the relative concentrations of oxyhaemoglobin and haemoglobin. The relative attenuation of different wavelengths reflects the ratio of haemoglobin to oxyhaemoglobin (Fig. 5). Early fibre-optic systems used two wavelengths of light and were unreliable, because the mathematical algorithms employed assumed incorrectly that the relationship between the reflected light ratio and oxygen saturation was linear. A three-wavelength system patented by Abbott Critical Care Systems (Chicago, USA) overcame these difficulties. The Oximetrix-3 system comprises a processor, optical module and fibre-optic catheter. Light of red and near infrared wavelengths is sent at 1 ms intervals down a transmitting fibre and reflected along a receiving fibre to a photoelectric sensor. Reflected signals are averaged over 5 s and updated every second. The system analyses raw optical data, uses a patented digital signal filter to reject vessel wall artefact and displays trended and current oxygen saturation. Reflected light intensity is also displayed and, when normal, appears as a bar between two dotted lines (Fig. 6). The reflected light intensity should always be examined before interpreting data. High reflected light intensity indicates vessel wall artefact (Fig. 7), while low intensity suggests catheter obstruction. By using three wavelengths, two independent reflected light intensity ratios, R_1 and R_2, are calculated. Oxygen

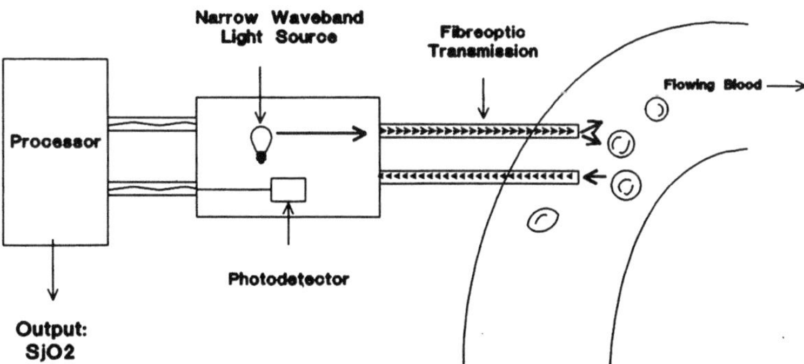

Fig. 5. The principles of reflection spectrophotometry; fibre-optic catheter in vivo jugular bulb oximetry. Light of selected wavelengths is transmitted down a fibre-optic cable, reflected from flowing red blood cells and returns through a receiving fibre-optic cable to a photodetector. This analyses the reflected light intensities and relays the information to a processor, which calculates the concentration of oxyhaemoglobin from the ratio of the reflected light intensities. SjO_2, jugular bulb venous oxygen saturation

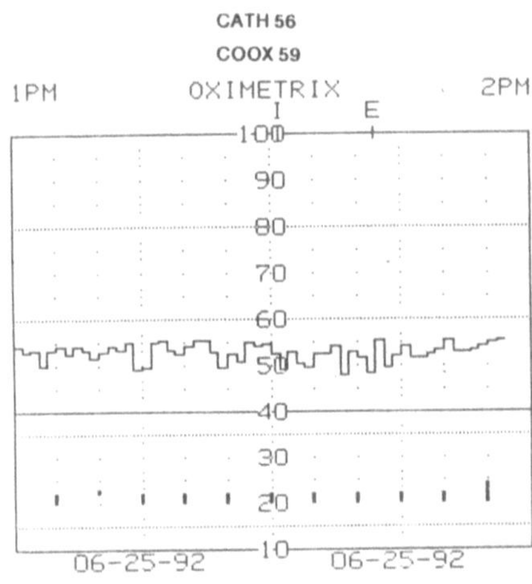

Fig. 6. Printout showing a normal recording from the Oximetrix-3 system during jugular bulb venous oxygen saturation (SjO_2) monitoring. The SjO_2 trace shows variability between 50% and 60%, and the light intensity display shows a series of *bars* between two *dotted lines* in the lower part of the display. *CATH*, catheter; *COOX*, laboratory co-oximeter

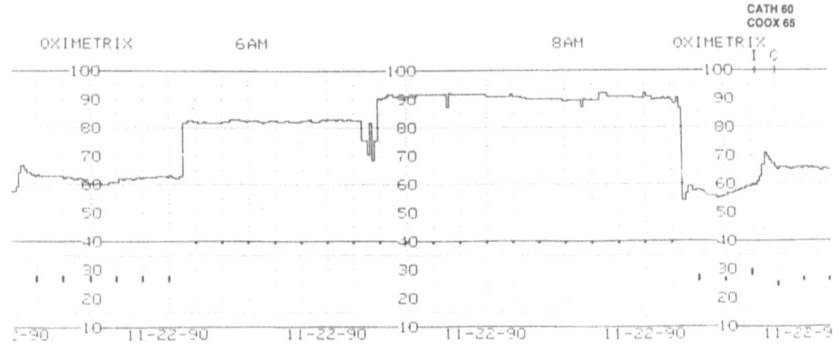

Fig. 7. Recording of normal jugular bulb venous oxygen saturation (SjO_2) with satisfactory reflected light intensity display until 5.45 am., when SjO_2 suddenly increases in association with high light intensity, suggesting the catheter tip is against the vessel wall. After 8.20 am., SjO_2 abruptly declines with return of acceptable light intensity readings. In vivo calibration at 8.40 am. confirms the catheter reading is within 5% of a sample analysed in the laboratory co-oximeter. *CATH*, catheter; *COOX*, laboratory co-oximeter

saturation (SO_2) can now be derived from two formulae:

$$SO_2 = A_1 + B_1R_1 + C_1[Hb] + D_1R_1[Hb]$$

and

$$SO_2 = A_2 + B_2R_2 + C_2[Hb] + D_2R_2[Hb]$$

(where A_1, B_1, C_1, D_1, A_2, B_2, C_2 and D_2 are constants; [Hb], concentration of haemoglobin). The two equations can be solved algebraically, eliminating the need for user input of [Hb]. Thus,

$$SO_2 = [(A_1C_1 - A_2C_1) + (B_1C_2 - A_2D_2)R_1 + (A_1D_2 - B_2C_1)R_2$$
$$+ (B_1D_2 - B_2D_1)R_1R_2]/C_2 - C_1 - D_1R_1 + D_2R_2$$

The use of two reflected light intensity ratios also virtually eliminates any contribution from carboxyhaemoglobin to the computed saturation.

The Oximetrix-3 System for Measurement of Jugular Bulb Venous Oxygen Saturation

Pre-Insertion calibration is done before introduction of the catheter into the jugular vein. Although this calibration, originally designed for use with pulmonary artery catheters, has proved unreliable for the 40-cm catheters used for SjO_2 monitoring [13], it should still be done, as it provides a check of the integrity of the whole system prior to catheter insertion. The jugular vein is then cannulated at the level of the thyroid cartilage in retrograde fashion and a 14-gauge introducer inserted until it abuts the skull base using the Seldinger technique. After pre-insertion calibration, the 40-cm Shaw Opticath is first flushed with heparinised saline and then advanced through this introducer into the jugular bulb.

Once inserted in the jugular vein, the catheter should undergo in vivo calibration by comparing and adjusting catheter recordings with values obtained by analysis of an aspirated sample of jugular bulb venous blood in a laboratory co-oximeter. It is necessary to repeat an in vivo calibration every 12 h in order to maintain the limits of agreement (LA) between catheter and co-oximeter below 5% (LA = mean difference $\pm$ 1.96 of the difference) [7].

Prognostic value of Monitoring Cerebral Blood Flow and Jugular Bulb Venous Oxygen Saturation after Severe Head Injury

Several studies suggest that the magnitude of reduction of $CMRO_2$ after severe head injury may serve as a prognostic indicator [8, 9]. In contrast, CBF correlates poorly with outcome, and both substantially elevated and extremely low flow states carry a poor prognosis [10, 11], although when hyperaemic patients with uncoupled flow and metabolism are excluded, good correlation exists between CBF and outcome [12].

Evidence for the prognostic value of continuous monitoring of SjO_2 comes from a recent report by Robertson [13], who studied 102 patients with severe head injuries during the first few days of intensive care with continuous SjO_2 measurements. A total of 143 episodes, of significant desaturation ($SjO_2 < 50\% > 10$ min) occurred; 61 patients had no episodes, 23 patients had one episode and 18 patients had more than one episode of desaturation. Desaturations were most frequent in the first 24 h after trauma and were from systemic and cerebral causes: raised ICP, Thirty; hypocarbia, Nineteen; hypotension, eight; hypoxia, six; anaemia one; and vasospasm, one. The average duration of desaturation was 1.1 h (range, 10 min–12 h) and so would probably not have been detected by intermittent measurements. Mortality rates for patients with no desaturations, one episode and more than one period of hypoperfusion were 17.9%, 45.5% and 70.6%, respectively. Reduction of SjO_2 was significantly related to poor outcome (p, 0.009), even when data were adjusted by logistic regression for all covariables.

Jugular Bulb Venous Oxygen Saturation During Treatment of Raised Intracranial Pressure

Therapy to reduce ICP after severe head injury may lead to significant cerebral ischaemia if applied injudiciously. Hyperventilation induces vasoconstriction and has been reported to lead to cerebral hypoperfusion or ischaemia [12, 2]. Similarly, reduction of CPP in an attempt to reduce ICP with barbiturates can induce cerebral ischaemia [12, 13]. As a general principle, ICP therapy should preserve or increase CPP. In our unit, SjO_2 monitoring has proved a useful adjunct both to define the critical CPP of a particular patient (see below) and to monitor the effects of ICP therapy during the intensive care of severely head-injured patients. Figure 8 shows an example of a patient with a CPP of 40 mm Hg during artificial ventilation on the intensive care unit, following evacuation of a subdural haematoma. Blood gas analysis revealed a $PaCO_2$ of 3 kPa, and his LOI was 0.17. Elevation of $PaCO_2$ to 4.1 kPa increased SjO_2 and LOI fell to 0.09, but ICP rose to 50 mm Hg and his left pupil developed a sluggish reaction to light. $PaCO_2$ was lowered and then raised to 3.7 kPa with a fall and then rise in SjO_2. ICP settled around 35 mm Hg and although pupil reactivity became equal, CPP remained at 48 mm Hg and LOI was still 0.09. Blood pressure elevation failed to increase CPP or SjO_2 and while administration of thiopentone resulted in a fall in ICP to 18 mm Hg, CPP declined to 40 mm Hg and SjO_2 was reduced further, although LOI remained at 0.09. Mannitol administration reduced ICP from 38 mm Hg to 18 mm Hg, and CPP rose from 42 mm Hg to 75 mm Hg. SjO_2 rose until CPP exceeded 60 mm Hg, at which point LOI was 0.07. Subsequent ICP/CPP therapy was continued with mannitol and frusemide for 6 days, ventilation was slowly adjusted to increase $PaCO_2$ to 4.5 kPa and the patient was weaned from the ventilator 11 days after injury. After 6 months he remained moderately disabled, but by 18 months he had made a good recovery.

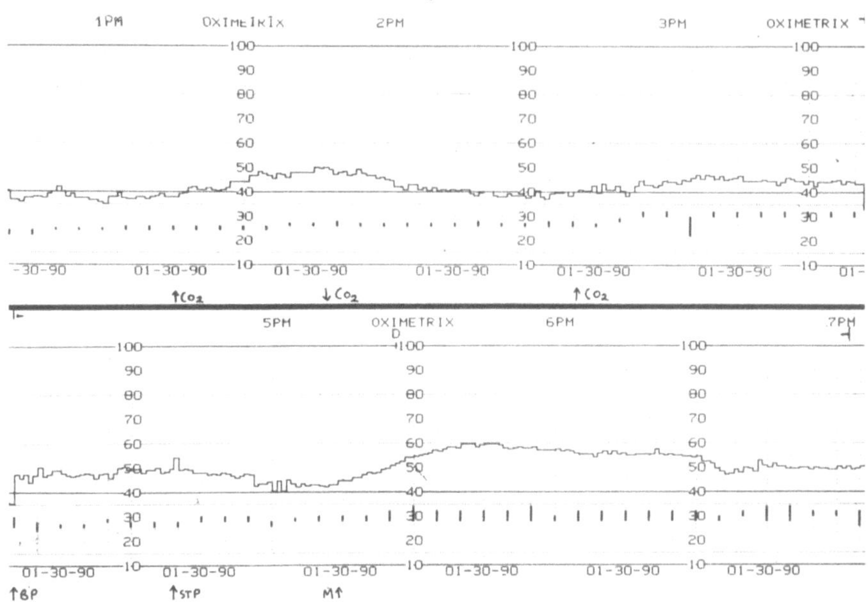

Fig. 8. Jugular bulb venous oxygen saturation (SjO_2) recording from a patient during intracranial pressure therapy following evacuation of a subdural haematoma after severe brain trauma. Hyperventilation and barbiturate therapy both reduced intracranial pressure, but compromised SjO_2. Blood pressure elevation had no effect on cerebral perfusion pressure (CPP) or SjO_2. Only mannitol improved CPP and corrected jugular desaturation associated with an increased lactate oxygen index (see text for details)

Jugular Bulb Venous Oxygen Saturation During Surgery Following Subarachnoid Haemorrhage

Our recent observations (unpublished) of SjO_2 in patients undergoing acute clipping of ruptured intracranial aneurysms, within 48 h of the bleed, suggest that systemic hypotension is poorly tolerated when compared to elective clipping of aneurysms. We have studied 18 patients–nine acute (first 48 h post-bleed) and nine elective (8–16 days post-bleed)–during aneurysm surgery with continuous SjO_2 monitoring on the side of angiographically derived predominant venous drainage of the aneurysm territory. In six of the nine acute patients, global hypoperfusion ($SjO_2 < 50\%$) was evident below a mean arterial pressure of 80 mm Hg, and this responded favorably to elevation of $PaCO_2$ above 4 kPa, to correction of low central venous or left atrial pressure with fluids and to blood pressure elevation with fluids, inotropes or vasoconstrictors, as appropriate. Cerebral hypoperfusion ($SjO_2 < 50\%$) only occurred in one of the elective patients and then only when mean arterial pressure fell below 50 mm Hg (Fig. 9).

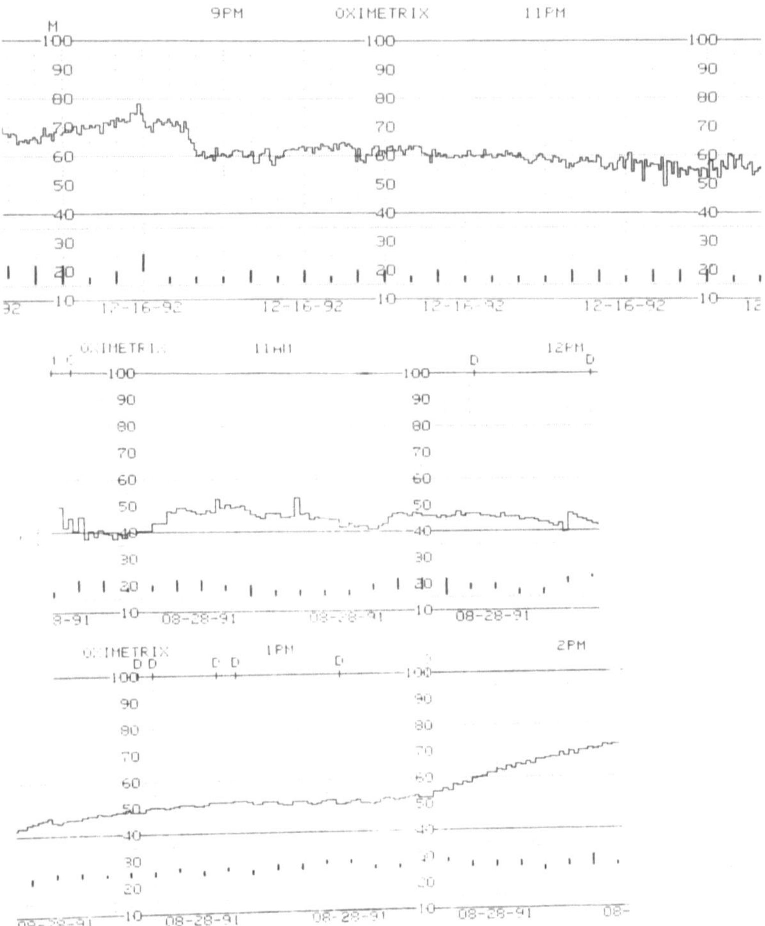

Fig. 9a,b. Jugular bulb venous oxygen saturation (SjO_2) recordings from two patients undergoing anesthesia for clipping of a middle cerebral artery aneurysm. The first patient (a) was fully awake without neurological deficit and was operated on electively 10 days after subarachnoid haemorrhage. (SAH). PaO_2 was maintained around 18 kPa, $PaCO_2$ at 4–4.5 kPa and mean arterial pressure was kept between 75 and 85 mm Hg during surgery. SjO_2 lactate oxygen index (LOI; both measured hourly) remained normal throughout. The second patient (b) was drowsy although orientated and obeying commands and had weakness of movement in the right arm and underwent surgery 24 h following SAH. Despite a PaO_2 of 19 kPa, $PaCO_2$ of 3.9 kPa and mean arterial pressure of 70–80 mm Hg, SjO_2 was below 40% with a LOI of 0.13. Elevation of $PaCO_2$ followed by intravenous colloids to raise the mean blood pressure to 95 mm Hg increased SjO_2 to around 50%, and LOI at 12 pm is 0.05. This level of blood pressure was maintained until after clipping of the aneurysm around 1.30 pm, when an adrenaline infusion was used to increase mean arterial pressure until SjO_2 was 60%–70%. On reversal of anesthesia, the patient's right arm had recovered

Following surgery, we have seen patients with focal neurological deficits following clipping of intracranial aneurysms with either reduced SjO_2 and elevation of LOI or elevated LOI despite normal or increased SjO_2. Furthermore, significant elevation in LOI may occur at mean arterial pressures well in excess of 80 mm Hg. The LOI is reduced and focal deficits are usually corrected by restoration of normovolamia, if appropriate, and thereafter further increase of CPP using fluids, vasoconstrictors or inotropes to raise blood pressure. In patients undergoing artificial ventilation following clipping of an intracerebral aneurysm, in whom neurological examination is compromised, SjO_2 and LOI may therefore prove a useful guide for management.

Transcranial Doppler Ultrasonography

Since the pioneering work of Aaslid in 1982 [14] the measurement of blood flow velocity in the cerebral basal arteries has become a routine technique in many neurosurgical centres. The technique involves directing a beam of sound waves at the basal cerebral vessels through 'cranial windows' in the skull where the bone is relatively thin. A relatively low frequency (2 MHz) is used to facilitate skull penetration and the signal is pulsed so that the piezo-electric ultrasound emitter also acts as a receiver for the sound beam. By convention, flow towards the probe is shown as a positive deflection and flow away as a negative deflection. Several cranial windows can be used. These are the transtemporal window above the zygomatic arch from just posterior to the orbit to the front of the ear, the transorbital window, where a low intensity ultrasound beam (10% of maximum) is directed through the superior orbital fissure to insonate the carotid syphon, and the suboccipital window, where the ultrasound wave is directed at the vertebral and basilar arteries through the foramen magnum. In the critically ill patient, the transtemporal window is usually chosen for insonation of the proximal segments of the anterior (ACA), middle (MCA) and posterior cerebral arteries (PCA). The low cost, mobility and non-invasive nature of this investigation makes it a desirable tool, although knowledge of its limitations is essential.

Theory of Transcranial Doppler Ultrasonography

Shift in the frequency of a sound wave when its source or receiver are moving with reference to the medium through which sound is propagated was first described in 1843 by Christian Doppler [15]. When a sound wave is directed at a basal cerebral artery, the change in frequency (dF) can be used to calculate the velocity of flowing blood according to the formula $dF = 2 \times Fe \times v \times \cos A/C$, where Fe is emitted ultrasound frequency, v is the real flow velocity of blood, C is velocity of sound in the transmission medium and, A is the angle between the ultrasound beam and the artery.

It should be noted that the true velocity v of flowing blood and the velocity V obtained from the Doppler machine may differ according to the angle of insonation, A. Since $V = v \times \cos A$, as the angle of insonation increases the perceived velocity V progressively underestimates true velocity v. Until recently, it was believed that A was between $0°$ and $30°$ and that the maximum underreading would be 15% since cos $30°$ is 0.86 [14]. However, more recent evidence suggests that A may be higher than $30°$ and accordingly v is underestimated by a greater amount [16]. It is envisaged that newer generation machines will measure A such that a true estimate of v can be made. Although the measured parameter is dF expressed in kHz, this is conventionally converted to a velocity scale (cm/s), and for the usual 2-MHz transducers, $V = d$F $\times 39$.

Cerebral Blood Flow and its Velocity

Transcranial Doppler ultrasonography (TCD) measures the velocity of flowing blood, not cerebral blood flow. The relationship between these two parameters depends on the diameter of the insonated vessel (which cannot be measured using TCD) and is non-linear, because velocity (cm/s) increases to the second power of the radius of the vessel, while flow (ml/min) rises in proportion to the fourth power of its radius. Without knowledge of the diameter of the insonated vessel (which may vary during the course of the examination), velocity cannot be converted to flow. TCD measurements can be made intermittently or continuously in the critically ill neurosurgical patient by using a headband to support a probe directed through the temporal window. Peak systolic velocity (SV) and end diastolic velocity (DV) of the termination of the internal carotid artery (ICA), the ACA (negative waveform), MCA and PCA (both positive waveforms) can be monitored. Normal SV in the ACA, MCA and PCA is around 80 cm/s, 90 cm/s and 55 cm/s, respectively [14, 17].

From the signal, several parameters can be derived. These include the time-averaged mean velocity (MV) [14], the dimensionless pulsatility index (PI) [18] and the assymetry index (ASI) [19]. The MV is the mean velocity over the cardiac cycle. Since velocity is proportional to the ratio of CBF to vessel diameter, it follows that this will be reduced during states of reduced blood flow or vessel dilatation and increased in the presence of raised cerebral blood flow or vasospasm. The ratio of the MCA MV (which rises with vasospasm) to the extracranial internal carotid artery (EICA) MV (which will fall with vasospasm) way help to differentiate these two states [14]. An MCA to EICA index of more than 3 correlates with angiographic evidence of vasospasm while severe spasm occurs with an index above 10 [20]. Normal MV for ACA, MCA and PCA is approximately 50 cm/s, 60 cm/s and 35 cm/s, respectively, and tends to decline with age [21]. The PI is calculated as (SV–DV)/MV and is related to distal resistance to flow. However, since PI is also related to systemic pulse pressure and pulse waveform itself is a function of vascular compliance, heart rate, temperature and $PaCO_2$, the relationship is complex [22]. ASI is used to differentiate normal variation in side to side velocity from significant pathological interhemispheric velocity assymetry: $ASI = [100(MV_a - MV_s]/(MVa + MV_s/2)\%$. The upper limits of normality for ACA, MCA, and PCA are 21%, 27% and 28%, respectively [19]. Values above these therefore suggest a pathological process.

Technique for Transcranial Doppler Ultrasonography in Critically Ill Neurosurgical Patients

The transtemporal window is normally chosen for insonation with the patient supine. Initially, the probe is placed over the posterior part of the window after application of ultrasound gel to the probe tip. The ultrasound beam can be range-gated, allowing 'focusing' of the beam on a particular cerebral vessel, and normally a depth of insonation of 55–60 mm is selected as the depth of the origins of MCA and ACA from ICA. By angling the probe slightly upwards and forwards, a flow signal is identified and then refined by small movements until a maximal signal is obtained, in an effort to minimise the angle of insonation. A bidirectional signal is usually sought as a reference point for the origins of MCA and ACA (Fig. 10). By reducing the depth of insonation from this point, the positive signal of the MCA can be followed until the vessel turns into the

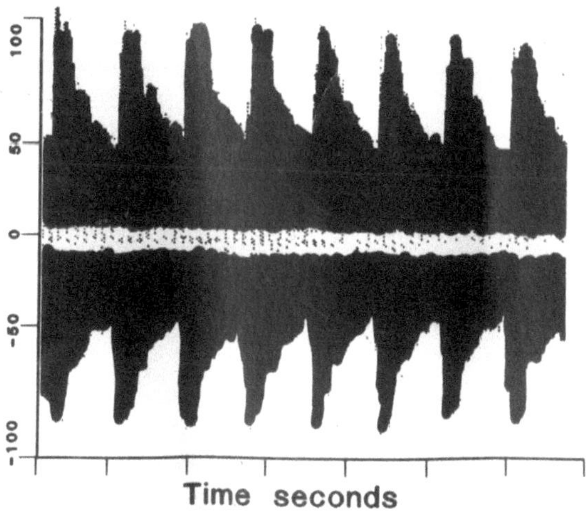

Fig. 10. Transcranial Doppler ultrasonography trace obtained through the left temporal transcranial window at 60 cm insonation depth identifying the origins of the anterior (negative deflection) and middle cerebral arteries (positive deflection) from the internal carotid artery

Sylvian fissure. Alternatively, by angling the probe anteriorly, the ACA can be followed and by turning more posteriorly and superiorly and increasing the depth setting by 5–15 mm, the PCA can be interrogated. Problems of insonation may be accounted for by thickening of the intracranial window with age [23] or variations in the anatomy of the circle of Willis [14].

Transcranial Doppler Ultrasonography After Head Injury

CBF may be increased, normal or reduced after brain trauma, while $CMRO_2$ tends to fall with increasing depth of coma [2, 12]. Increases in MV after brain trauma may therefore represent vasospasm or increased flow in the insonated vessel. The waveform of the Doppler signal may differ in these two states [24] (Fig. 11). The ratio of the MCA MV (which rises with vasospasm) to the EICA MV (which falls with vasospasm) may also be used to distinguish between these two states [14]. Patients with TCD evidence of vasospasm after brain trauma are more likely to develop regional infarction in the territory of the insonated vessel [25]. However, at high ICP and, more specifically, low CPP states, DV of the insonated vessel progressively reduces and a reverberant flow velocity pattern may emerge [26, 27] (Fig. 12). Under these circumstances, use of MV as an indicator of vasospasm is highly suspect, and in a recent study a MV in the MCA exceeding 100 cm/s was only seen if CPP exceeded 60 mm Hg [25].

After brain injury, persistent significantly reduced MV (30 cm/s) carries a poor prognosis, but there is no overall correlation between outcome and initial post-resuscitation TCD MV of the MCA. However, post-resuscitation MV is significantly lower after severe head injury, and good recovery or moderate disability 6 months after brain trauma is associated with a significant rise in TCD MCA MV between admission and discharge, while patients who remain severely disabled have persistently low velocities [28].

Transcranial Doppler Ultrasonography after Subarachnoid Haemorrhage

The major potential of this technique in the management of patients suffering SAH is the identification of patients liable to develop ischaemic neurological deficits. Cerebral vasospasm remains a major factor in the development of delayed ischaemia or infarction after SAH, is more prevelant with more severe grade of SAH and usually develops insidiously between 5 and 12 days after the initial event [29, 30]. It is generally considered that patients with good clinical grades after SAH (based on the World Federation of Neurosurgical Sciences

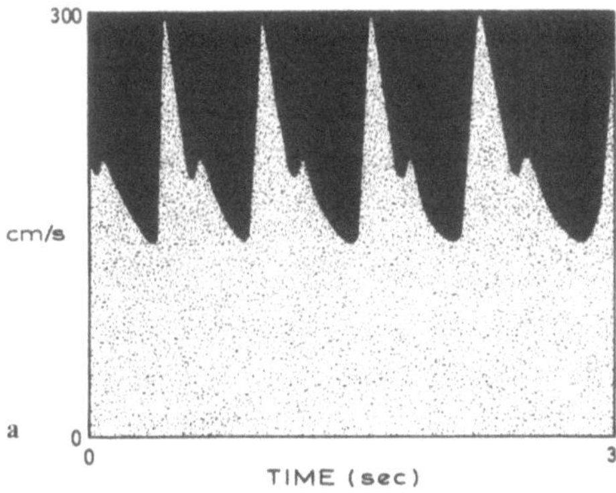

Fig. 11a, b. Transcranial Doppler ultrasonography signals from the middle cerebral artery (MCA) of two patients with increased time-averaged mean velocity (MV) following brain trauma. In the first (**a**) Cerebral perfusion pressure (CPP) was 88 mm Hg, jugular bulb venous oxygen saturation (SjO_2) was 65%, lactate oxygen index (LOI) was 0.07 and MV was normal in the opposite MCA. The patient subsequently developed an ischaemic infarction in this MCA territory following a period of low CPP secondary to treatment-resistant intracranial hypertension. It is suggested the trace indicates spasm of the insonated vessel. In the second patient (**b**), CPP was 80 mm Hg, SjO_2 was increased at 82%, LOI was 0.009 and MV was elevated in the MCA, anterior (ACA) and posterior cerebral arteries (PCA) on both sides, suggesting absolute hyperaemia. The two signals have different waveforms: trace **a** shows marked notching of the *downslope*, while in trace **b** the downslope is *smooth*

scale) should undergo early surgery to reduce the risk of rebleeding, but more severe grades receive later surgery because of the risk of exacerbating delayed ischaemia [31, 32]. If TCD could define those patients at greater risk of vasospasm-induced ischaemia, this would allow us to delay surgery beyond this period of increased susceptibility.

Following SAH, CBF is initially maintained and then progressively declines, according to clinical grade, for about 2 weeks, except for a brief period of increased flow after surgery [33]. This time course correlates with changes in

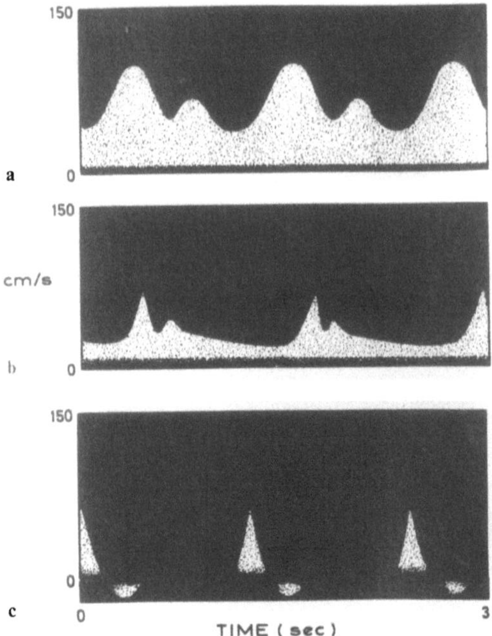

Fig. 12a–c. Series of transcranial Doppler ultrasonography traces from the middle cerebral artery (MCA) of a patient with progressively compromised cerebral perfusion pressure (CPP) due to intracranial hypertension associated with intracranial pressure 'A' waves. The initial trace (**a**) shows a normal pattern with systolic (SV) and diastolic waves (DV) at a CPP of 75 mm Hg. As intracranial pressure rises from 15 to 40 mm Hg, CPP falls to 45 mm Hg. The SV is reduced and DV is almost zero (**b**). At the peak of the 'A' wave (**c**), CPP is 12 mm Hg, and a reverberant pattern of flow velocity with systolic peaks and a negative diastolic phase is seen

TCD velocity [34] and with angiographic evidence of spasm [30], suggesting that increases in MV indicate vasospasm rather than increased flow through the insonated vessel. This hypothesis is further supported by the observation that flow velocities tend to be higher in patients with more severe clinical grades of SAH, except for the worst clinical grade V, who have such low CBF that MV is reduced [35]. Flow velocity is usually normal soon after SAH and then rises over the first few days [36]. Increased MV precedes ischaemic deficit in the insonated vessel territory, because of initial compensation by vasodilatation of the collateral circulation, and it is thought that very high mean flow velocities (> 120–200 cm/s) and rapid rates of increase in flow velocities indicate a heightened risk of developing ischaemic deficit [34, 35, 37]. However, it is not possible to accurately define threshold values of TCD MV below which ischaemic deficit will develop in the same way as can be achieved with CBF measurements. Because of the low specificity of TCD in defining patients with increased risk of developing focal ischaemia, use of TCD MV or rate of rise of velocity inevitably means applying prophylaxis to patients who would not develop ischaemia [34]. Autoregulatory responsivity to carbon dioxide is progressively impaired with increasing levels of vasospasm, and it has been suggested that the use of a carbon dioxide provocation test could refine the value of TCD as a predictor of severe vasospasm [38]. Inevitably, some patients will remain asymptomatic despite TCD evidence of vasospasm, because of adequate collateral circulation, while others will have spasm distal to the area of insonation that cannot be detected and will deteriorate. However, if neurological deficit has already developed, TCD may be useful for differentiation of deterioration due to vasospsam from other causes.

Combined Studies with Jugular Bulb Venous Oxygen Saturation Monitoring and Transcranial Doppler Ultrasonography

In critically ill neurosurgical patients, combined SjO_2 monitoring and TCD may be useful for determining the CPP below which venous desaturation starts because of exhausted cerebral autoregulation. In 16 brain-injured patients with SaO_2 above 95% and $PaCO_2$ between 2.7 and 4.7 kPa, measurements of SjO_2 and flow velocity in the MCA on the side of predominant pathology were made at the highest and subsequent lowest CPP recorded during a period of falling CPP due to either reduced mean arterial pressure or raised ICP. A biphasic response was recorded in SjO_2 and PI derived from TCD. Only when CPP fell below 71 mm Hg was there significant correlation between CPP and SjO_2 (r, 0.78, $p < 0.001$) or CPP and PI (r, 0.942, $p < 0.0001$) [26].

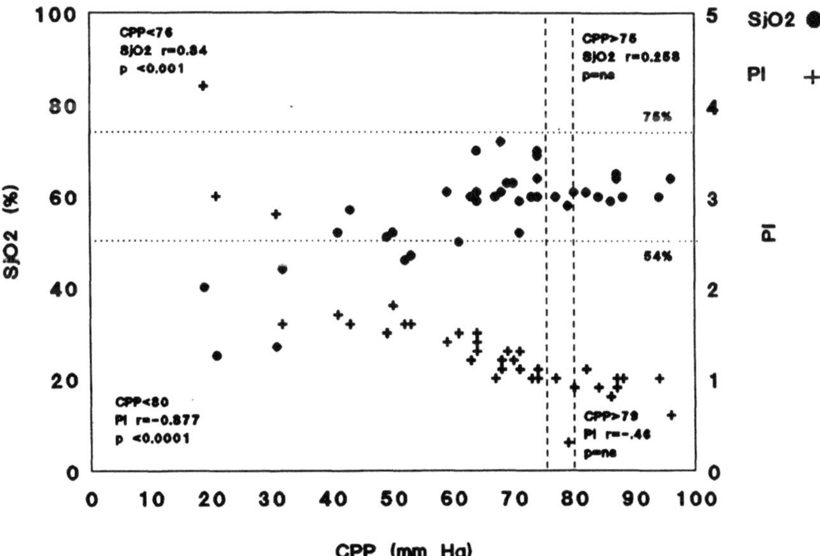

Fig. 13. Graph of cerebral perfusion pressure (CPP) against jugular bulb venous oxygen saturation (SjO_2) during 36 treatments of compromised CPP in 20 severely head-injured patients. For each event, two recordings were made, initially at the lowest CPP before therapy and again at the highest CPP achieved within 30 min of starting treatment. Critical CPP for SjO_2 is 68 mm Hg (see text for details). *PI*, pulsatility index

CPP, SjO_2 and TCD derived PI of the MCA were studied during the intensive care management of 20 severely head-injured patients, before and during 36 treatments of compromised CPP. Recordings were made at the lowest CPP before treatment and the highest CPP recorded during 30 min following therapy. Sequential linear correlation–regression analysis demonstrated that there were significant correlations between CPP and SjO_2 (r, 0.837, $p < 0.0001$; r, 0.941, $p < 0.0001$) and CPP and PI only below a CPP of 68 mm Hg and 70 mm Hg, respectively [39] (Fig. 13). Since during falling or rising CPP there is plateauing of SjO_2 and PI around a CPP of 70 mm Hg, it appears that after brain injury autoregulation remains at least partially preserved. Below this mean critical CPP of around 70 mm Hg, autoregulatory vasodilatation appears unable to sustain cerebral blood flow, and therefore increasing oxygen extraction maintains cerebral oxygen delivery associated with a fall in SjO_2 and rise in PI. This threshold CPP of 70 mm Hg is higher than previously considered as adequate to avoid cerebral hypoperfusion after adult brain injury and has considerable therapeutic implications. In patients with regional vasospasm, even higher CPP may be needed to avert ischaemia. The relationship between CPP and SjO_2 was examined in our department from four of 34 brain-injured patients with unilateral increased mean flow velocity associated

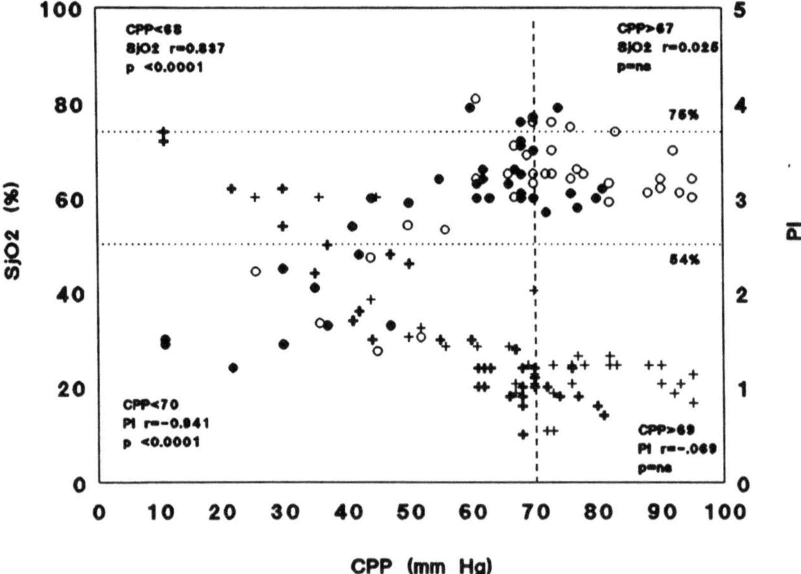

Fig. 14. Plot of consecutive 5-min recordings of cerebral perfusion pressure (*CPP*) and jugular bulb venous oxygen saturation (*SjO₂*) during episodes of decreasing CPP in four head-injured patients with transcranial Doppler evidence of vasospasm who subsequently developed regional, non-contusion-related cerebral infarctions. Critical CPP for the pulsatility index (*PI*) is 80 mm Hg and for SjO_2 76 mm Hg (see text for details)

with normal SjO_2 who subsequently developed non-contusion-related infarctions in the territories of the insonated vessel. Recordings of CPP and SjO_2 were made at 5-min intervals during periods of decreasing CPP. The CPP of 80 mm Hg and 76 mm Hg, below which PI began to rise and SjO_2 began to fall, were significantly higher than in patients without transcranial Doppler evidence of vasospasm, indicating a higher CPP was required to avoid cerebral hypoperfusion (Fig 14; K.H. Chan, N.M. Dearden and J.D. Miller, unpublished observations).

Conclusions

Monitoring of critically ill neurosurgical patients with continuous SjO_2 and TCD techniques offers potential for the early diagnosis and treatment of conditions likely to precede cerebral ischaemia. It is envisaged that further research and advances in technology will improve the sensitivity and specificity of these monitoring techniques in this regard.

References

1. Gentleman D, Jennett B (1990) Audit of transfer of unconscious patients to a neurosurgical unit. Lancet 335:330–334
2. Robertson CS, Grossman RG, Goodman JC, Narayan RK (1987) The predictive value of cerebral anaerobic metabolism with cerebral infarction after head injury. J Neurosurg 67:361–368
3. Robertson CS, Narayan RK, Gokaslan ZL, Pahwa R, Grossman RG, Caram P, Allen E (1989) Cerebral arteriovenous oxygen difference as an estimate of cerebral blood flow in comatose patients. J Neurosurg 70:222–230
4. Williams PL, Warwick R, Dyson M (eds) (1989) Grays anatomy, 37th ed New York, Churchill Livingstone, pp 793–805
5. Lassen NA (1959) Cerebral blood flow and oxygen consumption in man. Physiol Rev 39:183–238
6. Shenkin HA, Spitz EB, Grant FC (1948) Physiologic studies of arteriovenous anomalies of the brain. J Neurosurg 5:165–172
7. Andrews PJD, Dearden NM, Miller JD (1991) Jugular bulb cannulation:description of a technique and validation of a new continuous monitor. Br J Anaesth 67:553–558
8. Cold GE (1978) Cerebral metabolic rate of oxygen ($CMRO_2$) in the acute phase of brain injury. Acta Anaesthesiol Scand 22:249–256
9. Bruce DA, Langfitt TW, Miller JD, Schultz H, Vapalahti M, Stanek A, Goldberg HI (1973) Regional cerebral blood flow, intracranial pressure, and brain metabolism in comatose patients. J Neurosurg 38:131–144
10. Orbrist WD, Gennarelli TA, Segawa H (1979) Relation of cerebral blood flow to neurological status and outcome in head injured patients. J Neurosurg 51:292–300

11. Overgaard J, Tweed WA (1974) Cerebral circulation after head injury. 1. Cerebral blood flow and its regulation after closed head injury with emphasis on clinical correlations. J Neurosurg 41:531–541
12. Obrist WD, Langfitt TW, Jaggi JL, Cruz J, Gennarelli TA (1984) Cerebral blood flow and metabolism in comatose patients with acute head injury. Relationship to intracranial hypertension. J Neurosurg 61:241–253
13. Robertson CS (1992) Treatment of cerebral ischaemia in severely head injured patients. Br J Intens Care 1 (suppl):12–15
14. Aaslid R, Markwalderr TM, Nornes H (1982) Non-invasive transcranial Doppler ultrasound recording of flow velocity in basal cerebral arteries. J Neurosurg 57:769–774
15. Doppler CA (1843) Uber das farbige Licht der Doppelsterne und einiger anderer Gisterne des Himmels. Abhandl Konigl Bohm Ges Wiss Series 2:465–482
16. Yasaka M, Tsuchiya T, Yamaguchi T, Hasegawa Y, Kimura K, Omae T (1990) Transcranial real-time color-flow Doppler ultrasonography. 2. Reproducibility of measurements of blood flow velocity in the middle cerebral artery (abstract). Stroke 21:(suppl 1):113
17. Hennerici M, Rautenberg W, Sitzer G, Schwartz A (1987) Transcranial Doppler ultrasound for the assessment of intracranial arterial flow velocity 1. Examination technique and normal values. Surg Neurol 27:439–448
18. Gosling RG, King DH (1974) Arterial assessment by Doppler shift ultrasound. Proc R Soc Med 67:447–449
19. Zannette EM, Fieschi C, Bozzao L, Toni D, Argentino C, Lenzi GL (1989) Comparison of cerebral angiography and transcranial Doppler sonography in acute stroke. Stroke 20:899–903
20. Lindegaard KF, Nornes H, Bakke SJ, Sorteberg W, Makstad P (1988) Cerebral vasospasm after subarachnoid haemorrhage investigated by transcranial Doppler ultrasound. Acta Neurochirurg (Wein) 42(Suppl):81–84
21. Arnolds BJ, von Reutern GM (1986) Transcranial Doppler sonography. Examination technique and normal reference values Ultrasound Med Biol 12:115–123
22. Markwalder TM, Grolimund P, Seiler RW, Roth F, Aaslid R (1984) Dependency of blood flow velocity in the middle cerebral artery on end-tidal CO_2 partial pressure: a transcranial ultrasound Doppler study. J Cerebral Blood Flow Metabolism 4:368–372
23. Widder B, Paulat K, Hackspacker J, Mayr E (1986) Transcranial Doppler CO_2 test for the detection of haemodynamically critical carotid artery stenoses and occlusions. Eur Arch Psych Neurol Sci 236:162–168
24. Chan KH, Dearden NM, Miller JD, Midgley S, Piper IR (1992) Transcranial Doppler waveform differences in hyperemic and non-hyperemic patients after severe head injury. Surg Neurol 38:433–436
25. Chan KH, Dearden NM, Miller JD (1992) The significance of posttraumatic increase in cerebral blood flow velocity: a transcranial doppler ultrasound study. Neurosurgery 30:697–700
26. Chan KH, Miller JD, Dearden NM, Andrews PJD, Midgley S (1992) The effect of changes in cerebral perfusion pressure upon middle cerebral artery blood flow velocity and jugular bulb venous oxygen saturation after severe brain injury. J Neurosurg 77:55–61
27. Hassler W, Steinmetz H, Gawslowski J (1988) Transcranial Doppler ultrasonography in raised intracranial pressure and in intracranial circulatory arrest. J Neurosurg 69:745–751

28. Chan KH, Miller JD, Dearden NM (1992) Intracranial blood flow velocity after head injury: relationship of severity of injury, time, neurological status and outcome. J Neurol, Neurosurg Psychiatry 55:787–791
29. Wilkins RH (1986) Attempts at prevention and treatment of intracranial arterial spasm: an update. Neurosurgery 18:808–826
30. Weir B, Grace M, Hansen J, Rothberg C (1978) Time course of vasospasm in man. J Neurosurg 48:173–178
31. Chyatte D, Fode NC, Sundt TM (1988) Early versus late intracranial aneurysm surgery in subarachnoid haemorrhage. J Neurosurg 69:326–331
32. Saito I, Shigeno T, Arrotaki K, Tanishima T, Sano K (1979) Vasospasm assessed by angiography and computed tomography. J Neurosurg 51:466–475
33. Meyer CHA, Lowe D, Meyer M, Richardson PL, Neil-Dwyer G (1983) Progressive change in cerebral blood flow during the first three weeks after subarachnoid haemorrhage. Neurosurgery 12:58–76
34. Harders AG, Gilsbach JM, Hornyak ME (1988) Incidence of vasospasm in transcranial Doppler sonography and its clinical significance: a prospective study in 100 consecutive patients who were given intravenous nimodipine and who underwent early aneurysm surgery. In Wilkins R H (ed) Cerebral vasospasm. Raven New York, 32–52
35. Sekhar LM, Wechsler LR, Yonas H, Luyckx K, Obrist W (1988) Value of transcranial Doppler examination in the diagnosis of cerebral vasospasm after subarachnoid haemorrhage. Neurosurgery 22:813–821
36. Romner B, Ljunggren L, Brandt L, Saveland H (1989) Transcranial Doppler sonography within 12 hours after subarachnoid haemorrhage. J Neurosurg 70:732–736
37. Seiler RW, Grolimund P, Aaslid R, Huber P, Nornes H (1986) Cerebral vasospasm evaluated by transcranial Doppler ultrasound correlated with clinical grade and CT-visualised subarachnoid haemorrhage. J Neurosurg 64:594–600
38. Hassler W, Chioffi F (1989) CO_2 reactivity of cerebral vasospasm after aneurysmal subarachnoid haemorrhage Acta Neurochirurgica 98:167–175
39. Chan KH, Dearden NM, Miller JD, Andrews PJD, Midgley S (1993) Multimodality monitoring as a guide to the treatment of intracranial hypertension. Neurosurgery 32:547–553

Transcranial Doppler Sonography: Monitoring of Cerebral Perfusion

C. Werner

Introduction

Transcranial Doppler sonography (TCD) has been recently applied to anesthesia research as well as clinical anesthesia and critical care medicine. Interest in this method is derived from the unique capacity of TCD to measure cerebral hemodynamics noninvasively and continuously. Although TCD measures cerebral blood flow velocity rather than cerebral blood flow (CBF), the technique promises to generate new information concerning the dynamics of cerebral perfusion during administration of centrally active drugs such as anesthetics and narcotics. TCD also monitors the changes in cerebral perfusion associated with cerebral embolic events and during increases in intracranial pressure (ICP). This may increase our understanding of the time course and pathophysiology of focal and global cerebral ischemia and improve the quality of our clinical treatment.

Monitoring of Cerebral Blood Flow

Cerebral Blood Flow Autoregulation

Cerebral autoregulation maintains CBF within a wide range of cerebral perfusion pressure [22]. Drugs, trauma, tumor, or infection may impair or abolish the mechanisms of autoregulation. In order to avoid hypo- or hyperperfusion, monitoring of cerebral perfusion is of considerable clinical importance in patients where clinical evaluation of adequate CBF is impossible. Several studies have investigated the potential of TCD to assess cerebral autoregulation. Aaslid et al. [1, 2] studied the effects of acute arterial hypotension on cerebral blood flow velocity in volunteers. The deflation of cuffs around both thighs produced concurrent reduction of mean arterial blood pressure (MAP) and middle cerebral artery (MCA) blood flow velocity. The recovery time of MAP was significantly longer compared to MCA blood flow velocity. Since the return of blood flow velocity to baseline values was independent of arterial blood pressure changes, this was interpreted as intact CBF autoregulation. Giller [9] produced transient hyperemic responses in the MCA territory by common carotid artery occlusion for a period of 3 s. The hyperemic response was considered as autoregulatory cerebral vasodilatation and did not occur in patients with impaired cerebral autoregulation due to cerebral vasospasm or in brain-dead subjects. In dogs, Werner et al. [41] induced arterial hypertension beyond the upper limit of cerebral autoregulation (236 ± 11 mm Hg) using angiotensin infusion. Arterial hypertension resulted in concurrent increases in CBF and CBF velocity. However, the amount of CBF increase was underestimated by the TCD technique. It is likely that the sudden increase in transmural pressure produced by angiotensin injection dilated the insonated vessel segment. Thus, the increase in CBF during hypertension was a function of arterial dilation and increases in blood flow velocity. Barzó et al. [5] investigated the lower end of CBF autoregulation by stepwise increases in ICP using cisterna magna infusion in a rabbit model. The results by Barzó et al. [5] indicate a close correlation between CBF and TCD blood flow velocity within the lower range of CBF autoregulation (cerebral perfusion pressure CPP 40–80 mm Hg). This is consistent with experiments in rabbits where trimethaphan-induced arterial hypotension produced concurrent decreases CBF and TCD, indicating the lower limit of CBF autoregulation [18].

However, it is still unclear whether TCD measurements can reliably assess the entire range of cerebral autoregulation. Brooks et al. [7] have shown in patients with autonomic failure that orthostatic hypotension produces decreases in TCD blood flow velocity without changes in CBF. The changes in TCD without changes in CBF were discussed as autoregulatory vasodilation of the basal cerebral arteries. In contrast, Nelson et al. [18] found no close correlation between CBF and TCD during trimethaphan-induced arterial hypotension. In summary, these data indicate the potential of TCD to monitor changes in CBF associated with significant changes in CPP and to identify the upper and lower end of cerebral autoregulation. TCD at present does not provide a qualitative measure of CBF during moderate changes of arterial blood pressure.

Anesthetics and Narcotics

Anesthesia and long-term analgosedation produce drug-specific and dose-dependent changes in CBF and cerebral metabolism. It is clinically important to monitor the cerebral hemodynamic effects of these drugs in order to differentiate drug-induced hemodynamic changes from other etiologies. Several studies [28, 29, 34] have shown that volatile anesthetics such as halothane may increase CBF velocity, while narcotics or intravenous anesthetics such as alfentanil or propofol had no effect or decreased the TCD signal. This was interpreted as a reflection of drug response curves described in CBF studies. Recent experiments confirm that TCD measures the effects of anesthetics and narcotics and may be a clinically valuable tool to monitor these effects in the unconscious patient.

Werner et al. [38, 39] and Kochs et al. [16] investigated the effects of the intravenous anesthetic and sedative propofol, the volatile anesthetic isoflurane,

and the new narcotic sufentanil on CBF, MCA blood flow velocity electroencephalography (EEG), cerebral oxygen consumption ($CMRO_2$), and ICP in
dogs.

Propofol. Following recordings of baseline data, propofol (0.8 mg/kg per min)
was infused and a second series of measurements was performed at induction of
burst suppression. Propofol infusion was then discontinued and a third measure
was done following recovery of EEG to baseline levels. Propofol infusion
(0.8 mg/kg per min) was repeated and the last measurements were obtained after
induction of burst suppression. The results for CBF, EEG, and mean blood flow
velocity (V_{mean}) are shown in Fig. 1. Propofol significantly reduced CBF, V_{mean},
and $CMRO_2$ from baseline values and cerebral hemodynamics remained reduced following recovery (recovery time, 50–80 min). The second infusion of
propofol produced cerebral hemodynamic depression similar to the first burst
suppression period. Changes in CBF and V_{mean} were closely correlated (r, 0.86).
The reduction in CBF and V_{mean} was associated with decreases in ICP at each
respective propofol treatment. The close correlation between decreases in CBF
and V_{mean} indicates that TCD continuously measures changes in CBF following
administration of propofol. The significant decline in ICP indicates that
propofol-induced decreases in CBF are associated with reductions of cerebral
blood volume [39].

Isoflurane. Following baseline recordings, isoflurane was added to the inspiratory gas mixture, and cerebral hemodynamic and metabolic measurements were
repeated at end-tidal concentrations of 1%, 2% and 3% isoflurane, respectively.
Changes in CBF and V_{mean} during incremental isoflurane are shown in Fig. 1.
Isoflurane significantly increased CBF, V_{mean} and ICP at concentrations of 2%
and 3% compared to baseline values. Changes in CBF and V_{mean} were closely
correlated (r, 0.81, $p < 0.01$). The increase in CBF and V_{mean} was associated with
decreases in $CMRO_2$. The close correlation between increases in CBF and V_{mean}
indicates that TCD continuously measures dose-dependent changes in CBF
with isoflurane. The significant increase in ICP indicates that isoflurane increases both CBF and cerebral blood volume. The reductions of $CMRO_2$
parallel to increases in CBF indicates uncoupling of the ratio between cerebral
metabolism and CBF during isoflurane anesthesia [16].

Sufentanil. Following control measurements, 20 μg/kg sufentanil were injected
and data were obtained at 5, 15 and 30 min following sufentanil administration.
Figure 1 shows relative changes of CBF and V_{mean}. The correlation between
relative changes of CBF and V_{mean} was $r = 0.82$ for all measurements. ICP did
not change over time. The decreases in CBF following administration of sufentanil are related to decreased metabolic. TCD provides noninvasive and continuous monitoring of changes in cerebrovascular hemodynamics with a close
correlation to changes in CBF following sufentanil [38].

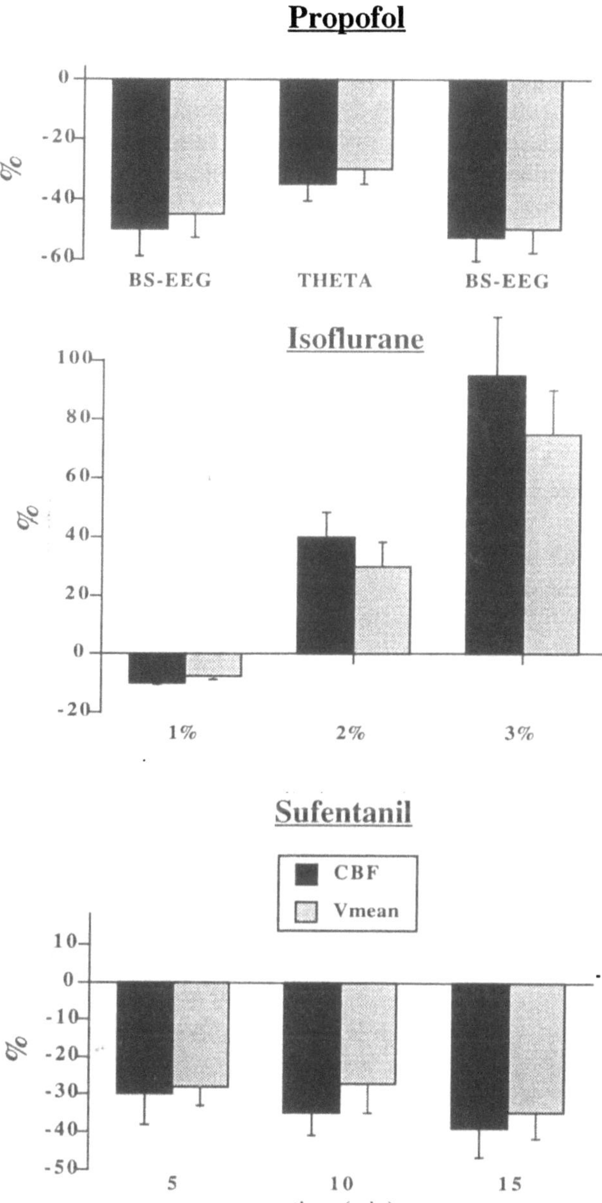

Fig. 1. Changes in middle cerebral artery mean blood flow velocity (V_{mean}) and cerebral blood flow (*CBF*) in percentage from baseline during infusion of propofol, administration of isoflurane, and infusion of sufentanil. *EEG*, electroencephalography; *BS*, burst suppression (mean ± SD)

These studies show that TCD is a noninvasive and continuous technique to measure relative changes in CBF during administration of anesthetics and narcotics. Drugs such as propofol and sufentanil decrease CBF as a function of decreased cerebral metabolism with or without decreasing intracranial volume. This suggests that these drugs can be safely used for sedation and analgesia during neurocritical care. Isoflurane is a cerebral metabolic depressant, but increases CBF and ICP at higher concentrations due to uncoupling of the flow to metabolism ratio.

Monitoring of Cerebral Ischemia

Cerebral ischemia may frequently occur in the perioperative period and during neurocritical care. Ischemic insults may be due to intraoperative occlusion of brain supplying arteries, implantation of vascular shunts, thrombus formation and reobliteration of endarterectomized vessels, elevated ICP, and generation of cerebral emboli. Several studies have tried to identify typical quantitative and qualitative changes in CBF velocity as sensitive parameters to indicate cerebral ischemia. Padayachee et al. [21] and Naylor et al. [17] have shown that temporary ligation of the internal carotid artery produced significant decreases in CBF velocity with a linear correlation to internal carotid stump pressure. In the study by Naylor et al. [17], a mean blood flow velocity of less than 30 cm/s was defined as a threshold indicative for cerebral ischemia. Halsey et al. [11] have shown that mean blood flow velocities of less than 15 cm/s correlate with critical CBF values of less than 20 ml/100 g per min. However, the authors suggest that the EEG may be more specific in indicating focal cerebral ischemia than TCD recordings. In a study by Thiel et al. [33], changes in TCD blood flow velocity were closely correlated with changes in amplitude and latency of somatosensory evoked potentials (SEP) during temporary unilateral internal carotid ligation, and reduced velocities of greater than 60% from baseline were defined as the ischemic threshold. The discrepancies between the results of these studies may be due to different sensitivities of TCD, EEG, and SEP in detecting regionally specific changes in cerebral perfusion and cerebral function.

More recently, a blood flow velocity pattern was identified that is associated with ischemic loss of neuronal function. Werner et al. [40] investigated the effects of hemorrhagic hypotension on MCA blood flow velocity in correlation to brain electrical activity (EEG) in fentanyl/N_2O-anesthetized dogs. Changes in blood flow velocity and EEG with decreasing MAP are shown in Fig. 2. V_{mean} and SEF did not change within the MAP range of 115 ± 7–49 ± 9 mm Hg. This suggests autoregulation of CBF and maintained cerebral function over this pressure range. These data are consistent with experiments in rabbits where CBF and CBF velocity were closely correlated at the lower end of CBF autoregulation (CPP range: 40–80 mm Hg) [5]. Below 49 ± 9 mm Hg, decreases in mean and diastolic blood flow velocity were associated with a shift of

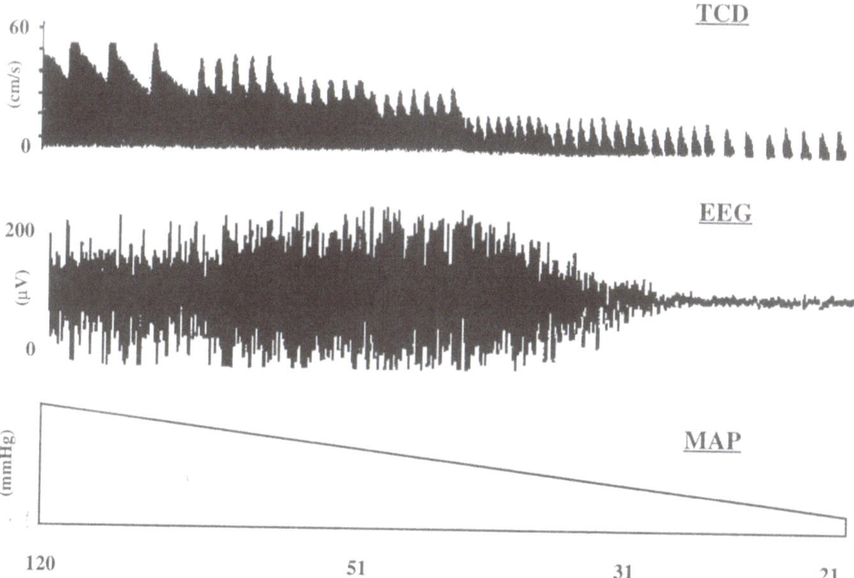

Fig. 2. Typical changes in transcranial Doppler sonography (*TCD*) blood flow velocity and electroencephalography (*EEG*) during hemorrhagic hypotension in dogs. TCD and EEG did not change within the mean arterial blood pressure (*MAP*) range of 115 ± 7–49 ± 9 mm Hg. Below 49 ± 9 mm Hg, decreases of the TCD signal were associated with a shift of the EEG to lower frequencies and higher amplitudes. Brain electrical silence occurred at a MAP of 31 ± 7 mm Hg, paralleled by a loss of the diastolic flow velocity pattern (mean $\pm$ SD)

the EEG to lower frequencies. Brain electrical silence occured at a MAP of 31 ± 7 mm Hg, paralleled by a loss of the diastolic flow velocity pattern. At this level of MAP, systolic and mean flow velocities were 35 ± 5 cm/s and 20 ± 6 cm/s, respectively. This suggests that the diastolic flow velocity pattern and neuronal function are closely correlated and that deterioration of the diastolic flow pattern is a more sensitive parameter in the detection of cerebral ischemia than absolute or relative changes in mean blood flow velocity. In summary, it has been shown that decreases in mean blood flow velocity to a level less than 15 cm/s or more than 60% from baseline are associated with cerebral ischemia and ischemic neuronal dysfunction. The diastolic flow velocity pattern may be even more specific in the detection of ischemic CBF. TCD provides information on adequate cerebral perfusion and neuronal viability in situations of maximal brain electrical suppression (burst suppression–EEG pattern induced by anesthetics, trauma, intoxication).

Besides arterial cross-clamping or arterial stenosis and obliteration, focal cerebral ischemia may be produced by arterial emboli. Recurrent embolization is not detectable in anesthetized or sedated patients unless EEG recordings can assess the ischemic challenge. However, ischemic EEG changes may escape in

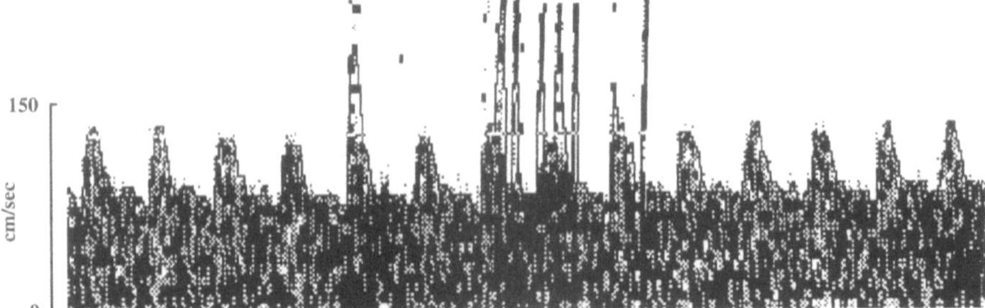

Fig. 3. High-amplitude flow disturbance signals in the middle cerebral artery during carotid endarterectomy

patients under anesthesia or long-term sedation. Recent studies suggest that TCD can detect cerebral emboli. TCD monitoring during carotid endarterectomy or during cardio pulmonary bypass detected the occurrence of high amplitude flow disturbance signals during implantation of shunts or cannulation of the aorta [20, 25, 32, 36] (Fig. 3). These high frequency signals were interpreted as ultrasonic reflexions of air or particulate matter. Animal research confirmed that these high amplitude flow disturbance signals are generated by cerebral emboli. Russell et al. [27] have shown that injection of air or particulate matter into the renal artery produces material-specific high-amplitude flow disturbance signals. Further studies [8, 27, 31] have suggested that the Doppler technique may even identify characteristics such as size and volume of emboli. Today, TCD is used as a sensitive detector of embolic sources during surgery and in neurocritical care. TCD monitoring is an early warning system for cerebral ischemia that provides insight into the mechanisms of embolic stroke and may guide early therapeutic interventions.

Increased Intracranial Pressure and Cerebral Circulatory Arrest

The requirements for an ideal and clinically reasonable monitoring system of cerebral hemodynamics are: (a) a technique with continuous and noninvasive data acquisition and (b) detection of cerebral ischemia. TCD may meet these requirements during elevated ICP and cerebral circulatory arrest. Several studies indicate that the diastolic flow velocity pattern is sensitive to changes in CPP [12, 13, 37, 42]. Tranquart et al. [35] have shown in rabbits that diastolic blood flow velocity decreases as ICP increases. Barzó et al. [5] found a close correlation between decreases in CBF and CBF velocity as a function of elevated ICP. Other experiments suggest a close correlation between cortical and brain stem electrical activity and the diastolic flow velocity pattern [30, 40]. During increases in ICP, the diastolic flow velocity signal decreases in response to decreases in CPP [12, 13, 37, 42]. This may occur with or without concomitant decreases in the peak flow velocity. Further increases in ICP beyond the level of diastolic arterial blood pressure reduce the diastolic flow velocity signal to zero (Fig. 4). The pathophysiological mechanisms of the loss of diastolic flow at this pressure level remain controversial. It is possible that the ICP-induced increase in transmural pressure generated consecutive capillary collapse [13]. However, it is more likely that increases in ICP reduce the pressure drop along the still patent vascular bed [4, 13]. The deterioration of the diastolic blood flow velocity pattern strongly suggests critical CBF and ischemic neuronal dysfunction due to decreases in cerebral perfusion pressure. Therefore, therapeutic interventions should be based on the analysis of continuous diastolic blood flow velocity recordings, since acute neurologic dysfunction may recover within the first few minutes following onset. TCD monitoring is particularly indicated in patients, where measurements of ICP are impossible (e.g., hepatic coma).

With persisting diastolic zero flow or further increases in ICP, cerebral circulatory arrest occurs (Fig. 5). Several studies have correlated TCD with classical techniques in the determination of brain death such as arteriography or measurements of CBF [15, 23, 24, 26]. These studies confirm that TCD is a reliable method to confirm cerebral circulatory arrest with a sensitivity of 91.3% and a specificity of 100% [19, 23]. The blood flow velocity patterns of brain-dead subjects exhibit five typical patterns (Fig. 4): (1) systolic flow without diastolic flow; (2) systolic flow with combined diastolic forward and reversed flow; (3) oscillating flow (biphasic flow) with reversed diastolic flow; (4) systolic peaks without diastolic flow; and (5) no detectable flow [5]. Lack of signals does not necessarily indicate impaired CBF or cerebral circulatory arrest, since the penetration of the ultrasonic beam may be impossible due to thick bones. Vessel

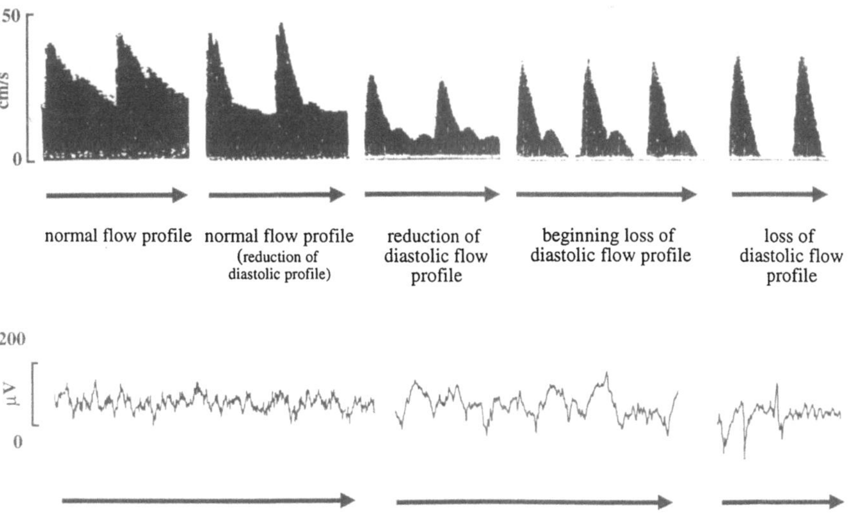

Fig. 4. Changes in the transcranial Doppler sonography flow velocity profile (*top*) and the electroencephalography pattern (*bottom*) during progressive increases in intracranial pressure

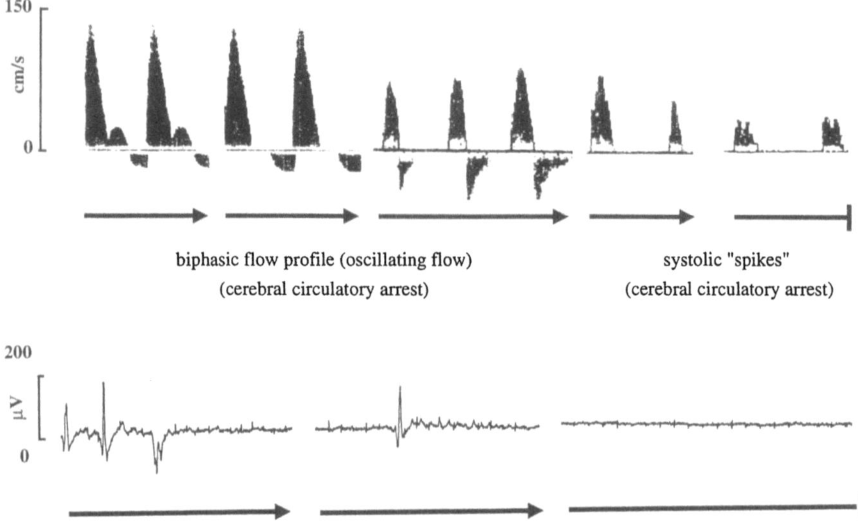

Fig. 5. Transcranial Doppler sonography flow velocity profile (*top*) and the electroencephalography pattern (*bottom*) with impeding and persisting cerebral circulatory arrest

identification may be also difficult due to traumatic or pathological displacement or anatomical variations.

The diagnosis of cerebral circulatory arrest should always involve monitoring of blood flow velocity in supra- and infratentorial brain regions, since flow may be present in supratentorial territories during infratentorial circulatory arrest and vice versa. Transient, reversible bidirectional TCD patterns may be present in situations that are not compatible with the diagnosis of brain death. Grote et al. [10] have shown that biphasic flow patterns may occur in the initial phase of subarachnoid hemorrhage or during the acute development of space-occupying lesions. Other studies indicate that transient biphasic flow patterns or zero flow may also occur during percutaneous transluminal aortic valvuloplasty [14] or in patterns with intra-aortic balloon pumps [16]. The confirmatory diagnosis of brain death should therefore be based on bidirectional flow patterns or systolic spikes of low amplitude and repeated measurements in supra- and infratentorial basal cerebral arteries for at least 30 min.

References

1. Aaslid R, Lindegaard K-F, Sorteberg W, Nornes H (1989) Cerebral autoregulation dynamics in humans. Stroke 20:45–52
2. Aaslid R, Newell DW, Stooss R, Sorteberg W, Lindegaard K-F (1991) Assessment of cerebral autoregulation dynamics from simultaneous arterial and venous transcranial Doppler recordings in humans. Stroke 22:1148–1154

3. Arnolds BJ, Von Reutern G-M, (1986) Transcranial Doppler sonography. Examination technique and normal reference values. Ultrasound in Med Biol 12:115–123

4. Auer LM, Ishiyama N, Hodde KC, Kleinert R, Pucher R (1987) Effect of intracranial pressure on bridging veins in rats. J Neurosurg 67:263–268

5. Barzó P, Dóczi T, Csete K, Buza Z, Bodosi M (1991) Measurements of regional cerebral blood flow and blood flow velocity in experimental intracranial hypertension: infusion via the cisterna magna in rabbits. Neurosurgery 28:821–825

6. Brass LM (1990) Reversed intracranial blood flow in patients with intra-aortic balloon pump. Stroke 21:484–487

7. Brooks DJ, Redmond S, Mathias CJ, Bannister R, Symon L (1989) The effect of orthostatic hypotension on cerebral blood flow and middle cerebral artery blood flow velocity in autonomic failure, with observations on the action of ephedrine. J Neurol Neurosurg Psychiatry 52:962–966

8. Bunegin L, Wahl D, Albin M (1991) Estimation of embolic air volume in the middle cerebral artery (MCA) using transcranial sonography. Anesthesiology 750:A 471

9. Giller CA (1991) A bedside test for cerebral autoregulation using transcranial Doppler ultrasound. Acta Neurochir (Wien) 108:7–14

10. Grote E, Hassler W (1988) The critical first minutes after subarachnoid hemorrhage. Neurosurgery 22:654–661

11. Halsey JH, McDowell HA, Gelmon S, Morawetz RB (1989) Blood velocity in the middle cerebral artery and regional cerebral blood flow during carotid endarterectomy. Stroke 20:53–58

12. Hassler W, Steinmetz H, Gawlowski J (1988) Transcranial Doppler ultrasound in raised intracranial pressure and in intracranial circulatory arrest. J Neurosurg 68:745–751

13. Hassler W, Steinmetz H, Pirschel J (1989) Transcranial Doppler study of intracranial circulatory arrest. J Neurosurg 71:195–201

14. Karnik R, Valentin A, Bonner G, Ziegler B, Slany J (1990) Transcranial Doppler monitoring during percutaneous transluminal aortic valvuloplasty. Angiology 41:106–111

15. Kirkham FJ, Levin SD, Padayachee TS, Kyme MC, Neville BGR, Gosling RG (1987) Transcranial pulsed Doppler ultrasound findings in brain stem death. J Neurol Neurosurg Psychiatry 50:1504–1513

16. Kochs E, Hoffman WE, Werner C, Albrecht RF, Schulte am Esch J (1993) Cerebral blood flow velocity in relation to cerebral blood flow, cerebral metabolic rate for oxygen, and EEG during isoflurane anesthesia in dogs. Anesth Analg 76:1222–1226

17. Naylor AR, Wildsmith JAW, McClure J, McL Jenkins A, Ruckley CV (1991) Transcranial Doppler monitoring during carotid endarterectomy. Br J Surg 78:1264–1268

18. Nelson RJ, Perry S, Hames TK, Pickard JD (1990) Transcranial Doppler ultrasound studies of cerebral autoregulation and subarachnoid hemorrhage in the rabbit. J Neurosurg 73:601–610

19. Newell DW, Grady MS, Sirotta P, Winn HR (1989) Evaluation of brain death using transcranial Doppler. Neurosurgery 24:509–513

20. Padayachee TS, Gosling RG, Bishop CC, Burnand K, Browse NL (1986) Monitoring middle cerebral artery emboli during carotid endarterectomy using transcranial Doppler ultrasonography. Br J Surg 73:98–100

21. Padayachee TS, Gosling RG, Bishop CC, Burnand K, Browse NL (1986) Monitoring middle cerebral artery blood flow velocity during carotid endarterectomy. Br J Surg 73:98–100

22. Paulson OB, Strandgaard S, Edvinsson L (1990) Cerebral autoregulation. Cerebrovasc Brain Metab Rev 2:161–191
23. Petty GW, Mohr JP, Pedley TA, Tatemichi TK, Lennihan L, Duterte DI, Sacco RL (1990) The role of transcranial Doppler in confirming brain death: sensitivity, specificity, and suggestions for performance and interpretation. Neurology 40:300–303
24. Powers AD, Graeber MC, Smith RR (1989) Transcranial Doppler ultrasonography in the determination of brain death. Neurosurgery 24:884–889
25. Pugsley W, Klinger L, Paschalis C, Aspey B, Newman S, Harrison M, Treasure T (1990) Microemboli and cerebral impairment during cardiac surgery. Vasc Surg 24:34–43
26. Ropper AH, Kehne SM, Wechsler L (1987) Transcranial Doppler in brain death. Neurology 37:1733–1735
27. Russell D, Madden KP, Clark WM, Sandset PM, Zivin JA (1991) Detection of arterial emboli using Doppler ultrasound in rabbits. Stroke 22:253–258
28. Schregel W, Beverungen M, Cunitz G (1988) Transkranielle Doppler-Sonographie: Halothan steigert die mittlere Blutflußgeschwindigkeit in der Arteria cerebri media. Anaesthesist 37:305–310
29. Schregel W, Schäfermeyer, Müller C, Geißler C, Bredenkötter U, Cuntiz G (1992) Einfluß von Halothan, Alfentanil und Propofol auf Flußgeschwindigkeiten, "Gefäßquerschnitt" und "Volumenfluß" in der A. cerebri media. Anaesthesist 41:21–26
30. Shiogai T, Sato E, Tokitsu M, Hara M, Takeuchi K (1990) Transcranial Doppler monitoring in severe brain damage: relationships between intracranial haemodynamics, brain dysfunction and outcome. Neurol Res 12:205–213
31. Spencer PM, Thomas GI, Nicholls SC, Sauvage LR (1990) Detection of middle cerebral artery emboli during carotid endarterectomy using transcranial Doppler ultrasonography. Stroke 21:415–423
32. Thiel A, Russ W, Kaps M, Marck GP, Hempelmann G (1988) Die transkranielle Dopplersonographie als intraoperatives Überwachungsverfahren. Anaesthesist 37:256–260
33. Thiel A, Russ W, Zeiler D, Dapper F, Hempelmann G (1990) Transcranial Doppler sonography and somatosensory evoked potential monitoring in carotid surgery. Eur J Vasc Surg 4:597–602
34. Thiel A, Zickmann B, Zimmermann R, Hempelmann G (1992) Transcranial Doppler sonography: effects of halothane, enflurane and isoflurane on blood flow velocity in the middle cerebral artery. Br J Anaesth 68:388–393
35. Tranquart F, Berson M, Bodard S, Roncin A, Pourcelot L (1991) Evaluation of cerebral blood flow in rabbits with transcranial Doppler sonography: first results. Ultrasound Med Biol 17:815–818
36. Van der Linden J, Casimir-Ahn H (1991) When do cerebral emboli appear during open heart operations? A transcranial Doppler study. Ann Thorac Surg 51:237–241
37. Werner C, Kochs E, Rau M, Schulte am Esch J (1990) Transcranial Doppler as a supplement in the detection of cerebral circulatory arrest. J Neurosurg Anesth 3:159–165
38. Werner C, Hoffman WE, Baughman VL, Albercht RF, Schulte am Esch J (1991) Effects of sufentanil on cerebral blood flow, cerebral blood flow velocity and metabolism in dogs. Anesth Analg 72:177–181

39. Werner C, Hoffman WE, Kochs E, Albrecht RD, Schulte am Esch J (1992) The effects of propofol on cerebral blood flow in correlation to cerebral blood flow velocity in dogs. J Neurosurg Anesth 4:41–46
40. Werner C, Hoffman WE, Kochs E, Albrecht RF, Schulte am Esch J (1992) Transcranial Doppler sonography indicates critical brain perfusion during hemorrhagic hypotension in dogs. Anesth Analg 74:S 347
41. Werner C, Kochs E, Hoffman WE, Blanc I, Schulte am Esch J (1993) Cerebral blood flow and cerebral blood flow velocity during angiotension-induced arterial hypertension in dogs. Can J Anaesth 40:755–760
42. Zurynski Y, Dorsch N, Pearson I, Smith F, Woods P (1989) The use of transcranial Doppler sonography in the diagnosis of brain death. J Neurosurg Anesth 1:323–327

BRAINDEX–An Expert System
for Supporting Brain Death Diagnosis

G. Rom, R. Grims, G. Schwarz, and G. Pfurtscheller

Introduction

Brain death is defined as the irreversible cessation of all functions of the entire brain, including the brain stem, during continued functioning of the cardiovascular system and other organs. Brain death is a necessary requirement for organ transplantation and since the number of organs needed is increasing, the determination of brain death is of special importance.

The determination of the irreversible loss of all brain functions presupposes investigations of the brain stem and the cerebral cortex. Located in the brain stem are the respiratory and cardiovascular centers and structures responsible for the control of the state of consciousness. The cerebral cortex is responsible for sensational, perceptual, motor behavioral, and cognitive processes and contains a variety of highly specialized regions responsible for speech, reading, hearing, etc.

The diagnosis of brain death is based primarily on the examination of cephalic reflexes mediated by the cranial nerves passing through the brain stem and measurements of bioelectrical activity of the brain (e.g., electroencephalography EEG; evoked potentials EP) and cerebral blood flow (angiography). Besides this, there are a great variety of examinations and tests which can be used to verify the irreversible cessation of all brain functions. The diagnosis of brain death is furthermore dependent on a number of prerequisites such as no shock, no hypothermia, no medication affecting the central nervous system and repeated investigations after intervals of several hours, whereby the time interval depends on the type of clinical and electrophysiological investigations. A further problem lies in the great variety of apparent divergence from the standard case, caused by various motor and vegetative actions or reactions of integrated spinal mechanisms.

From this, it can be seen that the diagnosis of brain death is a highly complex task for the physician, who must take various premises, facts, and measurements into account. The decision support system discussed in this paper should assist, but can and may not substitute, the physician in the difficult task of defining the irreversible cessation of all brain functions.

Features and Fields of Application of BRAINDEX

BRAINDEX (Pfurtscheller et al. 1989; Rom 1989) is intended to be used mainly by physicians occupied with the determination of brain death, especially in peripheral hospitals where the determination of brain death (in view of an organ transplantation) is an exceptional case. BRAINDEX can also be used in training students in the field of brain death. A further application is to use the system for documentation purposes.

The main tasks of the system for the determination of brain death can briefly be characterized as follows:

— Pointing out contraindications to brain death
— Timely detection of special constellations
— Indicating and explaining differing and untypical symptoms for the diagnosis brain death
— Suggesting or, in special cases if there is a contraindication, warning against additional tests, e.g., angiography, EEG, EP
— Calculating the necessary "monitoring period" depending on the anamnesis. This monitoring period is a waiting period between the first determination of the brain death syndrome and the final diagnosis of brain death
— Reaching a system diagnosis of brain death only with 100% certainty and justifying the diagnosis-finding process

Further features include:

— Making a coma grading about the patient's coma stadium (MBS1-MBS4, midbrain syndrome stages 1–4; BBS1-BBS2, bulbar syndrome stages 1 and 2) if the system says "not brain dead" or "no evaluation possible"
— Printing a consultation protocol of medical examinations for documentation of brain death
— Storing medical data for documentation and statistical purposes in a large data base
— Ability to use a medical lexicon which contains special terms pertaining to brain death as well as bibliographic data

Implementation with the Expert System Shell PC-PLUS

The version of the expert system called BRAINDEX-R is implemented with the expert system shell PC-PLUS (Texas Instruments), which is a successor of EMYCIN. Today, the knowledge base of BRAINDEX-R consists of about 450 rules organized in 15 frames. Every frame represents a special part of the medical knowledge (e.g., cephalic reflexes, spinal reactions, EP), but special frames for creating and displaying the diagnosis, for printer-output, etc., are implemented as well (Fig. 1). The block structure of the system is shown in Fig. 2.

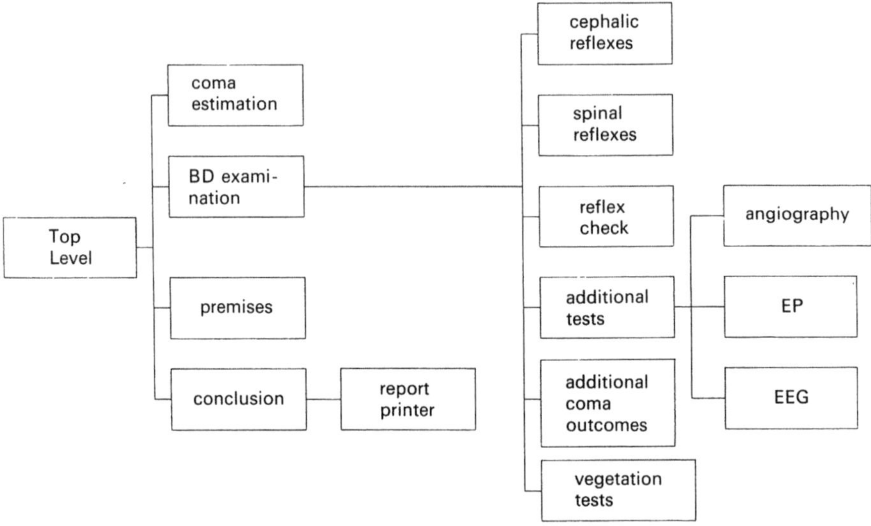

Fig. 1. Frame structure of BRAINDEX-R (Brain Death Expert System, version R). *BD*, brain death; *EP*, evoked potentials; EEG, electroencephalogram

For the diagnosis-finding process, the inference strategy known as backward chaining is used. In this process, the patient's data is checked on the assumption of "no brain death." This means that every relevant fact is tested to see whether it is a contraindication to brain death or not.

In this process, all the data are classified into confirmative data, contradictory data, untypical data, and not evaluated or not evaluable data. Based on this classification and the results of additional tests (EEG, angiography, EP), the system generates a diagnosis and justifies it.

Optional coma estimation was realized with forward chaining by using ten significant parameters (e.g., pulse, pupil dilation, blood pressure, etc.). This process yields percentage values for each coma stage (MBS1-MBS4, BBS1-BBS2) indicating the conformity of the parameters' values with each syndrome.

The explanatory text which appears on the screen during the consultation is not implemented as a part of the knowledge base, but is stored in a separate text archive. The advantage of this is that the text can be manipulated without changing the knowledge base, and thus it is not also necessary to create a new user version of the system if special medical text changes.

The rules of the knowledge base were implemented in ARL (abbreviated rule language), where PC-PLUS translates the ARL expressions into LISP form. For special purposes (e.g., reading and writing the text archive, printer control, use of windows, etc.), several functions were implemented directly in LISP. The user version was compiled with the integrated "build" function of PC-PLUS, which results in a considerable increase in speed.

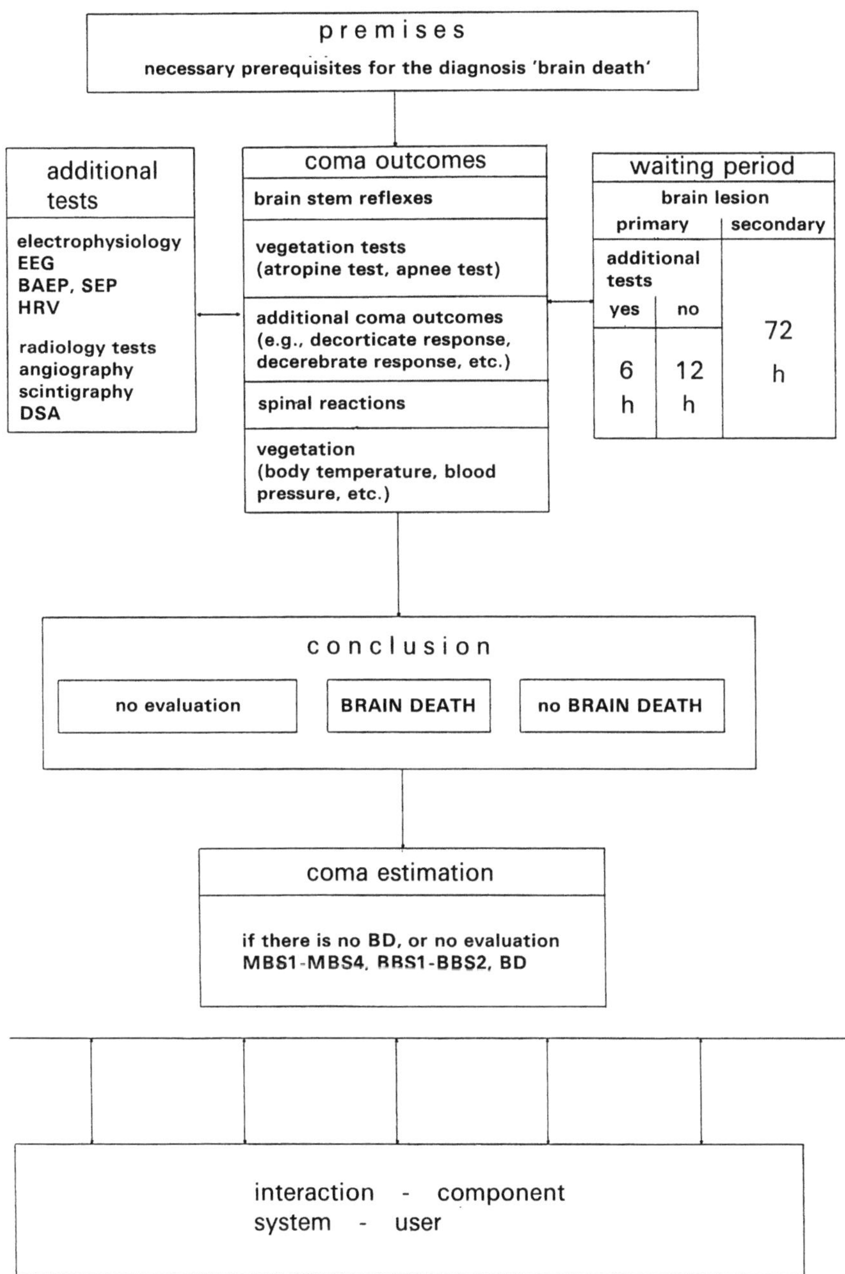

Fig. 2. Block structure of BRAINDEX-R. *EEG*, electroencephalogram; *BAEP*, brain stem auditory evoked potentials; *SEP*, somatosensory evoked potentials; *MBS1–MBS4*, midbrain syndrome stages 1–4; *BBS1–BBS2*, bulbar syndrome stages 1 and 2

Implementation with CLIPPER

The version BRAINDEX-G was not implemented with the use of artificial intelligence (AI) tools; instead, the expert system was developed in a conventional programming language (the data base language CLIPPER). BRAINDEX-G can be considered as the combination of a data base system with a rule-based expert system. The system consists of nine main data bases. They include a patient data base, the related examination data base, the integrated lexicon and reference list, and five data bases which represent the text archive of the knowledge base.

The expert system component works with an inference mechanism similar to backward chaining which was driven by the main goals "brain dead," "not brain dead," and "no evaluation." Based on the physician's judgement of the patient's status, the system starts by attempting to either prove or disprove this assessment (Fig. 3). To acknowledge a main goal, the corresponding list of attributes must be checked, whereby each attribute is characterized by rules implemented in the program code.

If rules are fired, the corresponding text from the text data base are activated and presented to the user. Intermediate results of the inference process and the pertaining rules are stored temporarily in an E&C data base (explanation and conclusion data base). This information is retrieved for the final diagnosis as well as for explanations and justifications. Thus, the system is capable of presenting the chain of reasoning, which explicates the diagnostic process to the user.

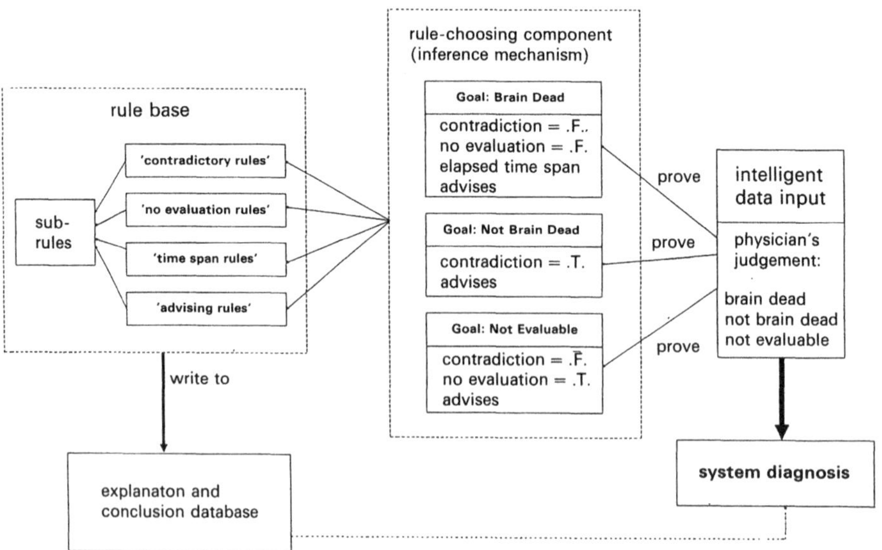

Fig. 3. Block structure of BRAINDEX-G

This version of BRAINDEX also allows optional coma grading. In this process, ten significant parameters are called upon and evaluated with statistical methods by using the same table as BRAINDEX-R.

Implementation with Neural Networks

During recent years, it has often been shown that neural network techniques provide very powerful learning paradigms for diagnostic problems. We chose a three-layer back propagation network to solve the specific problem of implementing coma grading for comatose patients.

Error propagation (Rumelhart and McClelland 1986) is a method of associative learning based on the generalized delta rule. A model designed to perform error propagation has to be built with at least three layers of elements: input, hidden, and output units. The learning procedure requires a set of paired input and output patterns. The input pattern is propagated through the network, and the generated output vector is compared to the desired output. The computed error is used to perform weight changes in direction of a better approximation. The goal of the back propagation algorithm is to train the system to form its own representations in the hidden layer.

One of the major problems of neural network systems is the encoding and interpretation of information in the units. Here, this question was solved in a very advanced way. There are many different medical parameters (50 in this case) which have to be checked by the physician for coma grading. Each parameter can take on several qualitative values (e.g., parameter "state of consciousness": clear, somnolent, soporous, unconscious, and not evaluable). In the net, each parameter is represented by a group of input units. A unit in a group stands for a specific value of the parameter and is active if the value is true. For this coding technique, we needed about 150 input units to represent all the possible values of the 50 medical parameters used. The seven units in the output layer represent the seven possible coma stages (or rather six coma stages and brain death; Fig. 4).

After a training phase, the system has learned to generalize with the set of training examples. It can now be used to grade other parameter constellations not yet presented. The parameter values are encoded into the input vector in the described way, and from these the output vector is generated. The output vector can then be interpreted as a determination of which syndrome stage corresponds best to the given symptoms. Although the implemented net was trained with a minimal set of data, the results obtained are quite promising.

A very large number of input parameters (50 compared to only ten with the expert system) was chosen here to additionally determine which parameters are absolutely necessary for an efficient coma evaluation. Analysis methods which allow the extraction of "microfeatures" were implemented. The medical information gleaned from this process about the importance of individual

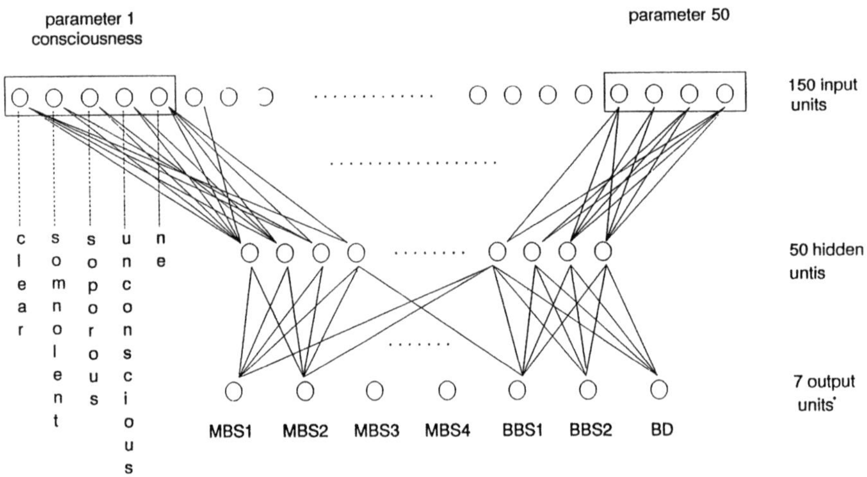

Fig. 4. Coma grading with a back propagation net *MBS1–MBS4*, midbrain syndrome stages 1–4; *BBS1-2*, bulbar syndrome stages 1 and 2; *BD*, brain death; *ne*, not evaluable

parameters can be implemented in the expert system as additional rules and can also be used to construct a neural network which only requires a smaller number of input parameters, namely, the most important ones.

By using the available data, it was shown that this method is viable. The next steps will be to train the already existing prototype with a larger number of data in order to substantiate the propositions.

Discussion

First of all, it should be pointed out that each method has its advantages and disadvantages. The use of an expert system shell allows the swift construction of a prototype and good structuring of the implemented knowledge. Since the inference mechanism is already present, only the knowledge must be entered in the proper form. However, this also has disadvantages, because the knowledge engineer is confronted with a relatively rigid structure.

The specific conditions involved in diagnosing brain death (a fuzzy logic is useless; intricately linked conclusions are not necessary) allowed the implementation of this system with a data base language (CLIPPER) within a reasonable time period. The advantage of this version is that the diagnostic process is substantially faster than when the shell is used. Furthermore, the design of the user interface could be greatly improved, as PC-PLUS is not always conducive to a good dialog with the user. One disadvantage encountered when using a conventional programming language is that the entire inference mechanism must be implemented; its maintenance, too, will certainly require more attention.

In order to store previous examinations for further consultations, statistical evaluations, and documentation purposes, an extensive data base (about 300 medical parameters per clinical check-up) was implemented using dBase and Clipper. It is almost superfluous to note that BRAINDEX-G (Clipper version) can optimally access the data base, whereas the PC-PLUS version can only utilize it indirectly, which results in prolonged running time.

After the completion of both systems, they were tested extensively and the results compared. Since both systems are based on the same medical knowledge, the test consultations produced the same results with only minimal deviations for a few comments. This process shows that the topic at hand can be treated not only with AI, but also with conventional methods.

The neural network was implemented only for the coma diagnosis, and not for brain death diagnosis, as were the previously described expert systems. Neural networks are less suited for making "100% statements", and a resulting diagnosis that a patient is 95% brain dead is useless; for this reason, only coma grading was performed with this paradigm.

The advantage of a neural network is that it is not necessary to implement a system of rules; instead, the network must be trained with examples. Thus, an adequately large amount of training data is a prerequisite, which can cause some practical problems. One disadvantage of neural networks is that interactive communication with the user (e.g., presentation of comments and warnings during the consultation) is almost impossible to realize; it is only possible to use it in the form of a batch diagnosis system.

Compared with the results of the expert system, this first attempt at constructing a coma diagnostic system using neural networks is very promising. As a last general comment, it should be noted that no one method alone is perfect: only a combination of all the available methods (conventional computer languages, AI languages and tools, and neural networks) can produce optimal solutions.

References

Pendl G (1986) Der Hirntod. Springer, Vienna New York
Pfurtscheller G et al. (1989) BRAINDEX – ein Expertensystem für die Hirntoddiagnostik. Biomed Tech (Berlin) 34:3–8
Rom G (1989) BRAINDEX – interaktives, wissenbasierendes System zur Unterstützung der Hirntod-Diagnostik. Diplomarbeit TU-Graz
Rummelhart DE, McClelland L (1986) Parallel distributed processing, MIT Press, Cambridge
Rumpl et al. (1979) The EEG at different stages of acute secondary traumatic midbrain and bulbar brain syndroms. Electroencephalogr Clin Neurophysiol 46:487–497
Schwarz G (1989) Dissoziierter Hirntod – computergestützte Verfahren in der Diagnostik und Dokumentation. Springer, Berlin Heidelberg New York
Walker AE (1985) Cerebral death. Urban and Schwarzenberg
Waterman DA (1986) A guide to expert systems. Addison Wesley, Reading

Subject Index